Chemical Sensors and Biosensors

Chemical Sensors and Biosensors

Edited by
René Lalauze

ISTE

WILEY

First published 2012 in Great Britain and the United States by ISTE Ltd and John Wiley & Sons, Inc.

ISTE Ltd
27-37 St George's Road
London SW19 4EU
UK

www.iste.co.uk

John Wiley & Sons, Inc.
111 River Street
Hoboken, NJ 07030
USA

www.wiley.com

Library of Congress Cataloging-in-Publication Data

Chemical sensors and biosensors / edited by Rene Lalauze.
p. cm.
Includes bibliographical references and index.
ISBN 978-1-84821-403-3
1. Electrochemical sensors. 2. Biosensors. I. Lalauze, Rene.
TP159.E37.C44 2012
543--dc23

2012012116

British Library Cataloguing-in-Publication Data
A CIP record for this book is available from the British Library
ISBN: 978-1-84821-403-3

Table of Contents

Preface

This book, written at the La Rochelle technical school, organized by Club Micro Capteurs Chimiques (the Chemical Microsensors Club), represents a major breakthrough in the detection of gaseous, ionic and biological environments.

Throughout this scientific work, fundamental and technological aspects concerning different types of sensors are examined by specialists from the world of research and academia.

These are sensitive elements, which are liable to interfere with gaseous spaces, semi-conductor materials, molecular crystals, ionic conductors and polymers all of which will be examined here.

In terms of detecting ionic environments in water solutions, these include silver salts, glasses and polymers which are the focus of detailed studies.

Biological recognition is approached using interactions between spaces for detection such as heavy metals, organophosphorus components or DNA and sensitive components such as enzymes, algae, bacteria or yeast.

Electrical, electrochemical, piezoelectric, mechanical, optical and biological transducers are commonly examined in terms of their implementation, production and performance in specific contexts and applications.

Microreactors, nanotechnology and flexible technologies also feature in this work, which focuses on the delicate problem of processing information over neural networks or electronic noses.

This book should therefore serve as a particularly useful and interesting tool for Masters and PhD students working in the field of chemical sensors, biosensors and biochips.

Chapter 1

Chemical and Biological Recognition

A (bio)chemical sensor allows the direct transduction of a concentration of (bio)chemical spaces into an electrical signal. This device includes a chemical or biological recognition feature coupled with a transduction system.

This chapter focuses on the main materials enabling the recognition of target spaces to be detected: gas molecules, ions, biological spaces.

1.1. Gas molecule recognition [MÉN 08]

Three types of recognition material are used: the exchange of electrons with inorganic or organic semi-conductor materials, the exchange of ions with ionic conductive materials and selective sorption in polymer matrices.

Chapter written by Nicole JAFFREZIC-RENAULT.

1.1.1. *Utilizing metallic oxides as semi-conductor materials for detecting gases*

It has been known for a long time (since the 1950s) that metallic oxides can present various semi-conductor properties, depending on the type of gaseous atmosphere they are in. We will examine different theories focusing on an example of the electronic catalyst theory or, more generally, the laws of adsorption. The use of metallic oxides as gas sensitive materials was introduced in the 1960s by Japanese researchers. As such, in this same period, Seiyma showed that variations in electric resistance of a ZnO film allowed the detection of reductor gases and Taguchi proposed the first gas sensor with porous SnO_2-based ceramics.

Metallic oxides with a general MO formula can present challenges to stoichiometry. The semi-conductive nature of these oxides comes from their ability to exceed or go under the stoichiometric amount. These challenges lead to more or less ionized punctual defects in crystalline networks in these oxides. This can consist of anionic gaps or interstitial cations which release electrons and make the semi-conductive material n-type, or even cationic gaps or interstitial anions which release holes of electrons and which make the semi-conductive material p-type.

Depending on the main type of defects in the solid, the four following types of oxides can be found: MO_{1+X}, $M_{1+X}O$, MO_{1-X}, $M_{1-X}O$ (M = metal, O = oxygen, x = discrepancy with stoichiometry). A type of solid MO_{1-x} is shown in Figure 1.1.

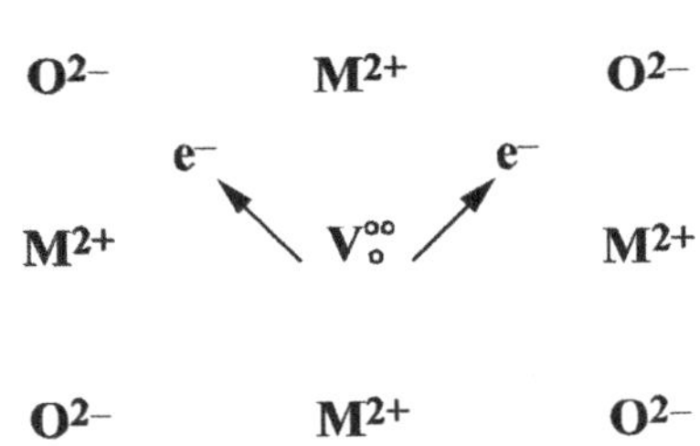

Figure 1.1. *A type of semi-conductive metallic oxide: oxygen gaps (n-type)*

For example SnO_2 has a tendency to be sub-stoichiometric in oxygen (written as SnO_{2-x}). According to punctual defect laws, oxygen gaps lead to free electrons in the oxide crystalline network, and therefore an increase in the concentration of free carriers, which provides the n character of the semiconductor.

For SnO_2, the band-gap-width is 3.5 eV and the donor levels are singly or doubly ionized oxygen gaps. As with every semi-conductor, an increase in temperature causes the electrons to move from these points towards the conduction bands. Next to the surface, the band diagram is modified by the electronic surface states, therefore creating a specific zone called the space charge zone. This can be the result of intrinsic faults in the material's surface and is especially due to the adsorption of atoms and foreign molecules. Variations in the space zone charge are directly linked to the extraction or injection of electrons by acceptor or donor spaces. This is the basis of gas detection properties by metallic oxides. The adsorption of oxygen on a n-type semi-conductor, such as SnO_2, leads to an electronic transfer towards the adsorbed molecules and as a result a decrease in the electric conductivity. This can be written as the following equation:

$$\frac{1}{2} O_2 + 2e^- = O^{2-}_{adsorbed}$$

The influence of other gases can therefore have two major effects. On the one hand, by direct transfer of charges following adsorption, as previously described, but also by chemical reaction with pre-adsorbed environments, in particular oxygenated environments. The reactions involved are essentially oxido-reduction reactions. Returning to the example of an n-type semi-conductor such as SnO_2, the reaction of the gases can be illustrated as follows.

With a reductive gas such as CO:

$$CO + O^{2-}_{adsorbed} = CO_{2\ adsorbed} + 2e^-$$

This involves an increase in the electrical conductivity and, in terms of the environment charge zone; this results in a decrease in the size of the electron free zone.

With an oxidizing gas such as NO_2:

$$NO_2 + 2e^- = O^{2-}{}_{adsorbed} + NO_{adsorbed}$$

which leads to a decrease in the electrical conductivity and, in terms of the environment charge zone; this results in an increase in the size of the electron free zone.

The function of a sensor with a metallic oxide base such as SnO_2 can therefore be schematized as follows: the measure of variations in the electrical resistance for a material depends on the gap at stoichiometry x, therefore the partial pressure in oxygen and the quantity of adsorbed oxygen, depends on all the oxidation-reduction reactions of the gases present.

1.1.2. *Molecular semiconductors*

A molecular crystal is shown in the form of a stack of molecules according to a precise crystallographic structure. Interactions between non-polar molecules are weak. They are of Van der Waals type and originate from the interactions between the dipoles caused by the molecules making up the crystal.

A number of molecules have been tested as the sensitive element in a semi-conductive gas sensor: polycenes, polyaniline, polyacethylene as well as other electroactive polymers, pophyrines, phtalocyanines, etc.

Among the different families, phtalocyanines (see Figure 1.2) are the most interesting for this application due to:

– their good thermal stability;

– their resistance to gases, which are also occasionally chemically aggressive (O_3, NO_2, etc.), that we want to potentially detect;

– their ability to be sublimated to form thin layers;

– the large quantity of phtalocyanines available (polymers, dimers, monomers substituted, or not, via various diverse chemical groups, etc.).

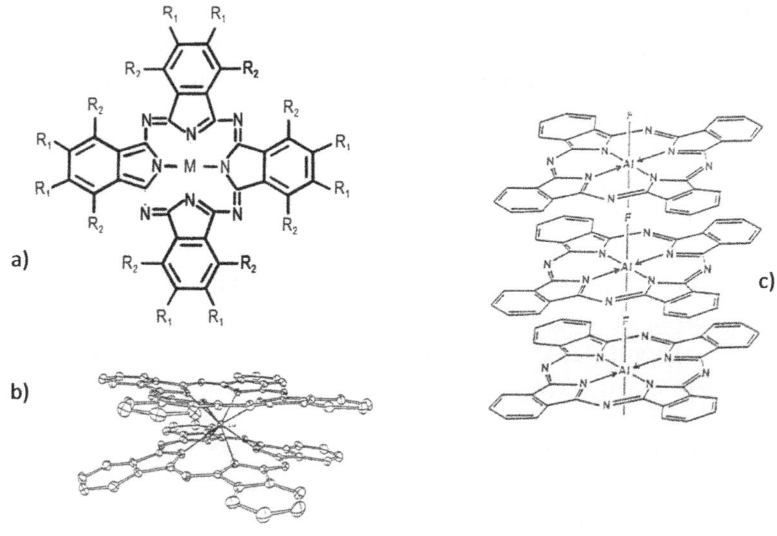

Figure 1.2. *Some phthalocyanines: a) The general form of a molecule. M is a metallic ion in a central position. R_1 and R_2 are two substitutive groups. When M = Cu and $R_1 = R_2 = H$ then the copper phthalocyanine molecule is written as PcCu. b) The diphthalocyanine lutetium molecule is written as Pc_2Lu. c) A molecule of a phthalocyanine polymer, aluminum poly-fluoro-phthalocyanine, is written as $(PcAlF)_n$*

The material's intrinsic semi-conductive property can be theoretically understood using an electrochemical model and can be summarized as follows:

The generation of charge carriers inside the crystal composed of neutral molecules A can be represented by the equation:

$$2A \Leftrightarrow A^+, A^-$$

The thermal energy ΔE_{int} required to generate, dissociate and delocalize the charges found in ionized molecules can be estimated using oxido-reduction energies from the A environments in solution $E^{ox}_{1/2,A}$ and $E^{red}_{1/2,A}$:

$$\Delta E_{int} = E^{ox}_{1/2,A} - E^{red}_{1/2,A}$$

Under the influence of an electrical field, the delocalized charges are susceptible to movement by jumping within the material from one ionized molecule to an initially neutral neighbor. They are intrinsically semi-conductive charge carriers.

In a semi-conductive gas sensor application, the simplest detection method involves measuring variations in conductivity caused by modifications in the gaseous environment of the sensitive material. It is therefore interesting to have a molecular material which we can shape into thin layers, by thermal sublimation in a vacuum, for example. Therefore, creating the sensitive structure on an isolating substrate in metal with electrodes is easy.

The potentially detectable gases are therefore likely to introduce reversible redox reactions with molecules being the sensitive element. As such, oxidization of the molecular material composed of phtalocyanine molecules PcM via electron accepting environments A can be expressed by the chemical equation:

$$PcM, A \Leftrightarrow PcM^+, A^-$$

This corresponds to the introduction of a positive charge in the thin phtalocyanine layer, which we can call a "hole" by analogy with mineral semi-conductors. This is therefore a "p-type" doping. Similarly, a reduction of the same molecular material via an electron providing space D can be expressed by the chemical equation:

$$PcM, D \Leftrightarrow PcM^-, D^+$$

corresponding to an "n-type" boosting.

All the experimental results show that thin layers of phthalocyanines are very sensitive to oxidizing gases (O_2) as well as highly oxidizing gases (O_3, NO_2, Cl_2, etc.). For example, some ppb of NO_2 in the air cause an important increase in conductivity. The experimental results also show that these same thin layers are almost unresponsive to reductive gases, such as NH_3 or CO, even if these are found in the sensor's environment in concentrations greater than 1%.

1.1.3. *Ionic conductor materials for detecting gas molecules*

The use of ionic conductors as basic elements of chemical sensors is not that recent, it dates back to the pH electrode (Cremer 1906). This was then developed with the discovery of electrodes sensitive to halides with silver based salts (Kolthoff, 1937) (see section 2.1) and more recently with the zirconia based oxygen gauge (Weissbart, 1961) which is now commonly used for controlling combustion.

To increase conductivity, these materials are generally doped by different valency elements, with electric compensation therefore being carried out by the creation of gaps. The most typical example is that given by stabilized zirconia (fluoride structure). When we dope ZrO_2 with CaO (or Y_2O_3), the doping cation substitutes Zr and, as such, anionic sites remain empty. The increase in concentration of mobile ions (O^{2-}) is limited by the solubility of the doping cation but the concentration can become high. In the case of zirconia doped to around 10%, we obtain a concentration in gaps of the order of 5 mol L^{-1}.

This doped zirconia based oxygen sensor, discovered in the 1960s and subsequently used in metal industries (to find the proportion of oxygen in cast steel) is today integrated into a number of vehicles for optimizing combustion by analyzing exhaust fumes (regulating stoichiometry in the injection of air/gas). The electrochemical chain is relatively simple:

$$\text{Pt, } O_2(P_1) \,/\, \mathbf{O^{2-}\ (zircon)} \,/\, O_2(P_2)\text{, Pt}$$

At each interface (porous platinum electrode) the following equation applies:

$$O_2 + 4\ e^-_{(Pt)} \Leftrightarrow 2\ O^{2-}_{(zircon)}$$

and the resulting voltage between the two platinum electrodes is equal to:

$$E = \frac{RT}{4F}\ln\left(\frac{P_1}{P_2}\right)$$

If one of the oxygen pressures is known (for example, in air), then the tension allows us to identify the other.

1.1.4. *Selective sorption in polymer matrices*

In balance with certain polymer matrices, gas molecules have a partition coefficient K_P:

$$K_P = \frac{C_P}{C_G}$$

where C_P is the concentration of the molecule in the polymer and C_G is the concentration of the molecule in the gas phase.

The sorption of the gas molecule by the polymer causes an inflation in the polymer matrix. The sensitive polymer must have a glass transition temperature below the sensor functioning temperature.

The sorption of the gas molecule can be detected either by the variation in induced mass (type QCM piezoelectric transducers or SAW), by varying the polymer refractive index or by varying resistivity when the polymer has been previously charged with conductive carbon particles.

The chemical interaction properties of polymers regarding gas molecules have been modeled using the linear solvation energy relationship (LSER). LSER models the sorption of gas molecules in terms of fundamental interactions: such as dispersion interactions, interactions involving dipoles, induced dipoles and hydrogen connections. The LSER equation expresses the sharing coefficient as well as the linear combination in terms representing the different types of interaction:

$$\log K_p = c + rR_2 + s\pi_2^H + a\sum\alpha_2^H + b\sum\beta_2^H + l\log L^{16}$$

The parameters R_2, π_2^H, $\sum\alpha_2^H$, $\sum\beta_2^H$ and $\log L^{16}$ characterize the properties of gas molecule solubility. The parameter R_2 is the molecular refraction parameter which gives a quantitative indication of the polarizable electrons n and π. π_2^H measures the ability of the molecule to stabilize a charge or neighboring dipole. $\sum \alpha_2^H$ and $\sum \beta_2^H$ measure the effective acidity or alkalinity respectively by hydrogen bonding and $logL^{16}$ is the liquid/gas sharing coefficient in the hexadecane at 298 K.

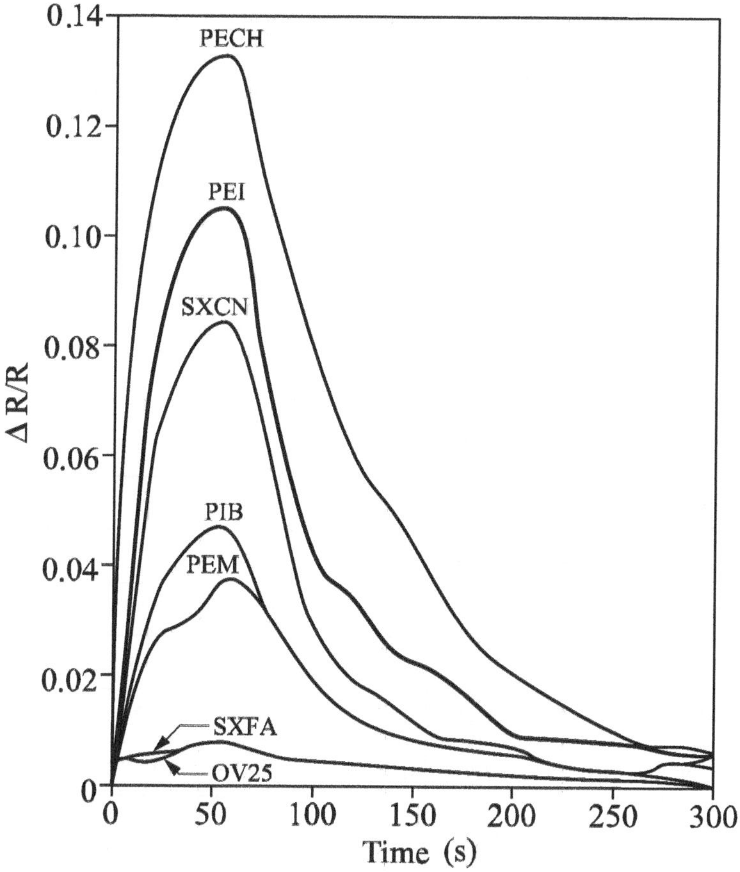

Figure 1.3. *The Response of sensors to resistors from different polymer films with the injection of chloroform vapors. PIB: poly(isobutylene); PECH: poly(epichlorohydrine); PEI: polyethyleneimine, OV25: 75% phenyl 25% methylpolysiloxane; SXCN: poly bis(cyanopropyl-siloxane; SXFA: hexafluoro2-propanol substituted polysiloxane; PEM: poly(ethylene maleate)*

The coefficients *s*, *r*, *a*, *b*, and *l* characterize the properties of the polymer as a sorbent. The coefficients *a* and *b* represent the alkalinity and acidity by the polymer's hydrogen bonding. The coefficient *s* is linked to the polymer's dipolarity/polarizability. The coefficient *l* is linked to dispersion interactions. The coefficient *r* represents the polymer's ability to interact with the pairs of electrons n and π of the gas molecule and provides an indication of polarizability. The constant *c* is an adjustment coefficient [GRA 99].

The sensor's responses to resistors from films from different polymers with the injection of chloroform vapors are shown in Figure 1.3.

The values of these different coefficients characteristic of the chloroform vapors are as follows:

$$R_2 = 0.425, \pi_2^H = 0.49, \sum \alpha_2^H = 0.15, \sum \beta_2^H = 0.02 \textit{ and } l \log L = 2.48$$

The values of the coefficients which characterize the different polymers are provided in Table 1.1.

Polymer	c	r	s	a	b	1
PIB	-0.77	-0.08	0.37	0.18	0	1.02
PECH	-0.75	0.1	1.63	1.45	0.71	0.83
OV25	-0.85	0.18	1.29	0.56	0.44	0.88
SXFA	-0.08	-0.42	0.6	0.7	4.25	0.72
SXCN	-1.63	0	2.28	3.03	0.52	0.77
PEI	-1.58	0.49	1.52	7.02	0	0.77
PEM	-1.65	-1.03	2.75	4.23	0	0.86

Table 1.1. *Solvation parameters for the different polymers. PIB: poly(isobutylene); PECH: poly(epichlorohydrine); PEI: polyethyleneimine; OV25: 75% phenyl 25% methylpolysiloxane; SXCN: poly bis(cyanopropyl-siloxane; SXFA: hexafluoro2-propanolsubstituted polysiloxane; PEM: poly(ethylene maleate)*

The strong response of the resistor found with PECH in relation to that of PIB can be explained by the high value of the coefficient *s*, the PECH polarizability. The polymer PEI compensates for the low value of 1 (dispersivity) with a strong value of *s* (polarizability) and *a* (alkalinity).

A network of six SAW sensors with different polymer films constituting an electronic nose has allowed the identification of six chemical warfare agent simulations [MAT 11].

1.2. Ionic recognition [FAB 03]

Ions, in water, are generally detected by selective electrodes (ion selective electrodes – ISEs) or by ion selective field effect transistors (ISFETS), which are based on a potentiometric measure. A difference in potential develops between the two sides of an ion exchange membrane, one side of which is in contact with a reference environment and the other side which is in contact with the environment to be analyzed.

The ion exchange membrane is a thin film (with less than a micrometer in thickness in the case of ISFETS and several hundred micrometers in the case of ISEs) made of an inorganic material or polymer material.

1.2.1. *Inorganic materials for ionic recognition*

1.2.1.1. *Insoluble silver salts*

These families of compounds (with an Ag_2S or Ag_2Se base) are not very soluble. Membranes composed of these components allow the detection of the anions S^{2-} and Se^{2-} by a simple exchange of ions. The detection limits are very low.

We can mix these salts with those of various anions (Cl^-, Br^-, I^-, CN^-, SCN^-) which therefore allow their detection. We can also mix these salts with those corresponding to various metal cations (copper, cadmium, lead), therefore enabling their detection.

Membranes with a mono-crystal base of LaF_3 allow the detection of fluoride ions, which are mobile in the crystalline network of lanthanide ions.

1.2.1.2. *Oxide or chalcogenide based glasses*

Oxide based glasses are used to detect protons (pH) and alkaline ions, mainly sodium. Their conductivity is nevertheless fairly mediocre (of the order of around 10^{-11} $S.cm^{-1}$ for electrodes in sodium), which requires thinning the membrane by about $\frac{1}{100}$ of a micrometer to lower impedance, creating a large amount of fragility. As such, the glasses' disordered structure is not very favorable to selectivity because the ions involved, whilst of a different size, diffuse across the conduction sites in the glass which are of a fairly random size.

More recently the discovery of glasses, which conduct via silver with a selenide or sulphuric base, have created new possibilities because their conductivity is better (nearly 10^{-3} S cm^{-1} at ambient temperature). These glasses allow us to analyze different spaces, particularly silver, as well as heavy metals which form sulphides or insoluble selenides.

Forming ISEs with silicate bases is achieved using relatively basic techniques (blowing the glass into a bulb shape), but these material's different deposition techniques into thin layers (evaporation, cathodic pulverization, CVD and other chemical procedures) lend themselves well to miniaturization.

1.2.1.3. *Ionic ceramic membranes*

These are oxide based materials which carry a frame composed of polyhedrons connected by covalent bonds. Ions (alkaline) can move through tunnels of 1D to 3D, whose size is uniform and can be adjusted by a chosen composition at the size of the ion to be detected. Membranes with a NASICON-type structure have, for example, been used for the selective detection of sodium and lithium. Their main advantages are their improvement and selectivity in relation to glass and their low impedance.

1.2.2. *Organic materials for ionic recognition*

Organic membranes are composed of a polymeric support containing a carrier and a complexing agent to obtain selectivity.

1.2.2.1. *Polymeric support*

In the literature there are currently a large number of polymers adapted to this role depending on the required usage for the electrode. One of the simplest solutions is the use of a plastisol carried out by mixing PVC (polyvinyl chloride) with a high or low molecular weight to a plasticizer such as dinonyl phthalate (DNP), 2-nitrophenyl octyl ether (NPOE), etc. To obtain good membranes, the plasticizer content must be composed of between 80 and 90% of the plasticizer mass. Polysiloxanes have a glass transition which is lower than that of PVC (of the order of around -50°C) and are more advantageous from this perspective. However, for industrial uses, membranes created using these materials have the disadvantage of being fragile and often being nonresistant to organic solvents. As such, membranes composed of cross-linked polymers are preferred.

1.2.2.2. *Carriers*

A simple solution is to include an ion salt in the membrane to be detected the insoluble in the water and which is soluble in the membrane layer, particularly a tetraphenylborate ($B\Phi_4^-$) from the cation to be detected. Selectivity is linked to the product of this salt solubility. Every other cation which could be linked to ($B\Phi_4^-$) will be excluded from the membrane. However, if other cations have solubility products of the same order or even more important than those of the cation to be detected, we lose the membrane's selectivity.

1.2.2.3. *Complexing agents*

In order to improve selectivity, we add a complexing agent to the membrane from the ion to be detected.

The formation of a complex between the complexing agent (ligand L) and the ion I to be detected is characterized by the complex formation equation:

$$L + I \Leftrightarrow LI$$

The formation constant of the complex K is linked to the variation in standard free energy $\Delta G°$ by the following relation:

$$\Delta G° = \Delta H° - T\Delta S° = - RT\ LnK$$

where R is the constant of perfect gases, T is the temperature expressed in Kelvin and $\Delta H°$ and $\Delta S°$ are the standard variations of enthalpy and entropy respectively.

If $\Delta G°$ is negative after the complex formation, but for values of K which are too high, (typically, $K > 10^4$, where $\Delta G° < -5.5$ kcal/mol), the reaction will be irreversible and the potential sensor will not be reversible and will function as a disometer.

The characteristics of the ion-molecule interaction recognition depend not only on the type of bonds between L and S which can differ in nature, but on electrostatic forces, hydrogen bonds, Van de Vaal bonds, hydrophobe bonds, etc., as well as complementarity in terms of size and form. As Table 1.2 shows, the stability constant of several complexes of the nickel ion strongly increases when we move from a simple sequestering agent to a chelate then to a macrocyclical receptor whose size adapts to that of the ion.

Type of ligand	Log K
hydroxide	4.6
Oxalic acid	5.3
Ethylenediaminetetraacetic acid (EDTA)	18.6
Cyclohexane diamine tetra (CDTA)	19.4
Tetraminemacrocycle 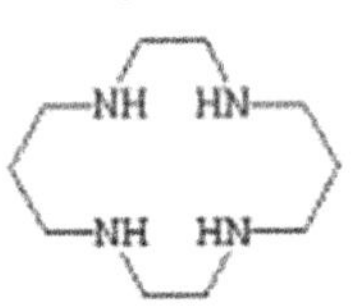	22.2

Table 1.2. *Complex stability constants for the nickel ion with different ligands*

The most common example concerning the geometry-selectivity ratio is that of crown ethers, which have very good selectivity for alkaline cation complexing.

The radius of the crown cavity adapts almost perfectly to that of the ion which will form the most stable complex (see Table 1.3).

Crown ether	Cavity radius/nm	Cation	Cation radius/nm
12-crown-4	0.06 – 0.075	Li^+	0.76
15-crown-5	0.086 – 0.092	Na^+	0.1
18-crown-5	0.134 – 0.143	K^+	0.14
21-crown-7	0.17 – 0.21	Cs^+	0.17

Table 1.3. *Comparison of crown ether cavity radii and alkaline ion radii showing the most stable complex. "15-crown-5" represents the component where the crown is composed of 15 atoms of which 5 are oxygen heteroatoms*

1.2.2.4. *Biologically derived sequestering agents*

Trans-membrane ion transporters in the cells have a large amount of ionic selectivity. This property has been used to create selective membranes. For example, valinomycin (the structure shown in Figure 1.4) a membrane transporter for potassium is extracted from a mushroom, *streptomycesfulvissimus*. It has excellent selectivity for the potassium ion (log K = 4.90) in relation to the sodium ion (log K = 0.67).

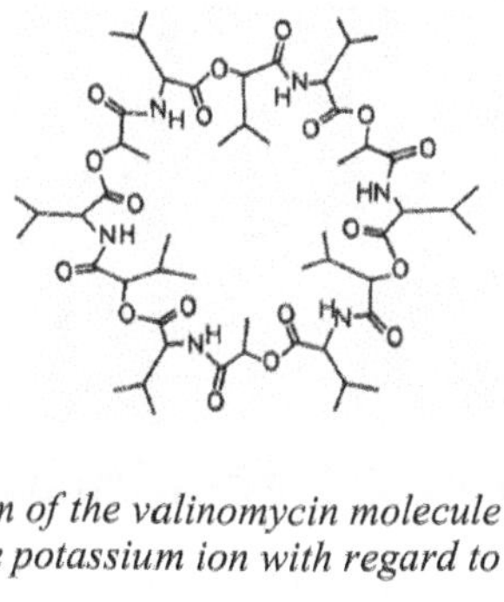

Figure 1.4. *Diagram of the valinomycin molecule showing excellent selectivity for the potassium ion with regard to the sodium ion*

1.3. Biological recognition [TRA 91]

1.3.1. *Enzyme biosensors*

Enzymes are classed according to families corresponding to a type of catalyzed reaction. Oxidases, reductases and hydrolases are the most commonly used enzymes in biosensors.

The hydrolysis reaction of butyrylcholine catalyzed by the butyrylcholinesterase enzyme (schematized in Figure 1.5) is shown here:

$$\underset{\text{butyrylcholine chloride}}{CH_3\text{-}(CH_2)_2\text{-}COO(CH_2)_2\text{-}N(CH_3)_3^+Cl^-} + H_20 \xrightarrow{BuChE} \underset{\text{butyric acid}}{CH_3\text{-}(CH_2)_2\text{-}COO^-} + H^+ + \underset{\text{choline}}{HO\text{-}(CH_2)_2\text{-}N(CH_3)_3^+Cl^-}$$

The enzyme is locked on a transducer which allows the detection of either the consumption of a co-factor entering into the enzyme reaction (example: oxygen in the case of oxydases) or the appearance of a product of the enzyme reaction. Generally, we bond a hydrolase to a potentiometric, a conductimetric transducer or an optic fiber, which allows us to detect local changes in pH or conductivity. Reductase or oxidase enzymes are bonded to an amperometric or conductimetric transducer, which allows us to record electronic transfers. These electronic transfers towards the electrode are favored by the use of redox mediators which enable us to reduce the imposed potential and which limit interference with other electroactive spaces.

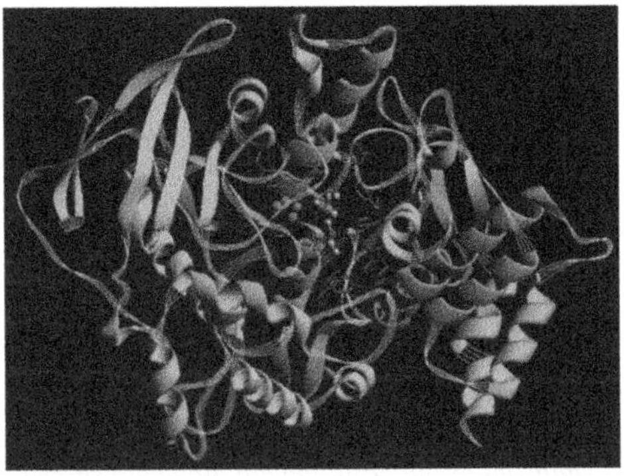

Figure 1.5. *Representation of the butyrylcholinesterase enzyme*

One of the main contributions of enzyme biosensors to ecotoxicological studies concerns the evaluation of aquatic neurotoxicity. This comes from organophosphoric components and carbamates, used as pesticides but also heavy metals and detergents from the inhibition of a particular group of enzymes, esterases. The neurotoxicity biosensors proposed in the literature are mainly examined on the basis of the two enzymes belonging to this family, acethylcholinesterase and butylcholinesterase.

Current developments aim to improve enzyme activity, either by genetic modification or by better immobilization on the transducer.

1.3.2. *Immunosensors*

Antibodies are the soldiers of the immune system. They are produced specifically to bind to the toxins attacking the living organism (poison, virus, bacteria, etc.), which we call antigens. A type G immunoglobule antibody is shown in Figure 1.6. The recognition part is found in the heavy chains. These antibodies are fixed on the transducer's surface which will allow the detection of the formation of the antigen-antibody complex.

The main methods used in immunosensors are:

– direct detection methods for the formation of the antigen-antibody complex (measures the modification of electric properties by electrochemical impedance spectroscopy, measures the modification of optic properties by surface plasmon resonance);

– methods of marking the fixed antigen with an antibody with a leading enzyme, either with colored or fluorescent products (optic transducer), or with electroactive products (electrochemical transducer).

A large number of immunosensor applications involve identifying pollutants or targeted groups of pollutants in the environment.

It is also possible to use them for the detection of substances called biomarkers in medical diagnoses as well as in the environment.

Vitellogenine, for example, is a phospholipo-glyco protein from the serum secreted by a fish exposed to large quantities of endocrine disruptors. Its presence is therefore an excellent biomarker enabling the identification of oestrogenotoxic identification of natural anthropogenic substances. It is detected using electrochemical, optical or piezoelectric biosensors with an anti-vitellogenine carp antibody base [BUL 07].

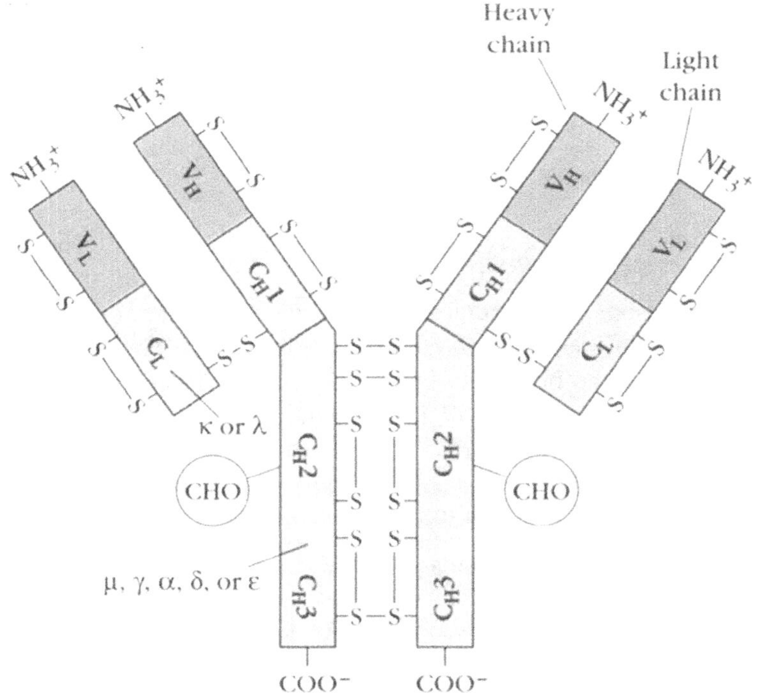

Figure 1.6. *Illustration of a type G immunoglobine (IgG)*

1.3.3. *Oestrogenic receptor biosensors*

Endocrine disruptors are chemical substances causing hormonal disorders and anomalies in the endocrine system, even the nervous system. Some affect the synthesis of endogen hormones or their receptors. Others have a neighboring structure for oestrogens linked to their receptors leading to inactivity or their anormal functioning.

A large number of molecules have these effects such as hormone synthesis or chemical substances such as phthalates, surfactants, PCBs, alkylphenols, parabens, HAPs, dioxins and some pesticides. To date, 320 priority substances likely to disturb the endocrine system have been listed by the European Union. Some of them (nonylphenol, di-2-ethylhexylphtalate and diphenyl-polybromide ethers) have been included in the list of priority substances by the European Union Water Framework Directive. Toxicity biosensors use the fixation abilities of these substances on immobilized oestrogenic receptors on a transducer's surface. The human oestrogenic receptor is the most commonly used. Different methods of transductions such as fluorescence, cyclical voltametry, SPR and electrophoretic mobility have been proposed.

1.3.4. *DNA and aptamer biosensors*

The structure of DNA is extremely sensitive to the influence of environmental pollutants such as heavy metals, polycyclical aromatic components and PCBs. These substances are known for having a strong relationship with DNA, which is at the basis of mutagenesis and carcinogenesis. Biosensors, based on measuring the interaction between these substances and simple DNA molecules or double immobilized strands on a transducer, have been developed and used for evaluating aquatic genotoxicity. Detection normally occurs when using electrochemical methods. DNA components are detected, either directly when there are electroactive spaces or by modifying the electrochemical DNA signal. Toxicity biosensors based on the detection of oxidization in DNA bases (essentially guanine but also guanosine and adenosine) or the breakdown of strands using an electrochemical bore have been developed and applied to the analysis of water containing different types of genotoxic aquatic contaminants (metals, pesticides, PCBs, aromatic amines, etc.).

Aptamers are specific sequences of ologonuclear acids which can be bonded closely to targeted molecules such as proteins. These are

NRA or DNA molecules (between 30 and 100 nucleotides) which are selected *in vitro* for a given molecule using the SELEX procedure from a large number of sequences selected at random. Aptamers are often called "synthetic antibodies". Aptamers have advantages in relation to antibodies, particularly in terms of stability, regeneration and production facility once the sequence has been defined. These aptamers have been used to develop biosensors and there are a number of reviews examining this topic [SAS 09].

1.3.5. *Cellular biosensors*

There is a large amount of literature and reviews focusing on cellular biosensors, which are used to identify analytes or specific families of analytes, or to evaluate aquatic toxicity.

The main organisms used to develop cellular bionsensors are bacteria, algae and yeast. They have the advantage of being much more resistant than their isolated components (enzymes, antibodies, DNA) and can be much more easily adapted to modifications in pH and temperature. There is an increased number of enzymes and decofactors in the cell which can react to a wide range of substrates, opening the way for the development of biosensors allowing the detection of a large number of analytes, sometimes to the detriment of selectivity. These organisms can nevertheless be easily genetically modified to limit their response to a given component or a family of components.

The transduction methods used for this type of biosensor are essentially electrochemical and optical. Bacteria, yeast, algae and fish cells have also been used to develop toxicity biosensors. The recorded response can result from a modification in cellular metabolism, altering within the cell and potentially resulting in its death or a change in the expression of certain genes in the case of modified organisms. Different types of toxicity also exist: neurotoxicity, genotoxicity, oestrotoxicity, acute toxicity [LAG 11].

1.3.6. *Immobilization methods for a bioreceptor*

An essential point for producing a biosensor is the immobilization of active biological entities (enzymes, antibodies, cells or tissue etc.) on the transducer's surface.

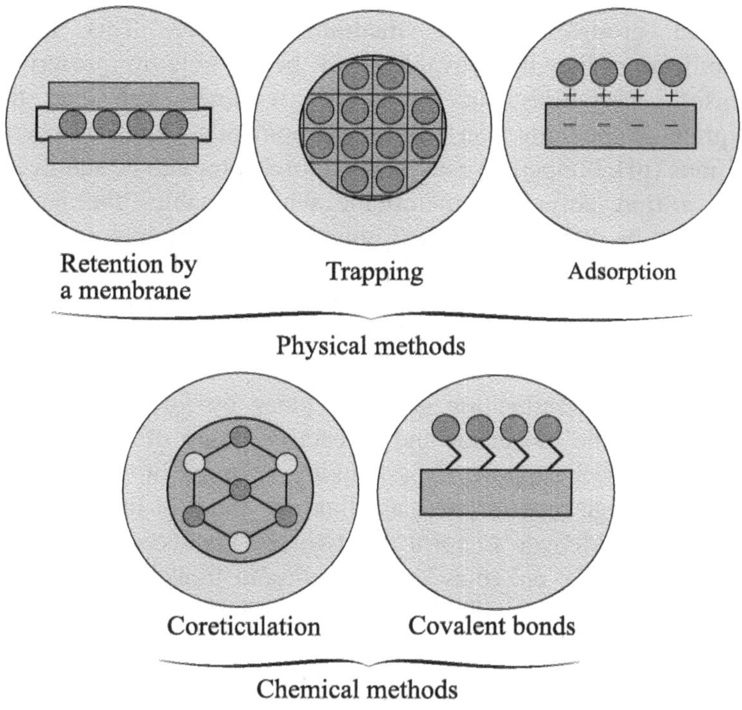

Figure 1.7. *Schematic representation of different immobilization methods for the biological metabolic element*

Different immobilization procedures for biosensors have been developed in order to obtain effective and stable immobilization, to keep the bioreceptor's biological properties intact and to guarantee maximal accessibility and reactivity. The selection of an appropriate immobilization method depends on the type of biological molecule, the type of transducer being used, the analyte's physical-chemical properties and the biosensor's operating conditions. There are several immobilization techniques that we can subdivide into physical and chemical procedures (Figure 1.7).

Physical immobilization can take place by adsorption. Adsorption is based on the establishment of ionic, polar or hydrophobic low energy interactions or even hydrogen bonds between the biosensor's functional groups and the medium's surface. This type of immobilization has the advantage of being able to preserve the biosensor's properties but the bond's weakness can lead to its desorption depending on the environment's physical-chemical conditions (pH, temperature etc.). The bioreceptor can be stabilized by adding a thin polymeric membrane which enables the particular molecule to be diffused. Physical immobilization by trapping consists of incorporating the biologically active molecule into an organic (naturally or synthetically derive hydrophone polymer) or inorganic (sol-gel) matrix.

Chemical immobilization can take place by forming a chemical bond between the biosensor and a reactive group of immobilization agents or a direct support. Cross linking is based on the interaction between a charge protein (such as albumin) and a bi-functional agent such as glutaraldehyde to form a network. This second technique is used mainly to fix enzymes or antibodies to transducers, it is used much more rarely for cells. Reaction conditions, which are often too aggressive, can cause modifications in the integrity of the cell's structure and lead to the loss of enzymes as well as a reduction in activity.

Covalent grafting is based on the reaction between a functional grouping of the active entity and a functional grouping of the previously activated medium. Functional groups available at the bioreceptor form lateral chains of protein amino acids (enzymes, antibodies, or surface cellular proteins), notably lysine ε-amino, aspartate and glutamate carboxyls, cysteine sulfhydryland tyrosine hydroxyphenyl groups.

To ensure the formation of covalent bonds with the transducer, we proceed to its functionalization. As such, for example, metallic surfaces such as gold and silver can be functionalized with amino, hydroxyl or carboxyl groups with aminoalkanethiols, hydoxyalcanethiols and carboxyalkanethiols simultaneously. Oxide surfaces are functionalized by organosilanes. More recently, metallic electrodes on which functionalized polymer conductor films are placed (polypurrole, polythiophene, polyaniline) have been used to immobilize active biomolecules by forming covalent bonds. Diazonium salts enable the electrofunctionalization of conductive surfaces (gold, diamond, etc.).

The use of new materials based on nanoparticles of gold, silver, iron or carbon nanotubes has enabled the significant improvement of the sensitivity of enzyme biosensors in terms of optical and electrochemical transduction. Methods for cell immobilization by trapping them inside a polymeric network or by grafting using functionalized support with antibodies by antigen-antibody interactions are the most common because they are mild and enable the micro-organisms' metabolic activity to be preserved.

1.3.7. *A biomimetic approach: molecularly imprinted polymers (MIPs)*

Molcular imprinting can be defined as the formation of specific recognition sites (with bonding or catalytic properties) in a material from its interaction with a "template" molecule, where it directs the positioning and orientation of the material's structural components by an auto-assembly mechanism (see Figure 1.8).

The material itself can be composed of a linear sequence or a polymerical sequence (organic or inorganic silicon gel-based imprinted MIPs) or a 2D surface network (grafted monolayers). MIPs have a number of advantages when compared with natural bio-molecules: they can be prepared for specific molecules (the best results have been obtained for molecules with a molecular weight of between 200 and 1,200 Da), they are stable for a large range of pH, pressure and temperature, can function as organic solvents, are insoluble in water, can easily create multisensor networks and are compatible with micro-production techniques.

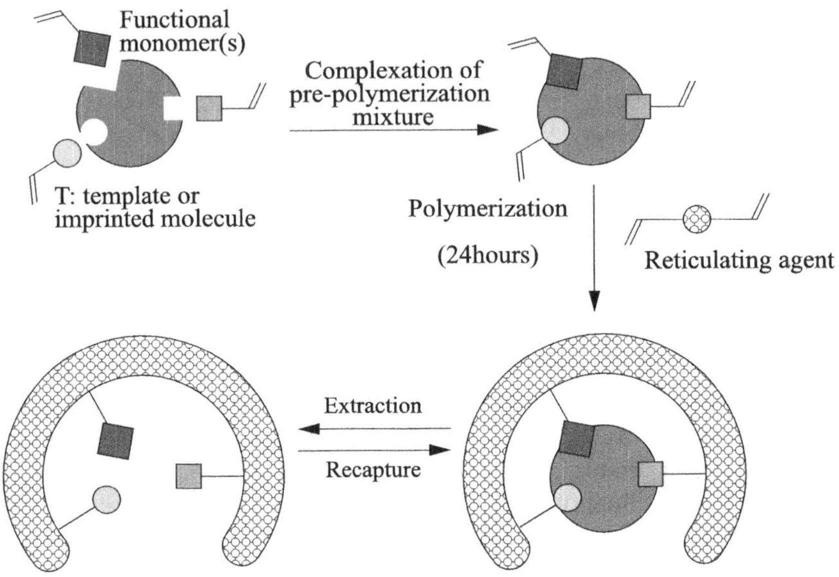

Figure 1.8. *The principal of molecular imprinted polymer fabrication*

The immobilization of MIPs on the transducer can be carried out either by *in situ* polymerization using a photochemical or thermal initiation or by surface grafting with chemical or UV initiation. Some polymers can be electropolymerized on the transducer's surface. For example, electropolymerization of phenylenediamine with glucose has been successfully used to construct an imprinted sensor for capacitive detection of this component which has been the focus of a number of recent reviews [WHI 11].

1.4. Bibliography

[BUL 07] E. BULUKIN, V. MEUCCI, M. MINUNNI, C. PRETTI, L. INTORRE, G. SOLDANI, M. MASCINI, *Talanta*, vol. 72, pp. 785-790, 2007.

[FAB 03] P. FABRY, J. FOULETIER, *Microcapteurs chimiques et biologiques: Applications en milieu liquide*, Hermès, Paris, 2003.

[GRA 99] J.W. GRATE, *Anal. Chem.*, vol. 71, pp. 4544-4553, 1999.

[LAG 11] F. LAGARDE, N. JAFFREZIC-RENAULT, *Analytical Bioanalytical Chemistry*, vol. 400, pp. 947-964, 2011.

[MÉN 08] F. MÉNIL, *Microcapteurs de gaz*, Hermès, Paris, 2008.

[MAT 11] D. MATATAGUI, J. MARTI, M.J. FERNANDEZ, J.L. FONTECHA, J. GUTIERREZ, I. GRACIA, C. CANÉ, M.C. HORRILLO, *Sensors and Actuators B*, vol. 154, pp. 199-205, 2011.

[SAS 09] A. SASSOLAS, L.J. BLUM, B.D. LECA-BOUVIER, *Electroanalysis*, vol. 21, pp. 1237-1250, 2009.

[TRA 91] C. TRAN MINH, *Les Biocapteurs: Principes, Construction et Applications*, Masson, Paris, 1991.

[WHI 11] M.J. WHITCOMBE, I. CHIANELLA, L. LACOMBE, S.A. PILETSKY, J. NOBLE, R. PORTER, A. HORGAN, *Chem. Soc. Rev.*, vol. 40, pp. 1547-1571, 2011.

Chapter 2

Adsorption Phenomena

2.1. The surface of a solid

The notion of a form which is normally associated with a surface is characteristic of a solid.

In crystals, each solid can be identified by the arrangement of atoms or molecules composing it. This arrangement, which is specific to each solid, constitutes a solid phase.

Generally, this kind of structure (the position of atoms, cohesion energy, etc.) is determined by the finite crystal hypothesis, which involves the same environment for all atoms. Next to the surface, this condition is no longer verified and there is a new local organization of atoms or charges.

In the case of ionic environments, in particular, and to respect local electroneutrality, it is often necessary to examine the solid's environment. A balance is established between the material and the different phases in contact with it.

Chapter written by René LALAUZE.

This therefore raises the notion of interface, a specific area of the solid from which all kinds of interactions between the solid and the different environmental components in contact with it originate.

Depending on the nature of these components, we talk about solid-solid, solid-liquid or solid-gas reactions.

To model the solid-gas interactions which we will examine in further detail, it is important to firstly imagine a simple bonding of a gas molecule on a solid. This fixed molecule could maintain a neutral character in relation to its support or it could react with it.

In our first hypothesis, the phenomena will be an adsorption phenomena which will constitute the overall reaction. This phenomenon has a reversible character. We therefore talk about adsorption-desorption (see Figure 2.1a, b and c).

In the second hypothesis, the adsorption phenomenon will be the first stage in a more complex process. This phenomenon therefore has a non-reversible character (see Figure 2.1c), causing the formation of a new GS compound, for example.

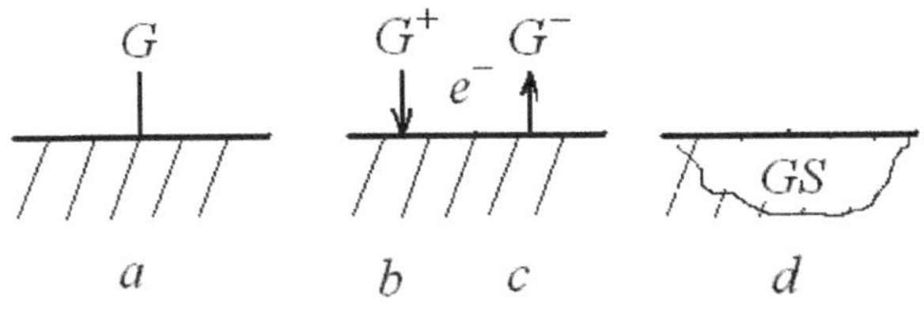

Figure 2.1. *Schema of different types of interactions between a gas and a solid: a) a physisorption phenomenon; b) a chemisorption phenomenon with electronic transfer from the gas to the solid; c) chemisorption phenomenon with electronic transfer from the solid to the gas; d) irreversible reaction*

The type of phenomenon observed will depend on the thermodynamic conditions (pressure, temperature etc.) and the chemical affinity of the elements present.

Regarding adsorption phenomenon, it is possible to further distinguish between physisorption (a) and chemisorption (b and c). Chemisorption is characterized by a simple transfer of electrons between the gas, the physisorbed state and solid, with the transfer occurring during a reversible chemical reaction between the two constituents (see Figure 2.1b and c). The appearance of chemisorption phenomena is again directly linked to thermodynamic conditions.

2.2. Examining the adsorption phenomenon

The most commonly used and explicit techniques for identifying and quantifying the adsorption phenomenon are manometry and thermogravimetry.

Manometry

In a closed system, the bonding of a gas molecule to a solid contributes to decreasing the partial pressure of this gas and a simple measure of the variation of this pressure in the reactor allows us to access the desired information.

In a test, we have two vessels, A and B (see Figure 2.2), where band A is linked to a device which allows us to measure the pressure in band A or band A+B when they are connected by a tap, R1. Gas is introduced into band A by a tap, R2, under a specific pressure in Pa. The solid sample is placed under band B. A simple release of the gas in band A+B allows us to measure the amount of pressure P(a+b).

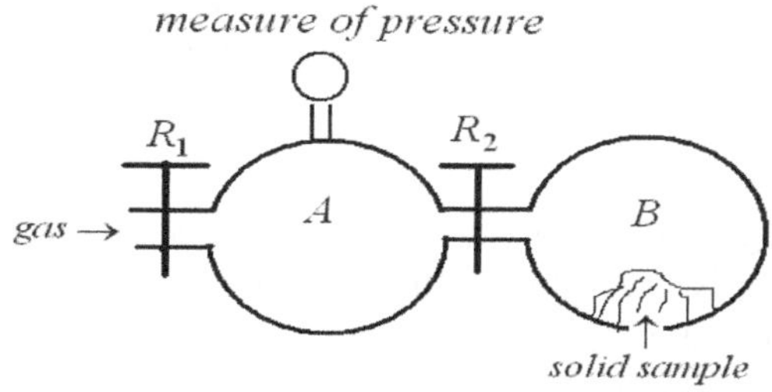

Figure 2.2. *Schema of a device measuring adsorption by manometry*

Generally, the number of gas molecules introduced N can be expressed:

– either by:

$$n_1 = P_A V_A$$

when band A is isolated from band B;

– or by:

$$n_2 = P_{A+B}(V_A + V_B)$$

after the gas is held in band A+B.

V_A and V_B indicate the respective volumes of the band A and B. In the absence of a solid sample in band B, we naturally find that:

$$n_1 = n_2$$

With the sample, we generally see that:

$$n_1 > n_2$$

The difference $n_1 - n_2$ represents the amount of gas fixed on the solid.

This experiment examining different gas pressures allows us to obtain adsorption isotherms for a given temperature $n = f(P)$.

Thermogravimetry

Bonding a gas molecule to a solid allows us to vary the solid's mass and simply weighing them allows us to directly obtain information on the fixed quantity, depending on the system's parameters. This method allows us to easily vary the phenomenon's reversible character (see Figure 2.3).

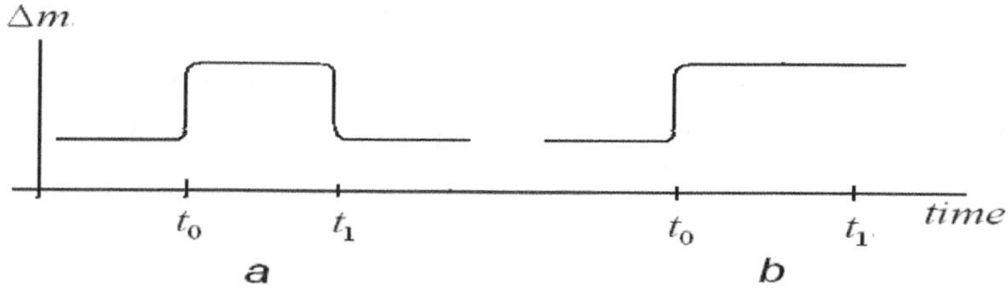

Figure 2.3. *Evolution of the increase in mass Δm for a solid sample subject to variations in pressure for $t < t_0$ $P = P_0$, for $t_0 < t < t_1$: $P = P_1 > P_0$, for $t > t_1$: $P = P_0$: a) example of a reversible phenomenon; b) an irreversible phenomenon*

2.3. Forces intervening between a gas molecule and the solid's surface

2.3.1. *Overview of van der Waals forces*

To interpret the origin of the physisorption phenomena observed at very low temperatures, we can, in parallel with research that has already been led on these intermolecular interactions, use van der Waals forces.

These forces, F, derive a potential φ:

$$\vec{F} = -\vec{grad}\,\varphi$$

These potentials are additive and the overall potential is the sum of the attraction potential φa and a repulsion potential φr, where:

$$\varphi = \varphi a + \varphi r$$

with:

$$\varphi_a = -\frac{C}{r^6} \text{ and } \varphi_R = \frac{B}{r^n}$$

r represents the distance between molecules.

The constant C is the sum between the three terms:

– Keesom's potential only applies to molecules which have a permanent dipolar moment, originating from the attraction between the molecules' permanent dipoles.

– Debye's induction potential, which comes from the molecule's polarizibility, is caused by the attraction between permanent dipoles and dipoles induced by the former's presence.

– London's dispersion potential comes from the molecule's instantaneous dipolar moment; this is often the most important.

In terms of the expression of φr, this is an empirical form for which we frequently adopt the following value:

$$n = 12$$

The global potential φ between two molecules can be expressed as:

$$\varphi = \frac{B}{r^{12}} - \frac{C}{r^6}$$

By expressing the fact that the potential has a minimal value φ_0 for the equation position r_0, we have: $\varphi = \varphi_0 \left[2\left(\frac{r_0}{r}\right)^6 - \left(\frac{r_0}{r}\right)^{12} \right]$

2.3.2. *Expression of potential between a molecule and a solid*

If we assume that all the molecules n of the supposed semi-infinite crystal interact with the gas molecule, and if we admit the additivity of these potentials, we have a total potential Φ for the solid which is expressed by:

$$\Phi_a = \sum_n \varphi_a \quad \text{for the attraction potential}$$

and:

$$\Phi_R = \sum_n \varphi_R \quad \text{for the repulsion potential}$$

These summations can be replaced by integrals of the following type:

$$\Phi_a = \int \varphi_a dn \text{ and } \Phi_R = \int \varphi_R dn$$

with:

$$dn = \mathrm{Ndv}$$

where N indicates the number of molecules per unit of volume.

The volume element taken for integration is between the radii of the spherical caps r and $r+dr$, (see Figure 2.4), where:

$$dv = S\,dr = \Omega\, r^2\, dr$$

Ω represents the solid angle and if α is the maximum angle of the sphere's radius with the normal on the solid's surface, we have:

$$\Omega = 2\,\pi\,(1 - \cos\,\alpha)$$

with $\cos\,\alpha = Z/r$; Z represents the distance of the molecule G from the solid.

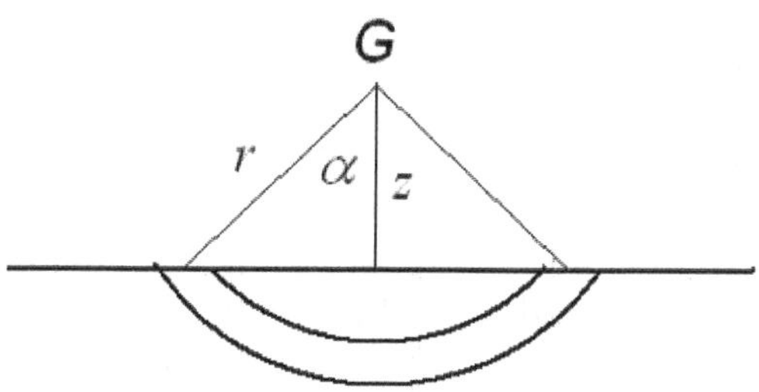

Figure 2.4. *Integration domain in the solid*

This occurs for the attraction and repulsion potentials respectively:

$$\Phi_a = -\frac{\pi NC}{6Z^1} \text{ and } \Phi_R = -\frac{\pi NB}{45Z^9}$$

As such, in the case of interaction between a gas molecule and a solid, a van der Waals potential in $1/r^3$ becomes an attraction potential in $1/r^6$ and a potential in $1/r^{12}$ becomes a repulsion potential in $1/r^9$. The solid appears to be a thousand times more attractive or repulsive than a single molecule.

2.3.3. *Chemical forces between a gas environment and the surface of a solid*

This physical chemical aspect, particularly important for explaining how chemical sensors function, will be further developed in subsequent chapters. For the moment, we will simply say that if the gas atom has single electrons there is the possibility of establishing a chemical bond between the gas and the solid and that in these two cases, the limits of polarization can be observed as:

$$G + e^- \leftrightarrow G^-$$

or:

$$G \leftrightarrow G^+ + e^-$$

2.3.4. *Distinction between physical and chemical adsorption*

The distinction between physical and chemical adsorption lies in the difference between the types of forces in the gas molecules on the solid's surface. If we study the curve $\Phi = f(r)$, we see that it has a minimum defined by Φ_0 and r_0. In the case of physical adsorption (see Figure 2.5), the value of Φ_0 is less than for chemical adsorption (1 instead of 5-6 Joules/moles); on the other hand, r_0 is weaker for chemisorption. Lastly, physical adsorption can be represented as a non-activated phenomenon and is, by extension, spontaneous and susceptible to occur at very low temperatures. In contrast, chemisorption is an activated phenomenon and the curve $\Phi = f(r)$ firstly reaches a maximum marked by an activation energy E_A. This activated character is linked to the fact that the transfer of electrons from the gas to the solid requires an input of energy; this therefore involves a kinetic process.

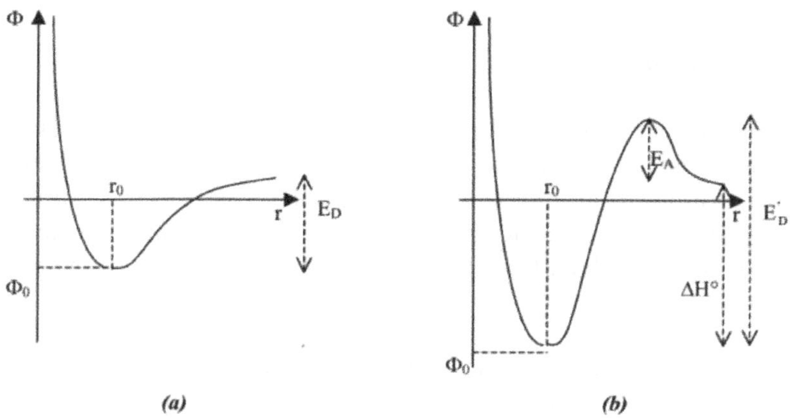

Figure 2.5. *Appearance of curves $\Phi = f(r)$: a) example of physisorption; b) example of chemisorption*

2.4. Thermodynamic study of physical adsorption

2.4.1. *Different adsorption models*

To build a thermodynamic model of physical adsorption, it is important to refer to experimental results. Isothermal adsorptions obtained by manometric or thermogravimetric methods allow us to verify that the quantity n of fixed molecules is a function of gas pressure and temperature. These results imply a divariant system. As such, all the proposed models should at least verify this essential condition.

Note that the adsorption phenomenon cannot in any case identify a simple condensation phenomenon. Condensation is in effect a monovariant phenomenon which results from the saturation effect of the gas phase. Saturation is reached for a pressure P which is greater or equal to the saturating vapor pressure of the gas P_0 which depends only on temperature. In contrast, the adsorption phenomenon is observed for gas pressures which are lower than P_0.

It is on the basis of such considerations that different thermodynamic models have been proposed.

2.4.2. *Hill's model*

In order to account for divariance, Hill considers that surface effects must be taken into account. As such, he supposes that the adsorption layer, i.e. the adsorbent + adsorbed set can be assimilated to a solution where the adsorbent is made up of free sites on the solid surface and for which the adsorbed set is constituted of gas molecules fixed on these sites.

In this case, the possible variables are pressure, temperature, the number of adsorbent moles n_s and the number of adsorbed moles n_a.

The number of independent constituents is 2 (adsorbent + adsorbed + gas – an equal relation between these three constituents).

The number of exterior parameters is 2 (*P* and *T*).

The number of phases is 2 (solid and gas).

With a variance v equal to:

$$v = 2 + 2 - 2 = 2$$

We can therefore indicate:

– isotherms $n_a = f(P)$ $T =$ constant;

– isobars $n_a = f(P)$ $P =$ constant;

– isosteres $P = f(T)$ $n_a =$ constant.

2.4.3. *Hill and Everett's models*

In contrast to the previous model which focuses on the surface and its surface spaces n_s, Hill and Everett convey these surface effects using a physical parameter, the solid's force field.

This model proposes that the adsorption phenomenon can be described using a localized condensation phenomenon, which tends to gradually cover the whole surface as gas pressure increases. A variation in the amount of coverage is conveyed by a variation in the contact energy between the two phases. This energetic term can be expressed by the product of an extensive size A_S, which is the contact surface between the phases and an intensive size P_S, which we can assimilate with a surface pressure. The term P_S is identified by the variation in the superficial tension coefficient γ before and after covering:

$$P_S = \gamma^0 - \gamma$$

In this case, there are three exterior parameters: pressure, temperature and surface pressure. If we consider that the number of independent constituents is equal to one with variance being expressed as:

$$v = 1 + 3 - 2 = 2$$

The number of components is identified at 1 as:

a solid + a gas – an equilibrium relationship

For $P = P_0$ (saturating vapor pressure), we can say that the whole surface is covered and that the solid's force field no longer varies. We therefore have the conditions for simple condensation where:

$$v = 1 + 2 - 2 = 1$$

2.5. Physical adsorption isotherms

2.5.1. *General remarks*

Physical adsorption isotherms can be obtained experimentally using either manometric or gravimetric methods. Before using the theoretical expressions taken from Hill and Hill and Everett's hypotheses, we should make a few remarks about the mobility of adsorbed layers on the solid's surface.

Two examples can therefore be envisaged:

– the first example concerns adsorbed molecules in the form of mobile layers;

– the second case concerns molecules which are perfectly localized on the solid's sites.

These limited examples are justified when we consider that the surface potential of a solid is not uniform and that it is often linked to the crystalline network's periodicity and the nature of the elements composing the solid.

This periodicity (see Figure 2.6) can be described by sites E_1 with weaker energies, separated by higher energy positions E_2. The probability that a molecule will move from one stable site to another is therefore proportional to $E_2 - E_1$. The difference $E_2 - E_1$, represents the energy barrier linked to the displacement of space on the surface.

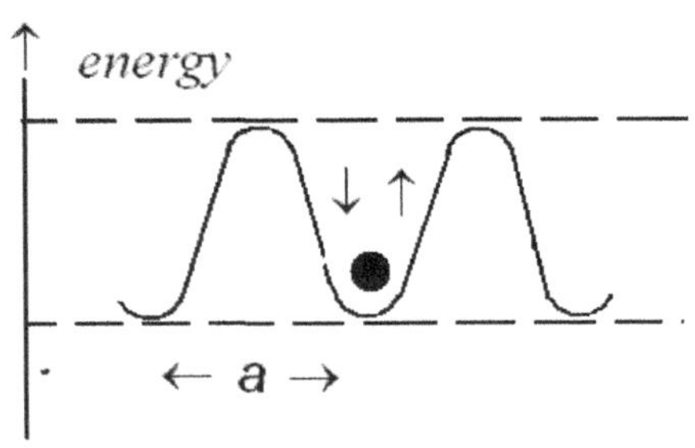

Figure 2.6. *Periodic variation in surface energy for a crystal for parameter a*

If $kT >> E_2 - E_1$, the jump probability is high and we say that the layer is mobile.

If $kT << E_2 - E_1$, the jump probability is weak and we say that the layer is localized.

With this classification, we can examine the examples of adsorption in mono-layers and the example of adsorption in multiple layers.

2.5.2. *Adsorption isotherms in mobile mono-layers*

In this case, we have a perfect analogy with the solutions. We can therefore apply the Hill model. The isotherm equation is that previously obtained:

$$n_S = \alpha P$$

However, this model does not account for interactions between adsorbed molecules. Generally, the hypothesis for diluted solutions will be applicable to adsorption phenomena; as such the thermodynamic model for solutions will be rarely used.

2.5.3. *Adsorption isotherms in localized mono-layers*

The adsorbed molecule is fixed on a weak energy position. This position is an active site. The solid's surface is therefore formed from a network of active and identical sites, which we will assimilate with the active spaces indicated by s.

Adsorption can therefore be formulated as a genuine chemical reaction for which we will use the following notation:

– $[A]$ expressed a compound A in the gas state;

– $<< A >>$ expresses that the component A is one of the constituents of a solid or liquid;

– $< A >$ expresses that the component A is pure in a solid or liquid phase.

In our example, the adsorption reaction is written as:

$$[G] + << s >> \quad \Leftrightarrow \quad << G - s >>$$

$<< G - s >>$ appears as the new space, we therefore have a solution G-s in s. This model is perfectly compatible with Hill's model.

To find the isotherm's equation, we will use the two complimentary methods.

2.5.3.1. *Thermodynamic method*

We apply the mass action law to the previous equation and obtain:

$$K = \frac{[<< G - s >>]}{P[<< s >>]}$$

K represents the equation's constant, which is expressed as:

$$K = K_0 \exp(-\frac{\Delta H^0}{RT})$$

In the case of adsorption, the phenomenon is exothermic: this implies that $\Delta H° < 0$. The value of $\Delta H° < 0$ implies that the equation is displaced towards the left if the temperature increases. The adsorption phenomenon is therefore favored at low temperatures.

In the hypothesis of an ideal solution, if S is the number of free sites and S_0 is the total number of sites, θ represents the fraction of occupied sites, with:

$$\theta = \frac{S_0 - S}{S_0} = [\ll G - s \gg]$$

therefore:

$$[\ll s \gg] = \frac{S}{S_0} = 1 - \theta$$

and we obtain:

$$K = \frac{\theta}{(1-\theta)P}$$

with:

$$\theta \frac{KP}{1+KP}$$

This relation constitutes the Langmuir isotherm, the degree of coverage is a homographic function of pressure. This function allows us to represent the majority of experimental results relating to adsorption.

We note that for low levels of coverage ($\theta \ll 1$), we have a Henry solubility-type proportionality law:

$$\theta = KP$$

2.5.3.2. *The kinetic model*

The equation can be obtained by considering the result of two dynamic phenomena as zero:

– adsorption which occurs at a speed of V_F;

– desorption which occurs at a speed of V_D.

The two phenomena can be considered as elementary acts which do not intervene in intermediary spaces.

In this case, the speed V_F is proportional both in terms of the number of impacts υ, i.e. the number of molecules hitting the surface per unit of time, and the number of spaces available on the surface.

Take:

$$V_F = \alpha\upsilon(1-\theta)$$

with:

$$\upsilon = P / \sqrt{2\pi mkT}$$

then:

$$V_F = \alpha' P (1 - \theta)$$

The speed Vd is proportional to the number of already adsorbed molecules:

$$V_D = \beta \theta$$

in equilibrium, $V_F = V_D$. We have:

$$\theta = \frac{KP}{1+KP}$$

with:

$$K = \frac{\alpha'}{\beta'}$$

2.6. Chemical adsorption isotherms

As previously stated, only chemical adsorption phenomena form the basis of electric perturbation. As such, these phenomena constitute the basis of how the resistive or potentiometric sensors, in which we are interested, function. Given the theoretical importance of this aspect, it will not be examined in this chapter. However, we will provide some information on this subject.

The form of chemisorption isotherms is asymptotic and Langmuir-type equations are perfectly compatible with the experimental results. However, in this case, the coefficient K, i.e. the equation's constant, can itself be a function of the covering degree. We have seen in effect that the equation's constant depends on molar adsorption heat ΔH_0. This is not necessarily constant for two main reasons:

– the first concerns a possible heterogeneity on the surface from an energetic point of view;

– the second concerns surface charge effects themselves, which are induced by chemisorption of the spaces concerned.

To resolve these different issues, we can adopt a distribution of sites characterized by their adsorption temperature. We therefore apply to the sites from the category I and the adsorption temperature ΔH_i the Langmuir theory, such that:

$$\theta_i = \frac{K_i P_i}{1 + K_i P_i}$$

We can therefore propose a type of energetic distribution for the sites.

We assume that adsorption heat varies with degree of coverage. This hypothesis is perfectly compatible where electric surface effects are involved. The rigorous relation, which exists between the degree of coverage and the adsorption temperature, is not simple. We can however, by averaging some simplified hypotheses, opt for simplified solutions. As such, laws of linear variation of type $\Delta H = \Delta H_0 (1 - \beta\theta)$ can be proposed. We therefore obtain an expression of the following kind:

$$Ln\ P = Ln\ (\theta_\infty/1 - \theta_\infty) - Ln\ A - \Delta H_0 \beta \theta_\infty / RT$$

For average coverage including $\theta_\infty = 0.5$, the first term is negligible and we have:

$$\theta_\infty = (RT/\Delta H_0 \beta)\ \mathrm{Ln\ AP}$$

This is the Temkin isotherm.

We have mentioned elsewhere that chemical adsorption phenomena are activated phenomena, leading us to highlight the kinetic aspect of such a process which is not necessarily an elementary act. We can therefore imagine that this phenomenon will involve two main stages:

– the first stage is identified as a simple physisorption phenomenon;

– the second stage conveys the transfer of a gas electron towards the solid or the solid towards the gas.

As such, in the direction of adsorption, only the second stage presents an activation energy E_A (Figure 2.5a). It is therefore the second stage which will control the kinetic process at adsorption.

For desorption, the two phenomena are activated. However, as is shown in Figure 2.5b, it is the chemisorption process which is imposed. This is due to the fact that the activation energy E_D from the chemisorption process is much larger than that of the physisorption process.

In the case of chemisorption, it therefore appears difficult to model the transfer of an electron from a stable diatomic molecule such as oxygen, for example. We can therefore envisage a preliminary dissociation stage in the molecule in either a gas or adsorbed state.

The complete process for the example of oxygen and for a dissociation of the gas state will be written as:

$$O_2 \Leftrightarrow 2O$$

with K_1 being the constant in this equation.

Subsequently:

$$O + 1 \text{ electron} + s \Leftrightarrow (O - s)^-$$

with K_2 being the constant in this equation.

Take:

$$K_1 = \frac{P_O^2}{P_{O_2}},\ K_2 = \frac{\theta}{P_O(1-\theta)} \text{ and } \theta = \frac{K_2\sqrt{K_1 P_{O_2}}}{1+\sqrt{K_1 P_{O_2}}}$$

In this example, the law of pressure for a diatomic molecule is no longer expressed as:

$$\sqrt{P_{O_2}}$$

More generally, for an atomicity gas, we have:

$$G_i \leftrightarrow i\,G$$

with:

$$\theta = \frac{K_2\sqrt{K_i P_{O_2}}}{1+\sqrt{K_i P_{O_2}}}$$

For an example of gas mixtures take a mixture of gases A and B, which can be adsorbed on a given solid. Each gas is characterized by its equation's adsorption constant K_A and K_B, and by its degree of coverage θ_A and θ_B.

In these conditions, the number of available sites on the surface will be equal to:

$$1-\theta_A-\theta_B$$

Therefore, for gas A we have:

$$K_AP_A = \theta_A / 1- \theta_A - \theta_B$$

which allows us to write $1- \theta_A - \theta_B = \theta_A / K_AP_A$

For gas B:

$$K_BP_B = \theta_B / 1- \theta_A - \theta_B$$

which allows us to write $1- \theta_A - \theta_B = \theta_B / K_BP_B$

We therefore have

$$\theta_A / \theta_B = K_AP_A / K_BP_B$$

By reporting this relation in one of the two previous expressions, we have:

$$\theta_A = K_AP_A/(1 - K_AP_A - K_BP_B)$$

Generally, and for a series of gases, we have:

$$\theta_A = K_AP_A/(1 - \Sigma K_iP_i)$$

2.7. Bibliography

[GER 59] J.E. GERMAIN, *Catalyse Hétérogène*, Dunod, Paris, 1959.

[FLO 67] E.A. FLOOD, *The Solid Gas Interface*, Marcel Dekker, Inc, New York, 1967.

[LAL 06] R. LALAUZE, *Physico-chimie des interfaces solides-gaz 1*, Hermes, 2006.

[LAL 08] R. LALAUZE, *Physical Chemistry of Solid-Gas Interfaces*, ISTE, London and John Wiley & Sons, New York, 2008.

[YOU 62] D.M. YOUNG, A.D. CROWELL, *Physical Adsorption of Gases*, Butterworths, London, 1962.

Chapter 3

Microcantilever Transduction

3.1. Introduction

The principle of atomic force microscopes (AFMs) is based on modifying the mechanical properties of a microcantilever placed several nanometers from a surface under test. These modifications are the result of interaction forces (van der Waal, Coulomb forces) between this mobile microstructure, commonly known as a microlever, and the surface of the sample being analyzed. In practice, this type of microscope is highly sensitive to changes in the environment (i.e. temperature, humidity, pressure, etc.). Two teams of researchers, one from Oak Ridge National Laboratory (USA) and the other from IBM Zurich (Switzerland) in 1994 first used parasite phenomena which were sensitive to the ambient environment, to develop a new kind of chemical microsensor [BAR 94, THU 94]. For around 15 years, a number of theoretical and practical studies have shown the potential of microcantilevers as chemical sensors [SEP 02, GOE 08, DUF 08a, DUF 08b, BOI 11].

The interest in the performance of these chemical microsensors is essentially due to the fact that when a mechanical device is reduced in size, the surface/volume ratio increases, therefore making the

---

Chapter written by Isabelle DUFOUR.

mechanical structure more sensitive to the surface effects which are key features of this type of microsensor.

In addition, from a technological point of view, these microsensors can be constructed using microelectromechanical system (MEMS) techniques derived from microelectronic fabrication technologies. A direct consequence of this type of fabrication is that it combines the advantages of miniaturization, collective fabrication and the integration of microelectronic commands and processing.

For (bio)chemical detection, the microcantilever is generally covered with a sensitive (bio)chemical layer designed to selectively sorb the species we want to detect. The sorbed species can therefore be detected either by measuring the microcantilever's resonant frequency shift (dynamic mode) or the microcantilever's quasi-static deflection (static mode). A simple way of explaining the basic principle of such sensors is to say that:

– in the dynamic mode, the change in mass due to species sorption in the sensitive layer causes a change in resonant frequency which is correlated with the concentration of the targeted species in the microcantilever's environment.

– in the static mode, the sorption of species causes the microcantilever to be deflected due to the layer's tendency to expand (or contract) when the targeted species is sorbed.

However, as illustrated by Figure 3.1, the behavior of microcantilevers is, in reality, much more complicated. As previously indicated, the overall effect of the dynamic mode lies in the change in the layer's mass whilst in the static mode it is due to a modification in the layer's stress and strain. However, secondary effects include changes in deflection due to the sorbed mass and the modification in resonant frequency due to the stress induced in the sensitive layer by sorption. Another effect which cannot always be ignored is the effect of the environment on the layer's sorption properties (e.g. the influence of temperature on the partition coefficient) or on the microcantilever's static or dynamic response (e.g. drift in the signal due to a thermal effect, which modifies both Young's modulus or the gas density and consequently the resonant frequency).

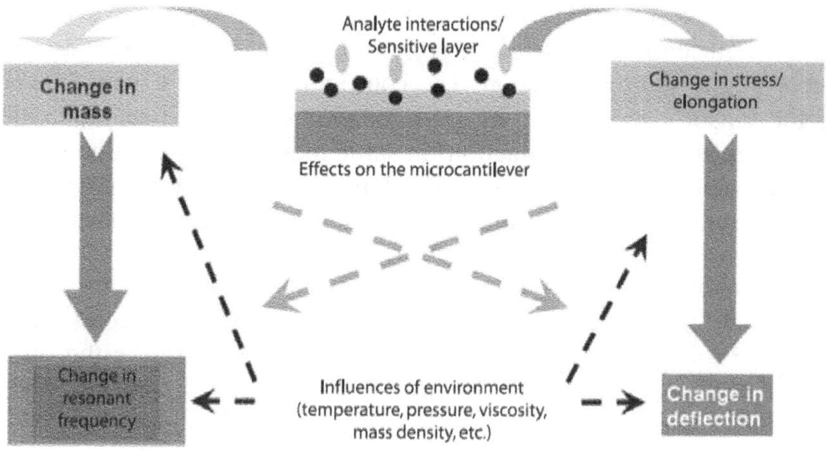

Figure 3.1. *Principle of (bio)chemical microcantilever based microsensors: dynamic and static modes*

3.2. Sensitive layers

The choice of a sensitive layer is important because it is an essential part of the sensor: the sensitive layer is involved in the sensor's sensitivity, selectivity and reversibility.

Reversibility is related to the type of interaction between the sensitive layer and the targeted species. The choice of sensitive layer is the result of a compromise between selectivity and reversibility. Non-specific interactions (or van der Waals forces) are reversible but are not sufficiently selective; whilst covalent interactions, with too much energy, render the sensor's response irreversible. These two types of interaction should therefore be avoided. As such it is important to favor non-covalent interactions as well as specific hydrogen bond type interactions relying on the capacity of molecules to accept or provide hydrogen bonds. On the other hand, the desired selectivity with regard to eventual interferents is directly linked to the sensitive layer's properties and can generally be improved by introducing specific functions favoring certain types of interactions. The sensitive layers used to create microcantilever chemical sensors are very often polymeric. This new tendency is due to the wide variety of materials available, their stability in time as well as their functionalization possibilities, which mean they can be adapted according to their application. The contribution of the sensitive layer to the sensor's sensitivity lies in its change in mass in dynamic mode and its expansion (or contraction) in static mode due to sorption of the target species. The layer's mechanical and physicochemical properties as well as its thickness therefore affect the sensor's overall sensitivity.

The choice of material and the microcantilever's geometry affect neither the sensor's reversibility nor its selectivity. In contrast, it influences sensitivity and noise measurement and therefore the limit of detection (LOD). The optimization of the microcantilever's materials and dimensions depends on the mode of operation.

3.3. Static mode

This mode is based on dissymmetry between the two opposite surfaces of the microstructure. Supposing, ideally, that the sensitive layer on one of the faces has a good affinity with the target species, it is the difference in behavior of these two surfaces regarding the target species which generates the structure's curvature and therefore its deflection (Figure 3.2). This phenomenon is comparable to the effect of a change in temperature on a mobile structure composed of two layers which have different dilation coefficients (bilayer effect).

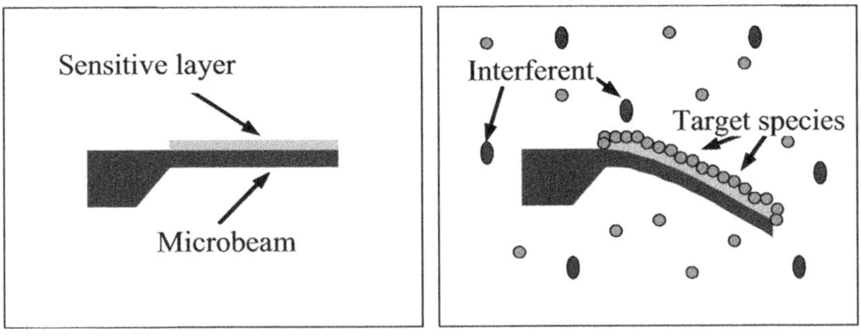

Figure 3.2. *Principle of static mode*

3.3.1. *Measurement principle*

Some of the most commonly used means of measuring deflection are:

– External optical measurement: a laser beam is reflected on the microcantilever's surface and is detected by photodiodes (method used in AFM).

– Capacitive measurement: the capacitance formed by one side of the microcantilever and an electrode facing it is modified by the movement of the microcantilever.

– Piezoresistive measurement: variations in piezoresistance placed where stress varies strongly (strain gauge).

3.3.2. *Equations*

For the static mode, Stoney's equation is normally used to express the link between the microstructure's curvature and the stress in the sensitive layer. It is preferable to use equation [3.1], which accounts for the Young's modulus of the sensitive layer and which is not limited to very low layer thickness compared to the microcantilever's thickness. This equation expresses the deflection along the beam [WEN 08]:

$$w(x) = -\frac{3(h_1 + h_2)}{h_2^2 K}\frac{\Delta\sigma}{\tilde{E}_2}x^2 \qquad [3.1]$$

with

$$K = 4 + 6\left(\frac{h_1}{h_2}\right) + 4\left(\frac{h_1}{h_2}\right)^2 + \left(\frac{\tilde{E}_1}{\tilde{E}_2}\right)\left(\frac{h_1}{h_2}\right)^3 + \left(\frac{\tilde{E}_2}{\tilde{E}_1}\right)\left(\frac{h_2}{h_1}\right) \qquad [3.2]$$

where $w(x)$ is the deflection of the microcantilever at the coordinate x, the symbols 1 and 2 represent the microcantilever and the sensitive layer respectively, where h is the thickness, $\tilde{E} = E/(1-\nu)$ is the bi-axial modulus, E is the Young's modulus, ν is Poisson's ratio and $\Delta\sigma$ is the stress induced in the sensitive layer during sorption.

The law governing the evolution of the stress in the layer according to the concentration C_g of the target gas is not well known and depends on a great number of parameters: it can be compressive or extensive.

In the case of a polymeric sensitive layer, the layer's viscoelastic properties can be influential and equation [3.1] therefore has to be modified. In this case, depending on the relationship between the time constant of the layer's sorption and mechanical relaxation time, an *overshoot* phenomenon can occur [HEI 09].

3.3.3. *Example of measurement*

To illustrate the static mode of microcantilever-based gas sensors, we will describe the research of the Oak Ridge National Laboratory on 2-mercaptoethanol detection [DAT 99]. The microstructure employed is a V-shaped silicon nitride microcantilever produced for AFM measurement heads (with a length of 200 µm and thickness of 0.6 µm) (Figure 3.3).

The sensitive layer detecting the sulfur compound is gold with a thickness of 50 nm placed on the microcantilever's surface. The deflection detection system is an external optical system composed of a laser diode and four quadrant photodiodes. So that the microcantilever's curvature is not modified by local increases in temperature caused by the laser beam (bilayer effect), an optical filter is placed between the objective lens serving to focus the beam and the laser diode.

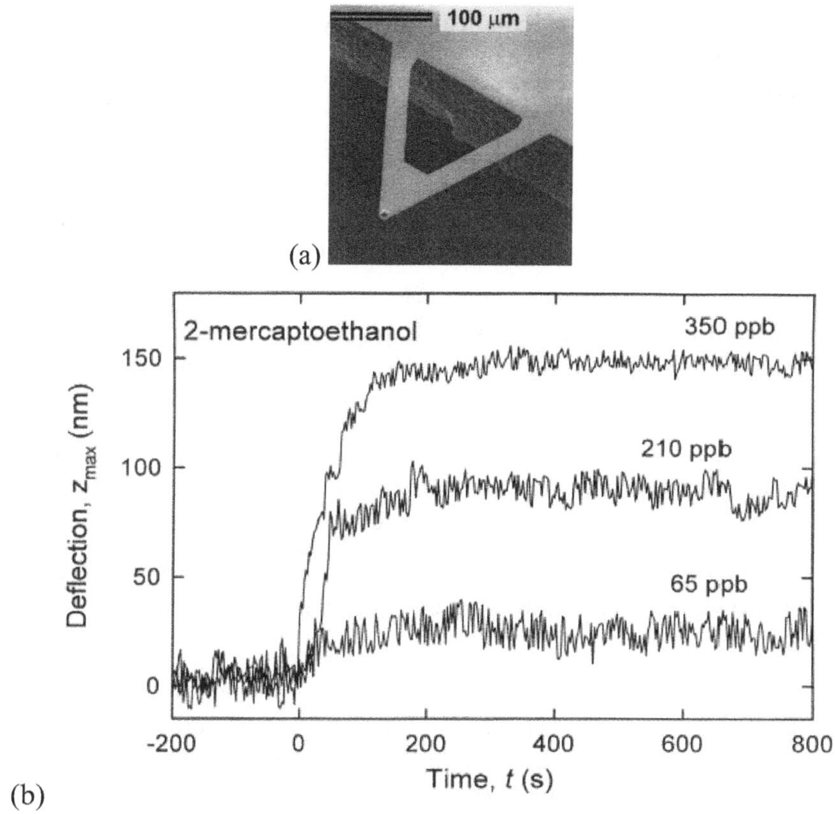

Figure 3.3. *(a) V-shaped microcantilever; (b) deflection measurement for different concentrations [DAT 99]*

3.4. Dynamic mode

The principle of this mode can be compared to that of a quartz crystal microbalance (QCM), which is frequently used in microelectronics for real-time thickness control during the deposition of thin layers. The addition of a mass due to the formation of a layer on the quartz surface causes a change in the resonant frequency. In this case, we can detect the resonant frequency shift via the variation in the quartz disc's impedance. In terms of microcantilever sensors, resonant frequency is measured by analyzing the vibration amplitude as a function of the excitation frequency. The structure must therefore be actuated; and the mechanical quantity to be measured is the vibration amplitude, which can be achieved using the principles listed in section 3.3.1.

3.4.1. *Actuation*

A number of methods exist for microcantilever actuation. Among the most common are:

– Electromagnetic actuation: a Laplace force is generated on the microcantilever by placing it in a magnetic field and by applying an electrical current on a conductive track placed on the microcantilever.

– Electrothermal actuation: mechanical stress induced by a local increase in temperature and controlled electrically causes a deflection similar to that caused by the bilayer effect.

– Actuation via a piezoelectric layer: a bilayer like effect is generated by a piezoelectric layer placed on the microcantilever's surface.

– Actuation via piezoelectric ceramic: the application of an electrical voltage to the electrodes of a piezoelectric ceramic results in the vibration of the microcantilever's support, thereby leading to deflection of the microcantilever by its inertia.

– Lastly, electrostatic actuation consists of an electrical voltage applied between two electrodes facing each other, one on a microcantilever's surface and the other on the support's fixed part, causing the microcantilever's deflection.

3.4.2. *Measurement of resonant frequency*

There are two main principles for measuring the resonant frequency: frequency sweep and microcantilever-based oscillator.

The principle of the frequency sweep method consists of sweeping the microlever's actuation frequency and measuring the electrical signal corresponding to the mechanical vibration. The study of the amplitude and phase of the transfer function of the actuation and measurement signals enables us to determine the microcantilever resonant frequency. The main advantage of this measurement is that it is possible to measure the microcantilever's quality factor, depending on the losses in the device, which impacts on the accuracy of the resonant frequency measurement and consequently the detection limit. The disadvantage of this method is the large amount of time required because a large number of frequencies have to be recorded with high accuracy to determine both the resonant frequency and quality factor precisely.

Nevertheless, there is a variant of the frequency sweep which does not involve any controlled actuation but simply the natural movement of the air around the microcantilever (random Brownian type movement, which actuates the microcantilever simultaneously at all frequencies). As such, by analyzing the measurement of the noise signal, it is possible to determine both the microcantilever resonant frequency and quality factor. However, due to the low effectiveness of air-based microcantilever actuation, the amplitude of the deflection is weak and therefore this method requires choosing an extremely accurate deflection measurement. Since the use of an external optical method is practically the only realistic solution in this case, this renders miniaturization a delicate procedure.

The principle of measurement using a microcantilever-based oscillator consists of amplifying the electrical signal of the deflection measurement in order to re-inject it as the microcantilever actuation signal to generate the vibration. We therefore create a closed-loop electrical circuit with an amplifier and the microcantilever as a resonator. The electrical signal obtained in this electrical circuit has the electrical frequency of the microcantilever's mechanical resonant frequency.

3.4.3. *Equation*

The resonant frequency f_0 for a microcantilever can be expressed as:

$$f_0 = \frac{1}{2\pi}\sqrt{\frac{k}{m_{eff}}} \qquad [3.3]$$

with k as the microstructure's stiffness, calculated by expressing the deflection at the end of the structure when a punctual force is applied at the free-end of the structure. Given the fact the sensitive layers used are generally polymer whilst microcantilevers are normally silicon ($E_1 >> E_2$), we can say that the microcantilever's stiffness is little modified in the presence of the sensitive layer and the microstructure's stiffness can therefore be expressed by:

$$k = \frac{E_1 b_1 h_1^3}{4L_1^3} \qquad [3.4]$$

m_{eff} is the effective mass depending on the resonance mode considered (mode 1: λ_n = 1.875; mode 2: λ_n = 4.694; mode 3: λ_n = 7.855, etc.) and the microstructure's total mass m:

$$m_{eff} = \frac{3m}{\lambda_n^4} \qquad [3.5]$$

The sensor's sensitivity is defined as the ratio of the output signal variation to the input variation. In the case of the examined sensors, we therefore define sensitivity to a gas concentration C_g by:

$$S_{C_g} = \left|\frac{df_0}{dC_g}\right| \qquad [3.6]$$

Given the expression of the microcantilever's resonant frequency, the variation in resonant frequency in the presence of gas can be due to the modification in both the structure's stiffness and mass. Generally, the modification of Young's modulus in the sensitive layer and therefore the microstructure's stiffness is not important in comparison to the variation in mass induced by the sorption of gas species into the sensitive layer. The variation in resonant frequency is therefore essentially the result of a modification in the structure's mass:

$$\frac{\Delta f_0}{f_0} = \frac{\Delta k}{2k} - \frac{\Delta m_{eff}}{2m_{eff}} \approx -\frac{\Delta m_{eff}}{2m_{eff}} = -\frac{\Delta m}{2m} \qquad [3.7]$$

In the presence of the target species, the structure's mass increases and therefore the resonant frequency decreases as a result.

As such, to express the sensor's sensitivity, we have to express the variation in mass Δm of the structure according to the concentration C_g of the gas. This is normally achieved using the partition coefficient K from the pair (sensitive layer, target gas), defined as the ratio between the concentration of gas in the sensitive layer and the concentration of gas in the surrounding fluid:

$$K = \frac{\Delta m / b_1 L_1 h_2}{C_g} \qquad [3.8]$$

where b_1 and L_1 are the width and length of the microcantilever respectively and h_2 is the sensitive layer's thickness.

By combining equations [3.6]–[3.8], the expression of the chemical microcantilever's sensitivity can be expressed as:

$$S_{C_g} = \frac{f_0}{2} \frac{K h_2}{\rho_1 h_1 + \rho_2 h_2} \qquad [3.9]$$

where ρ_1 and ρ_2 are the respective mass densities of the microcantilever and the sensitive layer.

Using this expression we see that sensitivity is proportional to the resonant frequency f_0, the partition coefficient K and the sensitive layer's thickness h_2 (provided that the sensitive layer's mass is negligible compared to the microcantilever's mass).

An important characteristic for chemical sensors is the limit of detection (LOD), which is the smallest detectable concentration of gas. This characteristic depends on both the sensor's sensitivity but also the measurement noise Δf_{noise}. Generally, we assume that we can measure this if it is three time higher than the measurement noise. As such, the limit of detection can be expressed as:

$$LOD = \frac{3\Delta f_{noise}}{S_{C_g}} \quad [3.10]$$

This expression allows us to demonstrate that a very low gas concentration measurement is possible by combining good sensitivity S_{Cg} with a low measurement noise Δf_{noise}. The measurement noise of resonant frequency weakens as the quality factor Q, associated with the resonant mode, increases.

3.5. Example of measurement

To illustrate the use of microcantilevers in the dynamic mode within the framework of chemical microsensors in gas environments, an example illustrating the notion of integration has been chosen. This is part of the work carried out by the Physical Electronics Laboratory at the ETH in Zurich.

Using CMOS production technology, which is commonly used in microelectronics, and using additional process steps to release the silicon microcantilever, the sensor and its electronic measurement system (oscillator) have been integrated into a single chip (Figure 3.4a).

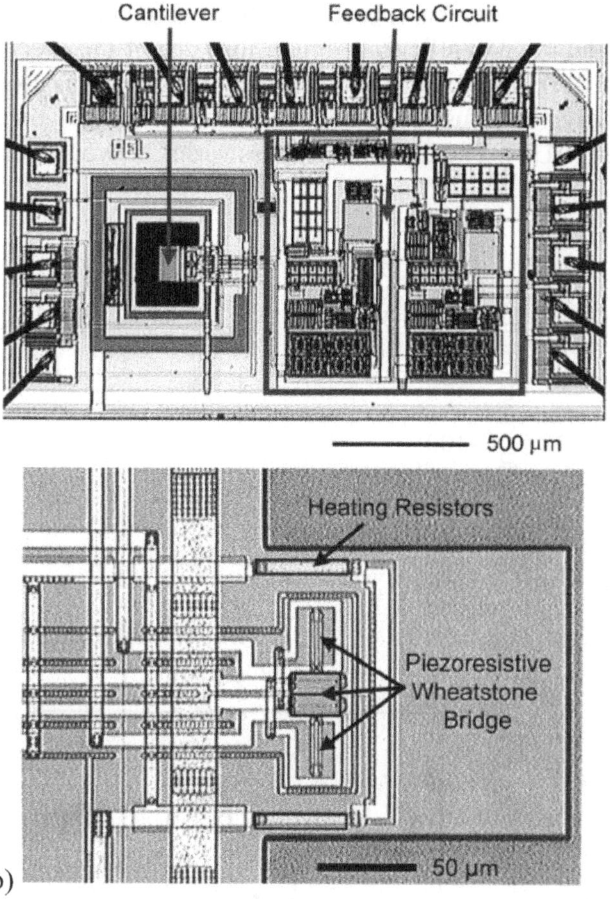

(a)

(b)

Figure 3.4. *(a) Microcantilever with integrated oscillator (feedback circuit). (b) Close-up of the microcantilever with its actuation system (electrothermal) with an integrated (piezoresistance) measurement [HIE 00]*

In this example, the microcantilever is moved in an integrated way with an electrothermal actuation. This actuation consists of an electrical current passing in a resistance placed in the cantilever's clamping (Figure 3.4b), and a local increase in temperature in the microstructure causes its vibration through a bilayer effect. This vibration is detected using piezoresistive detection (Figure 3.4b). Depending on the fabrication technique employed, the microcantilevers are composed of silicon, silicon oxide and silicon nitride. The microcantilever's dimensions are of the order of 150 μm in length, 150 μm in width and the overall thickness is around 10 μm. The sensitive layer used for alcohol detection is a polymer, polyether urethane (PEUT). The layer's thickness varies between 2 and 5 μm.

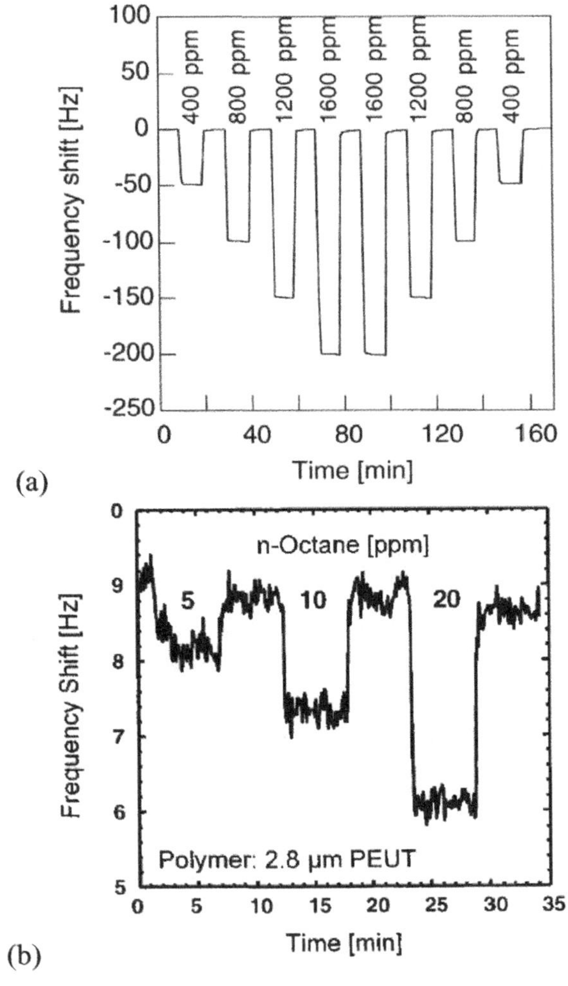

Figure 3.5. *Example of measurements: (a) toluene; and (b) n-octane [HIE 00, LAN 02]*

The measurement examples carried out using these devices are shown in Figure 3.5. Figure 3.5a shows the shift in resonant frequency (353 kHz) for different concentrations of toluene. Between each concentration tested, the test cell is purged with air. These measurements show the good linearity and reversibility of this type of sensor. Figure 3.5b, corresponding to n-octane measurements carried out at lower concentrations, enables us to understand the notions of noise and the sensor limit of detection. The resolution of the frequency measurement system is 0.03 Hz. If we suppose that the measurement provided by the sensor can only be considered valid for one higher or equal frequency variation, accounting for sensitivity, the sensor's limit of detection is 0.8 ppm for n-octane. Given the sensor's geometry, this corresponds to an increase in the structure's mass of 2 pg.

3.6. Other uses for dynamic mode

The basic equations presented in section 3.4 do not account for the microcantilever's environment. When a microcantilever vibrates in a viscous fluid (gas or liquid), the fluid resists the beam's vibration. The hydrodynamic force by unit length F_{fluid}, which is exerted on the microcantilever, can be written in the frequency domain [STO 51]:

$$F_{fluid} = -j\omega g_1(\omega)w + \omega^2 g_2(\omega)w \qquad [3.11]$$

where ω is the radial actuation frequency, w is the microcantilever's deflection and g_1 and g_2 are functions which depend on the fluid's properties, frequency and the microcantilever's section. These functions represent viscous losses and the effective fluid added mass.

The inertial and viscous effects influence the dynamic response of the microcantilever and, in particular, its resonant frequency f_r and its quality factor associated with viscous loss Q_{visc} [SAD 98, CHO 00]:

$$f_r = f_{0,vacuum} \frac{1}{\sqrt{1 + L_1 g_2/m}} \sqrt{1 - \frac{1}{2Q^2}} \qquad [3.12]$$

$$Q_{visc} = \frac{2\pi\sqrt{1+Lg_2/m}}{Lg_1/m} f_{0,vacuum} \quad [3.13]$$

$$Q = \frac{Q_{visc} Q_{\text{int}}}{Q_{visc} + Q_{\text{int}}} \quad [3.14]$$

where $f_{0,vacuum}$ is the microcantilever's natural frequency ([3.3]), Q is the total quality factor (including all losses) and Q_{int} is the quality factor associated with internal losses.

Equations [3.12]–[3.14] show that the resonant frequency depends on the terms g_1 and g_2 which are functions of the physical properties of the surrounding fluid. In the following sections, based on these equations, we will see that it is possible to carry out chemical detection in gas environments without a sensitive layer and that in-plane vibrations are preferable for detections in liquids as opposed to out-of-plane vibrations normally used in gas environments [DUF 12].

3.6.1. *Dynamic mode without a sensitive layer*

Using equations [3.12]–[3.14], it is possible to obtain the analytical expression for the shift in resonant frequency for a microcantilever without a sensitive layer corresponding to low variations in both the mass density and the viscosity of surrounding fluid. In the case of the binary mixes (He/N_2, CO_2/N_2 and H_2/N_2), the dominant effect is caused by the variation in the mass density of the gas mixture [TÉT 10].

Examples of detecting hydrogen in nitrogen or carbon dioxide in nitrogen are shown in Figure 3.6. Since hydrogen is lighter than nitrogen, the microcantilever's resonant frequency increases as the concentration of hydrogen increases, whilst it decreases when the amount of carbon dioxide in the nitrogen increases.

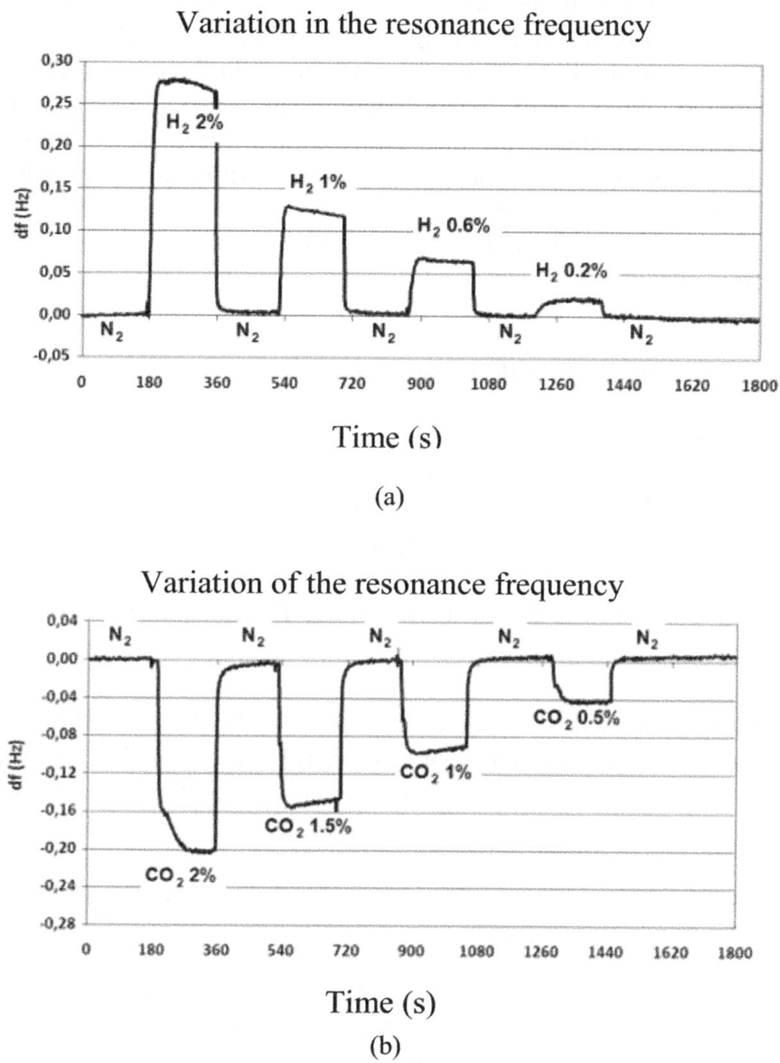

Figure 3.6. *Example of detection: (a) hydrogen in nitrogen; and (b) carbon dioxide in nitrogen with a silicon microcantilever (2000 x 400 x 15 μm^3) without a sensitive layer*

In comparison with detections with a sensitive layer, detections without a sensitive layer have weaker time constants because there is no sorption time in the sensitive layer. Furthermore, not using a sensitive layer results in a completely reversible and reliable behavior. Sensitivity and the limit of detection are better in the case of a binary mixture of gases with very different mass density. The main disadvantage of this type of sensor is that there is no intrinsic selectivity: as such, these sensors without a sensitive layer can only be used for specific applications where the concentration of a single component varies whilst the relative concentrations of other gases remain constant.

3.6.2. *Dynamic mode with in-plane vibrations*

Equation [3.12] shows that the decrease in resonant frequency when the microcantilever is immersed in liquid is essentially due to the fluid displaced mass effect (term due to g_2). On the other hand, the sharp decrease in the quality factor when the microcantilever is immersed in liquid comes from the greater influence of the viscous term g_1 compared to the influence of the fluid displaced mass (term due to g_2) in equation [3.13].

When immersed in liquid, in order to limit the decrease of both the resonant frequency (to maintain sensitivity) and the quality factor (to maintain the measurement noise), alternative vibration modes which shear the fluid rather than applying pressure have been studied and tested: these consist of in-plane vibration modes (lateral bending modes and elongation modes) instead of the classic out-of-plane vibration mode (transverse bending mode).

The measurements recorded for the first lateral bending mode for silicon microcantilever's with widths of 45, 60, 75 and 90 µm, and lengths of 200, 400, 600, 800 and 1,000 µm, and a thickness of 8 and 12 µm, have led to quality factors of up to 4,300 in air and 67 in water; with resonant frequencies in the water being only 5% to 10% weaker than those in air (generally for the first transverse bending mode, the quality factor does not exceed 20 in water and the reduction in resonance frequency is of the order of 50%) [BEA 10]. Piezoelectric cantilevers in PZT with dimensions of

$10 \times 2 \times 0.075$ mm^3, $8 \times 2 \times 0.075$ mm^3 and $6 \times 1 \times 0.075$ mm^3 have been produced using screen printing techniques and have allowed the actuation and read-out of in-plane elongation. Despite a relatively low quality factor in air (from 130 to 300) the quality factors obtained in dodecan (viscosity: 1.55 cP) are relatively high (from 70 to 107) and the decrease in resonance frequency is only 1% compared to the value in air [CAS 10].

3.7. Conclusion

Initial research into the use of microcantilevers as chemical sensors is relatively recent (1994). Since then, given the number of articles published on the subject, a number of research teams have focused on this new kind of transduction. As shown in this chapter, two modes, static or dynamic, can be used for chemical detection. Different materials (silicon, polymers, active materials etc.) and different microstructures (discs, membranes, bridges, etc.) can be used for the mobile structure. Note that microcantilevers can also be used as sensors for physical quantities (e.g. force [LAK 10] or fluid rheological properties [BEL 08]).

3.8. Acknowledgments

I would like to thank my colleagues from the MMM team (Microassembled Materials for Microsystems) at the IMS laboratory as well as the various researchers who have worked with us on the study of microcantilever-based sensors. I would also like to express my gratitude to Professors Fabien Josse and Stephen Heinrich from Marquette University in Milwaukee, USA and Professor Oliver Brand from GeorgiaTech, Atlanta, USA.

3.9. Bibliography

[BAR 94] J.R. BARNES, R.J. STEPHENSON, M.E. WELLAND, CH. GERBER, J.K. GIMZEWSKI, "Photothermal spectroscopy with femtojoule sensitivity using a micromechanical device", *Nature*, vol. 372, pp. 79-81, 1994.

[BEA 10] L.A. BEARDSLEE, A.M. ADDOUS, S. HEINRICH, F. JOSSE, I. DUFOUR, O. BRAND, "Thermal excitation and piezoresistive detection of cantilever in-plane resonance modes for sensing applications", *IEEE/ASME Journal of Microelectromechanical Systems*, vol. 19, pp. 1015-1017, 2010.

[BEL 08] N. BELMILOUD, I. DUFOUR, A. COLIN, N. LIVIU, "Rheological behaviour probed by vibrating microcantilevers", *Applied Physics Letters*, vol. 92, pp. 041907, 2008.

[BOI 11] A. BOISEN, S. DOHN, S. KELLER, S. SCHMID, M. TENJE, "Cantilever-like micromechanical sensors", *Reports on Progress in Physics*, vol. 74, pp. 036101, 2011.

[CAS 10] C. CASTILLE, I. DUFOUR, C. LUCAT, "Longitudinal vibration mode of piezoelectric thick-film cantilever-based sensors in liquid media", *Applied Physics Letters*, vol. 96, pp. 154102, 2010.

[CHO 00] J.W.M. CHON, P. MULVANEY, J.E. SADER, "Experimental validation of theoretical models for the frequency response of atomic force microscope cantilever beams immersed in fluids", *Journal of Applied Physics*, vol. 87, pp. 3978-3988, 2000.

[DAT 99] P.G. DATSKOS, I. SAUERS, "Detection of 2-mercaptoethanol using gold-coated micromachined cantilevers", *Sensors and Actuators B*, vol. 61, pp. 75-82, 1999.

[DUF 08a] I. DUFOUR, C. BERGAUD, "Capteurs à microstructures mobiles", Chapter 9 in F. MÉNIL, *Microcapteurs de gaz*, Hermès, 275-299, 2008.

[DUF 08b] I. DUFOUR, F. LOCHON, "Microcapteurs chimiques à base de micropoutres en silicium utilisées en régime dynamique", *Techniques de l'ingénieur*, vol. IN-84, 2008.

[DUF 12] I. DUFOUR, F. JOSSE, S. HEINRICH, C. LUCAT, C. AYELA, F. MÉNIL, O. BRAND, "Unconventiopnal uses of microcantilevers as chemical sensors in gas and liquid media", *Sensors and Actuators B*, 2012.

[GOE 08] K. GOEDERS, J. COLTON, L. BOTTOMLEY, "Microcantilevers: sensing chemical interactions via mechanical motion", *Chemical Reviews*, vol. 108, pp. 522-542, 2008.

[HEI 09] S. HEINRICH, M. WENZEL, F. JOSSE, I. DUFOUR, "An analytical model for transient deformation of viscoelastically coated beams: applications to static-mode microcantilever chemical sensors, *Journal of Applied Physics*, vol. 105, p. 124903, 2009.

[HIE 00] A. HIERLEMANN, D. LANGE, C. HAGLEITNER, N. KERNESS, A. KOLL, O. BRAND, H. BALTES, "Application-specific sensor systems based on CMOS chemical microsensors", *Sensors and Actuators B*, vol. 70, pp. 2-11, 2000.

[LAK 10] R. LAKHMI, H. DEBEDA, I. DUFOUR, C. LUCAT, "Force sensors based on screen-printed cantilevers", *IEEE Sensors Journal*, vol. 10, pp. 1133-1137, 2010.

[LAN 02] D. LANGE, C. HAGLEITNER, A. HIERLEMANN, O. BRAND, H. BALTES, "Complementary metal oxide semiconductor cantilever arrays on a single chip: mass-sensory detection of volatile organic compounds", *Analytical Chemistry*, vol. 74, pp. 3084-3095, 2002.

[SAD 98] J. SADER, "Frequency response of cantilever beams immersed in viscous fluids with applications to the atomic force microscope", *Journal of Applied Physics*, vol. 84, pp. 64-76, 1998.

[SEP 02] M. SEPANIAK, P. DATSKOS, N. LAVRIK, C. TIPPLE, "Microcantilever transducers: a new approach in sensor technology", *Analytical Chemistry*, vol. 74, pp. 568-575, 2002.

[STO 51] G. STOKES, "On the effects of the internal friction of fluids on the motion of pendulums", *Transactions of the Cambridge Philosophical Society*, vol. 9, pp. 8-106, 1851.

[TÉT 10] S. TÉTIN, B. CAILLARD, F. MÉNIL, H. DEBÉDA, C. LUCAT, C. PELLET, I. DUFOUR, "Modeling and performance of uncoated microcantilever-based chemical sensors", *Sensors and Actuators B*, vol. 143, pp. 555-560, 2010.

[THU 94] T. THUNDAT, R.J. WARMACK, G.Y. CHEN, D.P. ALLISON, "Thermal and ambient-induced deflections of scanning force microscope cantilevers", *Applied Physics Letters*, vol. 61, pp. 2894-2896, 1994.

[WEN 08] M.J. WENZEL, F. JOSSE, S.M. HEINRICH, E. YAZ, P.G. DATSKOS, "Sorption-induced static bending of microcantilevers coated with viscoelastic material", *Journal of Applied Physics*, vol. 103, pp. 064913, 2008.

Chapter 4

Piezoelectric Transduction (QCM)

4.1. General remarks and basic principles

Transducers which use acoustic waves can be used to form a range of sensors. Depending on the type of device being used and the type of wave generated, it is possible to carry out measurements in a vacuum, gas, liquid or on fine films. Due to their high sensitivity to changes in mass, quartz microbalances are often used for detection in chemical or biochemical spaces. However, they also allow us to identify a range of other properties in solids and liquids which are in contact with their surfaces, such as density, viscosity, conductivity and the shear module.

The use of acoustic waves in sensor and biosensor applications requires a device to generate them and an electronic surface or appropriate equipment to enable them to function and collect data.

4.1.1. *Piezoelectric effect*

In 1880, Jacques and Pierre Curie discovered that mechanical stress applied to some types of crystal, including quartz, leads to the appearance of an electric current through the crystal. They also

---

Chapter written by Hubert PERROT.

observed that the amplitude of this electrical current was proportional to the stress applied [CUR 80]. This has been termed a piezoelectric effect whose Greek root "*piezein*" means to "press". This property only exists in acentric materials, i.e. those which crystallize as a network without a center of symmetry. The charges generated in the quartz crystal originate from the formation of dipoles resulting in the displacement of atoms in the acentric material.

A short time after this discovery, the Curie brothers verified the effect of inverse piezoelectricity via further experiments: a difference in the piezoelectric current applied between the faces of a piezoelectric material causes a stress to occur inside it. This effect is the basis of how quartz microbalances function. The electrical field coming from polarization induces a reorientation of dipoles in the acentric material. It results in a deformation of the material. The direction of deformation depends on the direction of the electrical field applied whilst the extension of this deformation depends on the current's amplitude (Figure 4.1).

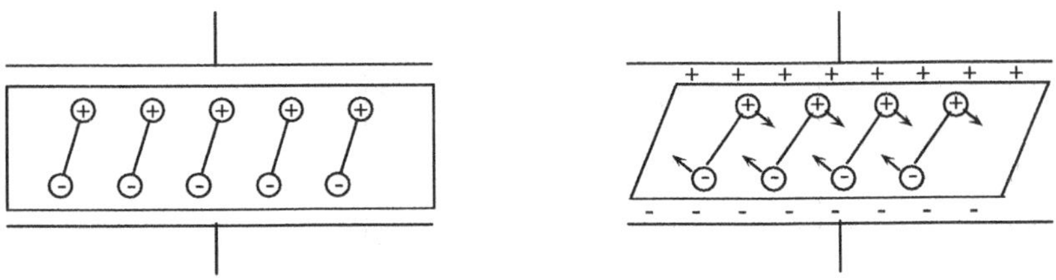

Figure 4.1. *Schematic representation of the deformation of polarized quartz [BUT 92]*

In the 1920s, Cady [CAD 64] demonstrated that the inverse piezoelectric effect could be used to create highly stable oscillator circuits.

The application of an alternating electrical current to the quartz crystal causes an equally alternative deformation. The vibrated (or oscillated) crystal therefore generates an acoustic wave inside the piezoelectric material. At the end of the 1950s, Sauerbrey [SAU 59] demonstrated that it was possible to use these piezoelectric materials to create a quartz microbalance which could function in gas environments. However, it would not be until the 1980s [KAN 85] that the use of piezoelectric materials in sensor or biosensor devices would be extended to include applications in liquid.

4.1.2. *Vibration modes*

The electromechanical coupling and stress from the applied electrical field depends on the crystal's symmetry, the angle of its cut and the configuration of exciting electrodes.

Vibration modes depend on the crystal's cut and the direction of the electrical field to which the quartz is subject. Different types of vibrations can appear separately or simultaneously. In the latter case, the quartz vibrates simultaneously at several frequencies. Since this is unhelpful in the majority of applications, these combinations of crystal cut/direction of electric field are avoided.

The main vibration modes include flexion vibration, longitudinal vibration, face shear vibration and thickness shear vibration. It is generally this last mode which is used for quartz microbalances because the crystalline structures glide in comparison to those shown in Figure 4.2 below.

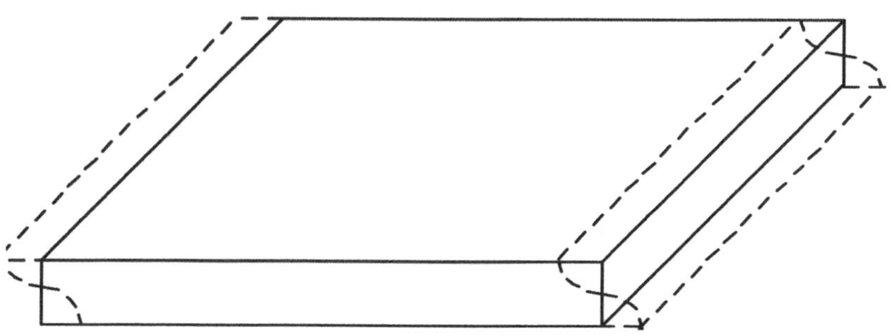

Figure 4.2. *Vibration in cutting thickness*

The quartz crystal resonates when the excitation potential frequency is such that the quartz's thickness is a multiple impair of half the length of the acoustic wave [MAR 91]. In this case, a stationary acoustic wave is generated throughout the quartz crystal. This is known as a quartz resonator.

4.1.3. *Different acoustic devices*

As we saw previously, the electromechanical coupling and the stress caused by the electrical field applied to a quartz crystal depend on the crystal's symmetry, the angle of the crystalline substrate's cut and the configuration of excitation electrodes. These different electromechanical coupling modes cause different types of acoustic waves and propagation modes for these waves. Several types of devices have therefore been developed from these different modes.

Firstly, thickness shear mode (TSM) devices are based on creating acoustic waves in thickness shear mode in the AT cut quartz. They are commonly known as quartz crystal microbalances (QCM). These devices are composed of a fine quartz blade which is sandwiched between two very thin metal electrodes (of the order of around 0.1 μm in thickness).

These electrodes create an alternating electrical current inside the quartz resonator via a circuit oscillator subject to resonance frequency. The latter is mainly sensitive to variations in the quartz's mass and its electrodes/films. The important parameter for these devices is the thickness of the piezoelectric substrate.

As we saw previously, the quartz crystal enters into resonance when the frequency of excitation tension is such that the quartz's thickness is a multiple impair of the acoustic wave's half length. As such, a reduction in the crystal's thickness leads to an increase in its resonance frequency and, as we will see, the microbalance's sensitivity.

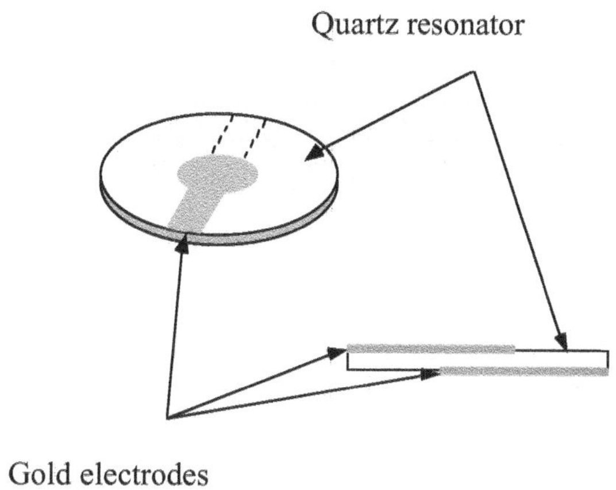

Figure 4.3. *View of a TSM quartz resonator and its two gold electrodes*

Other devices, generally with inter-digitized electrodes, allow us to carry out gravimetric measures such as (surface acoustic waves – SAW), SH-APM (shear horizontal acoustic plate mode) or even Love waves [ARN 08].

For example, SAW devices are based on the study of the prevention of an acoustic wave's propagation speed, called a Rayleigh wave, propagating between two pairs of inter-digitized electrodes on the surface of the quartz plate (typically with ST cut). The propagation speed of the wave, v, is characteristic of the material used and the crystal's cut. The wavelength propagated between pairs of electrodes is equal to the periodicity for a pair of electrodes, p, defined as the distance between two consecutive teeth of one of the IDT combs. As a result, the device's frequency, $f = v/p$, is simply the function of the gap between the teeth in the comb for a given piezoelectric material.

An acoustic wave surface device configured in a line is represented by Figure 4.4. Two pairs of combs are placed on the quartz's active face opposite each other. A sensory layer is placed between the two pairs of electrodes. When this layer gets heavier (by absorbing a gas, for example), the wave's propagation speed decreases in proportion to the added mass.

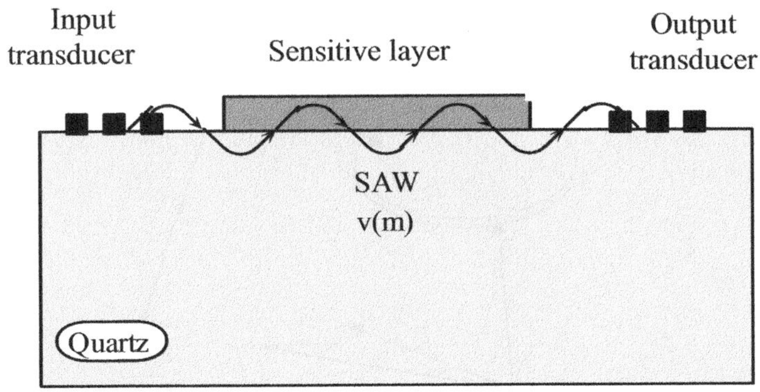

Figure 4.4. *SAW device where v(m) is the wave's propagation speed [BOU 96]*

Frequencies typically achieved with this type of device are of the order of 30 to 300 MHz. Sensitivity is of the order of 10 pg cm^{-2} for a device whose resonance frequency in fundamental mode is 200 MHz.

4.1.4. *Measuring methodology*

4.1.4.1. *Active mode: classic quartz microbalance*

As we have seen, the quartz crystal and its electrodes make up a piezoelectric resonator: surface changes (mass, mechanical stress, etc.) are transformed into measurable electrical signals. The quartz resonator, linked to an electrical circuit that is strong enough to create an oscillator sees its resonance frequency vary under the effect of surface changes. This frequency is generally measured using a frequency counter and it is possible to follow the microbalance's frequency over time with precision, which depends on the measuring parameters and the microbalance's stability [ROD 05].

The electronics in a number of oscillators are based on a single transistor acting as an amplifying element accompanied by a reaction loop [EHA 02]. This is illustrated by the diagram in Figure 4.5.

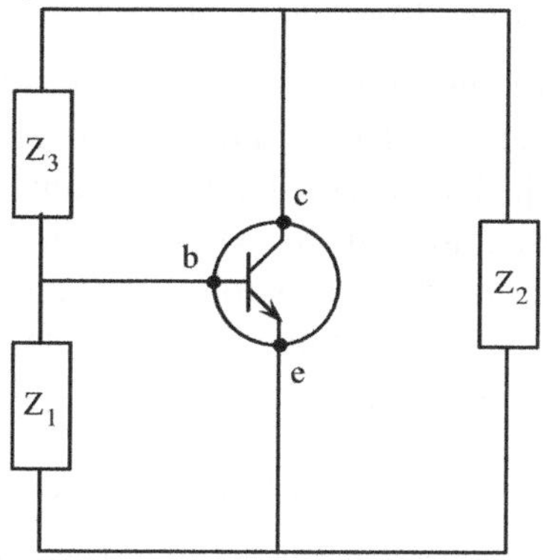

Figure 4.5. *Oscillator based on a single transistor. Z_1, Z_2 and Z_3 are impedance to passive electronic elements*

In order to find the correct values for the different electronic components, we use Stockhausen's conditions. The choice of oscillator depends on a number of criteria: serial or parallel resonance frequency, circuit bandwidth and charge resistance. There are several models available for this, e.g. Pierce, Miller (Figure 4.6) or Colpitts [MAT 83].

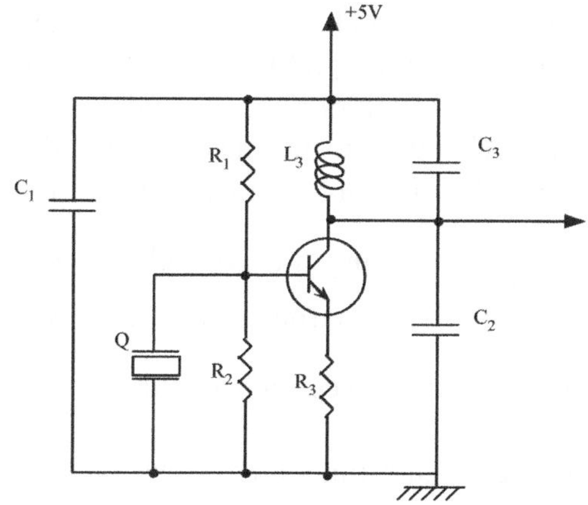

Figure 4.6. *Diagram of a Miller oscillator*

4.1.4.2. *Passive mode: electroacoustic measures*

The measuring methodology is based on identifying a quartz resonator's impedance of electric admittance around its resonance frequency [BOU 95]. To do so, a network analyzer is used and a model of this response allows us to extract the important data which is characteristic of the quartz resonator. This approach can also be employed to confirm results obtained using a classic microbalance in terms of gravimetry [GAR 00].

4.1.4.3. *Coupled mode: cyclical and dynamic electrogravimetry*

A very interesting approach for microbalance is to couple it with electrochemical measures. In cyclical mode measures are carried out using a combination of three electrodes, the principle being to carry out two inverse successive current scans, from which the term cyclical originates, between the working electrode and the reference electrode following the variation in current in the resulting active electrode. Using the gold electrode from the microbalance as the active electrode, it is possible to follow the variation in microbalance frequency, Δf_m, according to the current and, as such, evaluate variations in mass in the electrochemical system at the same time.

On the other hand, in dynamic mode, by measuring a function of electrogravimetric transfer, ac-electrogravimetry is an electrochemical impedance analogue technique [GAB 02, GAB 07, SCH 09]. A low amplitude sinusoidal perturbation ΔE, applied to an electrochemical system, can lead to changes in the mass of the film placed on the microbalance, Δm, allowing us to measure the electrogravimetric transfer function $\frac{\Delta m}{\Delta E}(\omega)$. Using the gold electrode from the quartz microbalance as a working electrode, we can carry out two distinct measures simultaneously: an electrochemical impedance $\frac{\Delta E}{\Delta I}(\omega)$ and an electrogravimetric transfer function $\frac{\Delta m}{\Delta E}(\omega)$.

4.2. Theoretical aspects of the quartz microbalance

4.2.1. *Measures in vacuum or gas environments*

The first quantitative study of this subject was carried out by Sauerbrey, expressing the difference in frequency Δf_m (in Hz) according to the added mass ΔM_f (in g) with the following equation:

$$\Delta f_m = -\frac{2f_0^2}{\sqrt{\mu_q \rho_q}} \frac{\Delta M_f}{A} \qquad [4.1]$$

where f_0 is the fundamental resonance frequency of the non-charged crystal, μ_q is the shear module of a shear AT quartz (2.947×10^{11} g cm^{-1}s^{-2}), ρ_q is the quartz's density (2,651 g cm^{-3}), ΔM_f is the total mass applied to two sides of the crystal and A is the area under the electrode of one side of the crystal.

By introducing the mass added by a surface unit, $\Delta m_f = \frac{\Delta M_f}{A}$, we obtain the ratio: $\Delta f_m = -C_f \; \Delta m_f$ with $C_f = 2.26 \times 10^{-6}$ f_o^2 Hz cm^2 g^{-1} and Δm_f expressed in g cm^{-2}.

Let us take quartz with a fundamental frequency of 6 MHz as an example, C_f is therefore equal to 8.14×10^7 Hz cm^2 g^{-1}. Note that Sauerbrey has shown that a thin film place on the surface of a sensor device functioning in thickness shear mode can be treated as an increase in the quartz's mass. C_f has been termed, by Sauerbrey, as the mass sensitivity coefficient and it increases quadratically with an increase in the quartz's resonance frequency, f_0. This property is particularly interesting in terms of the increase of the quartz microbalance sensitivity [BIZ 00]. According to the study carried out by Lu and Czanderna [LU 84] the limitation of Sauerbrey's model's validity is given by $\Delta m/m_q < 2\%$, where m_q is the mass by the surface unit mass of the non-charged quartz.

Since a crystal's fundamental frequency is inversely proportional to its thickness, the crystal being used needs to be as thin as possible.

For example, an AT shear crystal vibrating in a shear thickness of 6 MHz in a fundamental model has a thickness equal to 280 μm. This last parameter technically limit's the microbalance's sensitivity because blades which are too thin are fragile.

4.2.2. *Measurements in liquid environments*

4.2.2.1. *The effect of a liquid*

In the 1980s, it was common opinion that shear thickness wave sensors could not be used in liquids due to a destruction of the crystal's vibration by depreciation in viscosity [GUI 84]. This view was expressed as a study showed that a shear thickness can be used in liquid chromatography.

In 1985, two simple physical models were developed by Bruckenstein and Shay [BRU 85], and Kanazawa and Gordon [KAN 85]. In the latter case, the quartz is treated as an electronic solid without loss and the liquid as a purely viscous fluid. The variation in the crystal's resonance frequency is attributed to the coupling between the stationary shear wave in the quartz and the depreciating wave in the liquid. They therefore obtained a simple ratio expressing the variation in resonance frequency for the piezoelectric crystal depending on the characteristic parameters of the crystal and the liquid in the contact with the resonator's face:

$$\Delta f_m = -f_0^{3/2}\left(\frac{\rho_L \eta_L}{\pi \mu_q \rho_q}\right) \qquad [4.2]$$

where ρ_q and ρ_L are the quartz and liquid's respective densities, η_L is the liquid's dynamic viscosity, μ_q is the quartz's shear module and f_0 is the resonance frequency of the quartz only.

For quartz vibrating at 6 MHz with one side in contact with water, the theory predicts a difference of around 5,000 Hz between the quartz's frequency in air and the frequency of this same quartz when in water.

According to Yao and Zhou [YAO 88], the frequency response of a frequency acoustic wave sensor in liquid also depends on the liquid's dielectric and conductive effects. From their experimental results, Yao and Zhou deduced the following empirical formula:

$$\Delta f_m = C_1 + C_2\rho_L^{1/2} + C_3\eta_L^{1/2} - C_4\varepsilon_L \qquad [4.3]$$

where C_1, C_2, C_3 and C_4 are numerical constants, ρ_L is the liquid's density, η_L is the liquid's dynamic viscosity and ε_L is the liquid's dielectric constant.

4.2.2.2. *Effect of rugosity*

With the aim of better understanding the behavior of acoustic wave sensors in liquids, it seems important to consider other interfacial phenomena (rugosity, pressure, etc.). Schumacher *et al.* [SCH 85] have, for example, shown that the rugosity on the crystal's surface considerably affects the device's resonance frequency. They examined a rugose surface composed of a series of half-cylinders, liquid traps resulting in a rigidly fixed liquid layer (Figure 4.7).

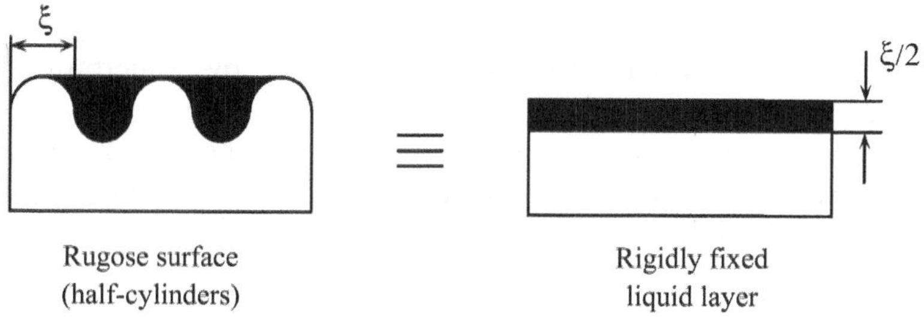

Figure 4.7. *Model of rugosity by a quartz microbalance*

In these conditions, the variation in frequency must contain the parameter $\Delta m_L = \frac{\rho_L \xi}{2}$, which represents the liquid's mass by surface unit, held in rugosity cavities ξ is the half cylinders' half-cylinders.

Schumacher has therefore developed a formula containing this parameter:

$$\Delta f_m = -\frac{2f_0^2 \Delta m_L}{(\mu_q \rho_q)^{1/2}} \qquad [4.4]$$

where f_0 is the resonance frequency for only the quartz, μ_q is the quartz's shear module, ρ_q is the quartz's density and Δm_L is the variation in the liquid's mass by surface unit.

4.2.2.3. *Combined influence of a liquid and a film*

In the case of a deposit of weak masses placed in a Newtonian liquid, variations in the quartz's frequency due to the displaced mass are added to those from the viscous coupling with the liquid. This has been demonstrated by Martin *et al.* [MAR 91].

The theory proposed by Martin *et al.* is applied to an electromagnetic analysis leading to an electric modeling of a resonator simultaneously charged by a thin film and a Newtonian liquid.

This liquid has a semi-infinite thickness. They furthermore examined the following hypotheses proposed by Sauerbrey: the thickness of less important electrodes, the deposed mass rigidly linked to the quartz, the continuity of displacement at the quartz/film and film/quartz interfaces and stress at the quartz/film interface. They therefore obtained the following equation:

$$\Delta f_m = -\frac{2f_0^2}{N\sqrt{C_{66}\rho_q}} \left[\rho_f h_f + \left(\frac{\rho_L \eta_L}{4\pi f_0}\right)^{1/2} \right] \qquad [4.5]$$

where f_0 is the quartz's serial resonance frequency, N is the order of the vibration partial, C_{66} is the quartz's complex elastic constant, ρ_L is the liquid's density, ρ_q is the quartz's density, ρ_f is the film's density, η_L is the liquid's dynamic viscosity and h_f is the film's thickness.

4.2.3. *Electroacoustic modeling*

The most common model relies on an equivalent electrical circuit (Figure 4.8).

The electrical characteristics of the quartz resonator can be evaluated using a physical quantity called quartz electric admittance, defined as the relationship between the current in the quartz and the excitation tension applied between its electrodes (electric admittance, Y, is the inverse quantity of the electrical impedance, Z). This parameter contains information on the stored energy and the power decreasing both in the quartz and the environment in contact with it.

In this model, C_o is the static capacity representing the flat condensator's capacity for the two electrodes separated by the quartz crystal. The fact that the quartz is piezoelectric introduces a dynamic branch series (R, L, C) in parallel with the capacity C_o. The quartz's capacitative behavior is dominant outside of resonance, whilst its dynamic behavior predominates resonance. R characterizes the losses in the resonator's energy, L the gain/loss in mass and C the device's piezoelectric properties.

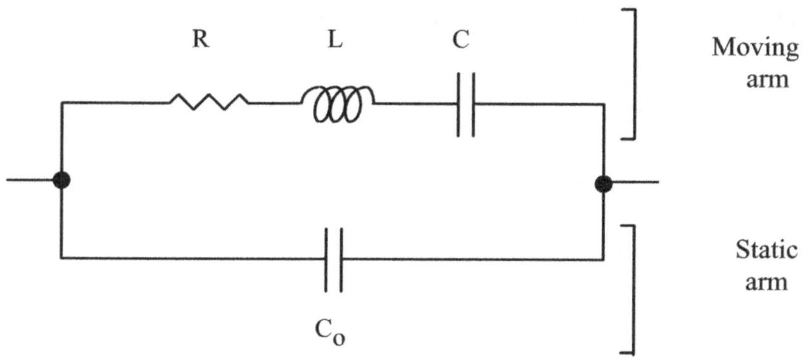

Figure 4.8. *Butterworth–Van Dyke circuit modeling the electrical behavior of a single piece of quartz*

Unfortunately, this approach is not always easily correlated with the different parameters of the film being studied. As such, models based on transmission lines have been developed [ROS 88].

It is possible both to correlate the films mechanical properties and validate the gravimetric result for the quartz microbalances by separating the mechanical and mass contributions.

The charged quartz resonator's electrical impedance can be expressed as follows:

$$Z_{th}^{V}(\omega) = \frac{h_q^{\ 2}}{4e_q^{\ 2}A} Z_f \frac{Z_s + Z_f \times \tanh(\theta_f h_f)}{Z_f + Z_s \times \tanh(\theta_f h_f)} \qquad [4.6]$$

where h_q is the quartz's thickness, e_q is the quartz's piezoelectric coefficient, A is the resonator's active surface (the area underneath the gold electrodes), Z_f is the film's acoustic impedance, Z_s is the liquid's acoustic impedance, θ_f is the film's complex propagation coefficient, h_f is its thickness and ω is the pulsation around the quartz's resonance frequency. More precisely, we obtain:

$$Z_f = \sqrt{\rho_f G_f} \qquad [4.7]$$

and $\theta_f = \sqrt{\dfrac{\rho_f}{G_f}}$ [4.8]

where ρ_f is the film's density and G_f is the film's shearing coefficient which can be written as a complex number, $G_f = G_1 + jG_2$, where G_1 is the film's rigidity coefficient and G_2 is the loss term.

4.3. Applications of quartz microbalances

4.3.1. *Electrochemistry*

This example allows us to illustrate the potential uses for quartz microbalances in electrochemical sensors and specifically in corrosion. A microbalance provides a simple and direct measure of corrosion speeds and, as such, an estimation of the effectiveness of organic compounds as inhibitors of corrosion.

The study presented relies on characterizing pure iron discs (99.99%) with a thickness 50 µm applied to a microbalance [EHA 01a]. This is immersed in a solution of H_2SO_4 1 M. Figure 4.9 shows the variations in the microbalance frequency over time, losses in mass and the iron disc.

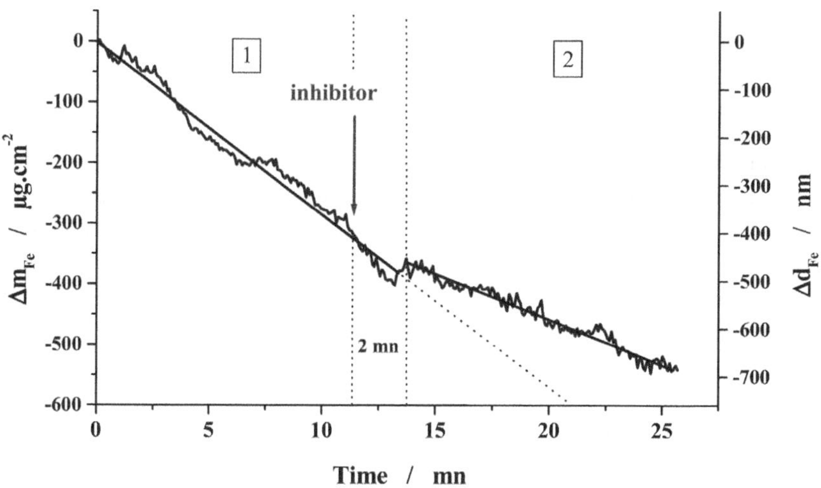

Figure 4.9. *Changes in the iron disc's mass and thickness [EHA 01b]*

The curve is shown between the two domains of globally linear variation, associated with the free corrosion of the iron without an inhibitor [1], followed by free corrosion with propargylic alcohol [2].

This second domain, which has a weaker gradient, confirms the inhibiting action of the organic compound. The curve also indicates that the influence of the inhibitor is only sensitive after two minutes following its introduction to the agitated solution. This gap represents the time required by the inhibitor to reach the sample's surface and to absorb in sufficient quantity. We have therefore verified in water that the introduction of the inhibitor does not modify the physical properties of the solution to the extent that it decreases corrosion speed due to the propargylic alcohol.

The corrosion speeds V_{corr} and $V_{\overline{corr}}$ measured before and after the inhibited action of the propargylic alcohol respectively, corresponding to the gradient of the curves in Figure 4.9, enable us to calculate the inhibiting ability of the propargylic alcohol. In this case studying the relationship R, the two speeds have been estimated as 1.9.

4.3.2. *Biosensors*

4.3.2.1. *Immunosensors*

Sensitivity in the quartz microbalance's surface or more precisely its gold electrode allows us to create an immunosensor [BIZ 99]. An example is provided here with the staphylococus enterotoxine B (SEB)/ anti-SEB antibody pair. This is isolated by direct adsorption on the microbalance's electrode in order to form the sensory layer. The interaction is characterized by a fall in frequency as shown in Figure 4.10.

Direct detection (without additive) *in situ* is carried out in a few minutes. The detection limit is of the order of 10^{-9} mole l^{-1} and it is possible to quantify the SEB mass fixed on the sensory layer via the microbalance's response since the electroacoustic measure has allowed us to validate Sauerbrey's relation for these monomolecular films. The more interesting fact concerns the large amount of accuracy with the sensory layer: the use of a swab containing another antigen (PO) does not lead to a variation in frequency as can be seen in Figure 4.10.

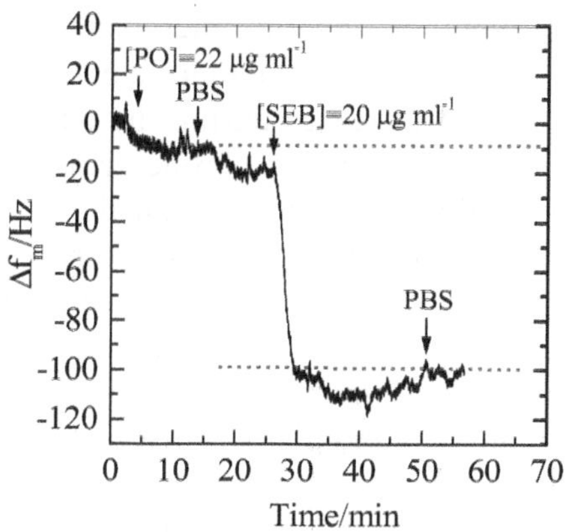

Figure 4.10. *Detection of toxins using a 27 MHz microbalance*

4.3.2.2. *DNA sensors*

The development of a rapid, sensitive and selective detection method for toxic elements is a current issue. For example, some toxic algae, such as Alexandrium Minutum (one of the most harmful algae), produces paralyzing substances which are ingested by shellfish such as mussels, oysters, clams, cockles and other molluscs, making them inedible during blooming of seasonal efflorescent algae. These paralyzing substances results in illness for those consuming them, with (potentially fatal) effects appearing as quickly as thirty minutes afterwards.

For this problem, the methodology used for detection is based on isolating a 20 base oligonecleotide corresponding to the complimentary sequence of a sequence of 20 bases characteristic of the genome DNA of the targeted toxic algae, Alexandrium Minutum. There are three methods for isolating this 20 base oligonucleotide, which is the DNA biosensor probe on a solid transducer: the adsorption of a probe marked with thioalcohol on a gold surface, the adsorption of a probe marked by a disulphide group on a gold surface and complexing a probe marked by a bitine group on a avidine biofilm. An example of isolation is provided in Figure 4.11 for an isolated disulphide probe [LAZ 05, LAZ 06].

For these three methods, quartz microbalances functioning at 27 MHz have been used for the different stages (Figure 4.12). The microbalance's frequency decreases during the circulation of a saline solution from a disulphide DNA probe corresponding to the chemisorption of a disulphide monolayer on the quartz's uncovered surface. The biosensor is selective because there is no variation in frequency during the circulation of a non-complimentary DNA solution. Lastly, the biosensor is sensitive because there is a reduction in frequency when a DNA complimentary DNA solution characteristic of Alexandrium Minutum is circulated, this decrease is attributed to the hybridization reaction of this complimentary DNA on the biosensor's surface. As such, it has been possible to characterize in detail the biomolecular structure in terms of both kinetics and thermodynamics, notably by calculating the kinetic constants of interaction and hybridization.

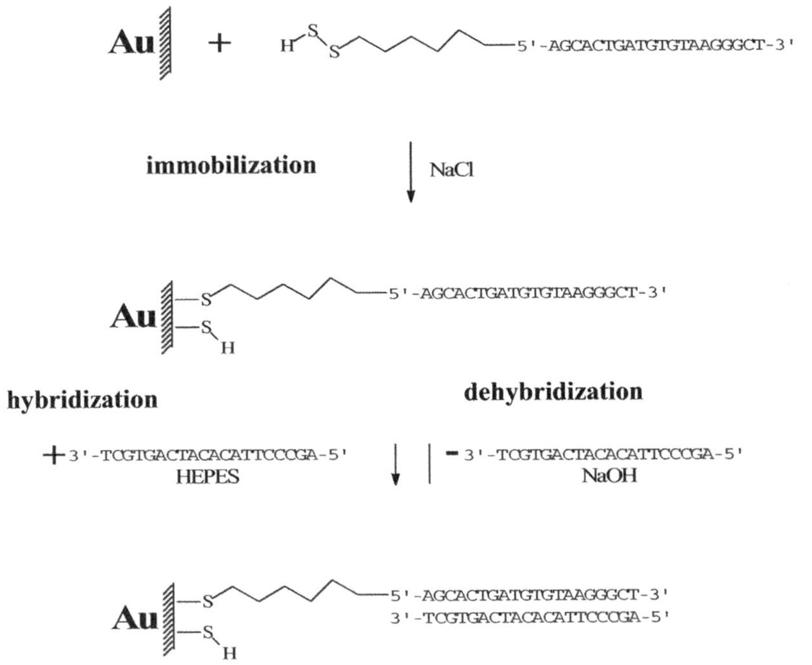

Figure 4.11. *Immobilization of a DNA probe marked with a diulphite group and the principle of a complimentary strand*

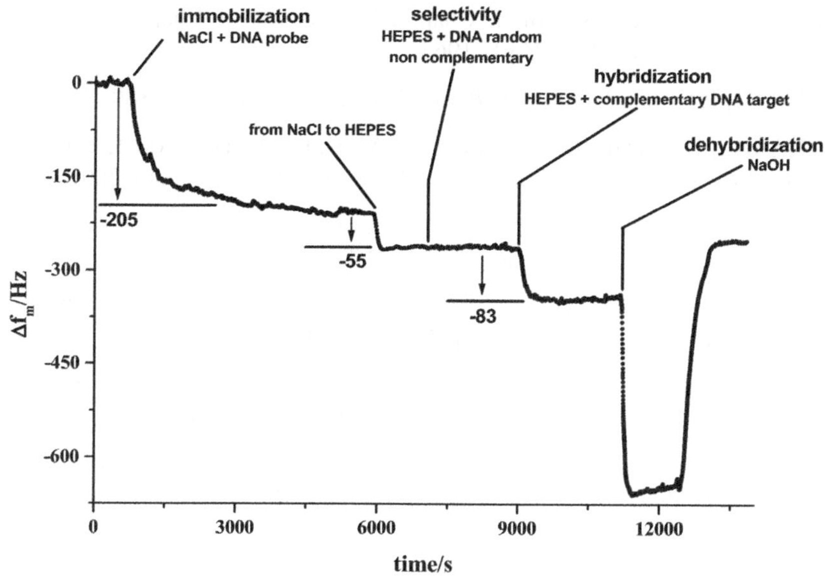

Figure 4.12. *Followed by a quartz microbalance of 27 MHz of the different stages of immobilization and detection of DNA strands*

4.4. Bibliography

[ARN 08] A. ARNAU (ed.), *Piezoelectric Transducers and Applications*, 2nd edition, Springer, 2008.

[BIZ 99] K. BIZET, C. GABRIELLI, H. PERROT, *Analysis EurJAC*, vol. 27, no. 7, pp. 609-616, 1999.

[BIZ 00] K. BIZET, C. GABRIELLI, H. PERROT, *Applied Biochemistry and Biotechnology*, vol. 89, pp. 139-150, 2000.

[BOU 95] D. BOUCHÉ-PILLON, C. GABRIELLI, H. PERROT, *Sensors and Actuators B: Chemical*, vol. 24, no. 1-3, pp. 257-259, 1995.

[BOU 96] D. BOUCHE-PILLON, Etude et mise au point de transducteurs ultra-sensibles fonctionnant en milieu liquide, PhD Thesis, Paris VI, November 1996.

[BRU 85] S. BRUCKENSTEIN, M. SHAY, *Electrochim. Acta*, vol. 30, p. 1295, 1985.

[BUT 92] D.A. BUTTRY, M.D. WARD, *Chemical Review*, 1992.

[CAD 64] W.G. CADY, vol. I, Chap. XIII and vol. II Chap. XIX, Dover Publications, Inc, New York, 1964.

[CUR 80] P. CURIE, J. CURIE, *C.R. Acad. Sci.*, vol. 91, p. 294, 1880.

[EHA 01a] H. EHAHOUN, C. GABRIELLI, M. KEDDAM, H. PERROT, Y. CÈTRE, L. DIGUET, *J. Electrochem. Soc.*, vol. 148, no. 9, pp. 333-336, 2001.

[EHA 01b] E. EHAHOUN, Etude et mise au point d'un capteur de corrosion basé sur une microblance à quartz, PhD Thesis, Paris VI, January 2001.

[EHA 02] H. EHAHOUN C. GABRIELLI, M. KEDDAM, H. PERROT, P. ROUSSEAU, *Anal. Chem.*, vol. 74, pp. 1119-1127, 2002.

[GAB 02] C. GABRIELLI, J.J. GARCÍA-JAREÑO, M. KEDDAM, H. PERROT, F. VICENTE, *J. Phys. Chem. B*, vol. 106, pp. 3182-3191, 2002.

[GAB 07] C. GABRIELLI, H. PERROT, D. ROSE, A. RUBIN, J.P. TOQUÉ, M.C. PHAM, B. PIRO, *Review of Scientific Instruments*, vol. 78, no. 074103, pp. 1-6, 2007.

[GAR 00] J.J. GARCIA-JARENO, C. GABRIELLI, H. PERROT, *Electrochemistry Communications*, vol. 2, pp. 195-200, 2000.

[GUI 84] GUILBAULT G.G., in C.S. LU, A.W. CZANDERNA (eds), *Applications of Piezoelectric Quartz Crystal Microbalance*, Elsevier, New York, vol. 7, p. 251, 1984.

[KAN 85] K.K. KANASAWA, J.G. GORDON, *Anal. Chim. Acta*, vol. 175, p. 99, 1985.

[LAZ 05] M. LAZERGES, H. PERROT, N. RABEHAGASOA, E. ANTOINE, C. COMPERE, *Chem. Com.*, vol. 48, pp. 6020-6022, 2005.

[LAZ 06] M. LAZERGES, H. PERROT, E. ANTOINE, A. DEFONTAINE, C. COMPERE, *Biosensors and Bioelectronics*, vol. 21, no. 7, pp. 1355-1358, 2006.

[LU 84] LU C.-S., CZANDERNA A.W., Elsevier, Amsterdam, 1984.

[MAR 91] S.J. MARTIN, V.E. GRANSTAFF, G.C. FRYE, *Anal. Chem.*, vol. 63, p. 2272, 1991.

[MAT 83] R.J. MATTHYS (ed.), *Crystal Oscillators Circuits*, John Wiley & Sons, 1983.

[ROD 05] L. RODRIGUEZ-PARDO, J. FARINA, C. GABRIELLI, H. PERROT, R. BRENDEL, *IEEE Sensors*, vol. 5, no. 6, pp. 1251-1257, 2005.

[ROS 88] J.F. ROSENBAUM (ed.), *Bulk Acoustic Wave Theory and Devices*, Artech House, 1988.

[SAU 59] G. SAUERBREY, *Z. Phys.*, vol. 155, p. 206, 1959.

[SCH 09] M. SCHLESINGER (ed.), *AC-electrogravimetry Investigation in Electroactive Thin Films*, Springer, 2009.

[SCH 85] R. SCHUMACHER, G. BORGES, K.K. KANASAWA, *Surf. Sci.*, vol. 163, p. L621, 1985.

[YAO 88] S.-Z. YAO, T.-A. ZHOU, *Anal Chim. Acta.*, vol. 212, p. 61, 1988.

Chapter 5

Metal Oxide Gas Sensors

5.1. Introduction: gas detection and micro-sensors [MOS 87]

Over the past 20 years, there has been growing development in gas detection systems using micro-sensors. In contrast to classic analysis methods, these systems must be simple, small and cheap. Whilst they may not perform as well as classic analysis methods, they perform well enough for a number of applications, such as pollution control alarms or process automization.

Sensor performance is often measured against the "3S", i.e. sensitivity, selectivity and stability. Sensitivity is the easiest aspect of performance to satisfy as sensors are generally sensitive enough for the majority of applications. However, the same cannot be said for selectivity and stability. The latter affects all information processing inside the sensor and we obviously cannot process a temporally derived signal, as is often the case with "chemical" sensors. The lack of selectivity is the current main limitation of industrially developed micro-sensors. In contrast to purely physical methods of analysis, the physicochemical principles used in the majority of micro-sensors allow only partial selectivity: we can detect a specific gas in relation to one or other family of gases but this still provides a more or less

---

Chapter written by Christophe PIJOLAT.

significant result. In the majority of cases, this governs the use of micro-sensors. On the one hand, different performances are often interdependent and in some cases it is preferable to choose working points that decrease sensitivity in exchange for increased stability and selectivity. However, the areas of application for micro-sensors change according to improvements in selectivity: with almost zero selectivity, only applications based on major gases are possible (detection of a threshold level in process automization), whilst sensors with partial selectivity are currently successfully used for "selective" detection of a number of gases (CO, CH_4, hydrocarbons, etc.) to control pollution in domestic and atmospheric environments.

Another element to consider is the cost of the sensor. In order to be used several times in a measure network or a range of individual devices, these sensors must be effectively produced at low costs. Successfully marketed gas sensors are simple and cheap to produce (e.g. catalytic-type detectors) with their production relying primarily on microelectronics, a technique which has become increasingly competitive due to recent developments in micro-informatics, allowing extensive automization of production processes and therefore a significant reduction in costs.

Micro-electronics has also enabled the miniaturization of sensor elements with all its accompanying advantages (decrease in consumption and new areas of application), guaranteeing reproducibility on important batches.

These micro-electronic techniques also have other advantages. Generally, a sensor is made from a sensor element, which initiates a signal and a transducer ensuring its exploitation. In the majority of modern detection devices, these two functions are combined in a miniaturized system. Transistor-based MOS-FET detectors are a perfect illustration of this kind of integration.

Today the methods used, physical or physicochemical concepts involved in detecting or analyzing chemical spaces, are numerous. The chosen reaction mechanism often relies on an electrical disturbance. Other physical phenomena are used with important developments currently involving new gas sensors based, for example, on changes in mass (piezoelectrical sensors) and acoustic (SAW sensors) or optical properties. The latter certainly have significant potential and, as in micro-electronics, recent progress in fiber and integrated optics has enabled the development of optical sensors in recent years. However, these detection systems only differ in terms of the mode of information transduction. By pushing reasoning to the extreme, the same gas sensitive element can be used in a piezoelectric system where we track the increase in mass created by the adsorption of gas on the sensor material; or in an optical system where we record a variation in the material's indicator during this same adsorption. In fact, the performance of these systems depends as much on the transducer as it does on the sensor element itself. It is therefore important in each case to ensure the performance (the 3S' as well as reversibility) of the sensor material placed in interaction with a gas environment.

Taking into account the diversity of these possible transduction devices (piezo, SAW, optical, etc.) and the planned areas of application, this chapter will concentrate on gas micro-sensors involving electrical concepts on solid materials (excluding organic conductors). Furthermore, whilst this chapter will focus on metal oxide-based semi-conductors, we will introduce the two other types of gas detectors using solid materials which have led to the development of industrial sensors, including catalytic detectors often called pellistors and potentiometric detectors. Given the growing number of gas sensors developed industrially and their manufacturers' more or less successful attempts to commercially break into the market, we will not provide an exhaustive overview of these devices in this chapter, instead we will look at significant examples for the three types of sensor. It is important to provide an overview of the technologies involved as well as the performance expected for these products. In relation to these "conventional" technologies, we are beginning to see sensors using micro-electronic technology appear on the market. Again, it is important to examine the importance of these technologies with regard to these different types of sensors.

5.2. Catalytic sensors

A pellistor is a miniature calorimeter used to measure the energy created when a combustible gas is oxygenized. Here, we are directly exploiting the exothermic reaction created when a gas comes into contact with a catalyst. The sensor element is made from an oxide with a large specific surface area supporting a metal catalyst fitted with a platinum wire. This wire has two uses. On the one hand it heats the catalyst to an adequate working temperature (between 500 and 800°C) and on the other it has a transducer allowing us to measure changes in temperature (measure of electrical resistance) caused the by oxidization of flammable gases. Note that these detectors require oxygen and cannot detect gases in non-oxidizing environments, such as nitrogen. As such, this detector has low selectivity. In terms of hydrocarbons, systems using several elements which are sensitive to different temperatures allow selective detection of particular compounds, specifically methane: a high temperature filament enables all hydrocarbons to be cracked with the exception of methane which, since it is the most energetically stable gas, requires selective detection by difference. Catalytic detectors are also used in detection systems for other reductor gases such as carbon monoxide or hydrogen. Again, selectivity is a significant limitation. Technologically, these problems of lifespan and poisoning are the main limitations of commercialized products which have governed their industrial development over the last few years.

Historically, "pellistors", at the beginning of the 1960s, were the first industrially developed gas sensors. These sensors are used to detect hydrocarbons in air, specifically methane (explosivity) but are also used for other gases. Early versions simply included a platinum wire heated to 800 or 1,000°C. Due to the rapid evaporation of platinum, these sensors only have a lifespan of around a few hours. In addition, the systems that are currently being developed include a platinum wire cased in porous aluminum containing metallic catalysts. The increase in catalytic activity due to the large specific surface area allows it to operate at temperatures reaching 500°C and, by extension, reduce the sensor's consumption and increase lifespan.

The characteristics of the pellistor developed by the EEV, in the UK are a perfect example of an industrial product (Figure 5.1). The diameter of the platinum wires used is generally around 0.5 mm. The energy consumed is around 0.8 W but can be reduced by decreasing the wire's diameter. This wire is covered with an aluminum bead which contains the platinum or palladium based catalyst. The detection system comprises two detectors placed in a Wheatstone bridge. One is active (containing the catalyst) and the other is a compensation element where the aluminum bead has been treated to prevent oxidization of the combustible gas. In the absence of catalyst poison, these elements can last up to 10 years. The decrease in catalytic activity after exposure to specific products is one of the main problems with this type of sensor. For example, it can be completely destroyed after being used for 24 hours when there is 25 ppm of tetramethyl lead. The effects of HMDS (hexamethyl-disoloxan) are even more significant. To solve this problem, versions incorporating filters (sieving and active carbons) have been recently developed. This allows us to limit poisoning (lower sensitivity after several hours of 10 ppm HMDS) but not completely. An important aspect of research currently being carried out into pellistors consists of avoiding this poisoning and reducing consumption, either by reducing the working temperature or miniaturizing systems.

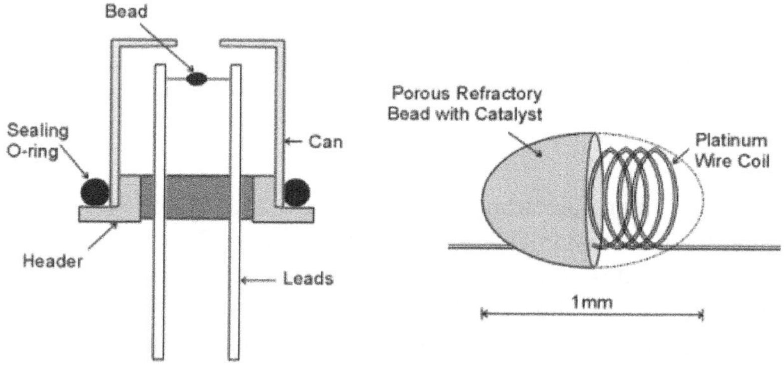

Figure 5.1. *Principle of how a catalytic sensor works (EV – GB)*

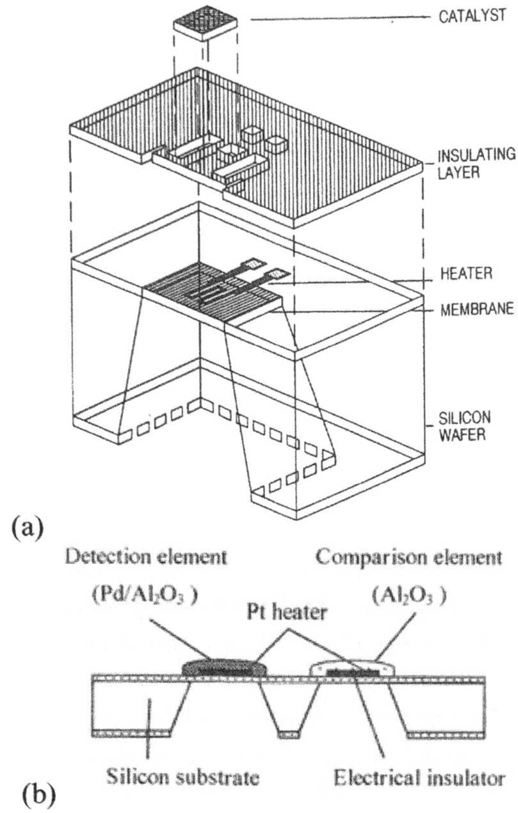

Figure 5.2. *a) Catalytic sensor developed by Microsens [KRE 93]] (Neuchatel – Switzerland). b) Catalyst with two elements on a silicon structure (Tokyo Gas – Japan)*

Today micro-electronic technologies are used to maximize miniaturization and enable industrial production. The first to introduce micro-hot plates on silicon were the researchers at Microsens (Figure 5.2a) in the 1980s. Since then, a number of similar technological developments have been proposed [VAU 91]. An example is provided in Figure 5.2b with the Tokyo Gas sensor (Japan). In pulsated mode, these types of sensor can have a consumption of several mW and result in more selective detections by combining signal processing and filtering superficial layers (detection of CH_4 and CO).

5.3. Potentiometric sensors [FOU 82]

In these devices, the transport properties of specific materials' electrical charges, called solid electrolytes, are used. In contrast to semi-conductor materials, which have electronic conductivity, transport of charges is ensured by the migration of ions (ionic conduction) or more generally by point defects in the solid. Whilst there is a modification in the concentration of mobile species on the surface of the material (for example, consumption of oxygen for an oxide), there is a displacement of charges creating an electrical field between the surface and the volume. Equilibrium is therefore reached when the force of chemical diffusion due to the spaces' gradients are exactly opposite to the resulting electrical force. If the effects are symmetrical on the solid sample's different faces, there is an exact compensation of fields and no electric effect is detectable. Inversely, all dissymmetry is the basis of an electromotor force, E, which is measurable and the result of Nernst's law: E = a Log P1/P2, with P1 and P2 representing the concentrations of gases at the two extremities of the ionic conductor.

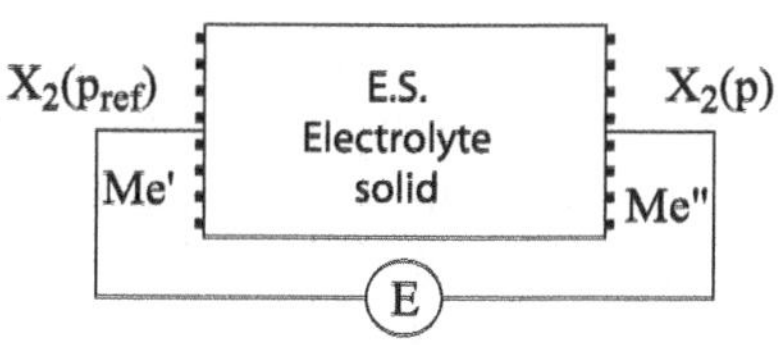

$$\underset{\text{Gas}}{1/2\, X_2} + \underset{\text{Metal}}{ne^-} = \underset{\text{E.S}}{X^{n-}}$$

$$E = \frac{RT}{2nF} \log \frac{P_{X_2}}{P_{X_2}^{ref}} \quad \text{Nernst's law}$$

Figure 5.3. *Principle of an electrochemical sensor and Nernst's law*

The most common example is the oxygen sensor. The system is composed of two gas cells tightly separated by a metallic oxide used as a solid electrolyte. If we fix the partial pressure of oxygen in a compartment, we can measure the partial pressure of oxygen in the other compartment. These detectors' performances depend on the type of electrolytes and their working conditions. In particular, sufficient electrical conductivities are required. This is obtained by doping the material and/or working at high temperatures, with temperature also influencing kinetics and reactions, and therefore the reversibility of such systems. We will see later that the industrial development of oxygen sensors uses zirconium dioxide doped either with yttrium (mixture of ZrO_2-Y_2O_3), or calcium (CSZ).

The above use of oxygen constitutes the "concentration cell" mode but this same sensor can also be used in "combustion cell" mode. We therefore obtain an electrochemical system in which a combustible is directly converted into an electrical current. For example, in the case of carbon monoxide, the oxidization of CO at the anode leads to the formation of electrons which are consumed at the cathode, reducing the amount of oxygen present. The measure of the current within the exterior circuit between the anode and the cathode therefore allows us to measure the concentration of carbon monoxide. Such oxidation reduction reactions are similarly used for detecting various gases such as hydrogen.

Aside from oxygen, the possibilities for detecting other chemical environments are numerous since there only needs to be an ionic conductor whose mobile space is identical to the species being analyzed or which can react with it. Some of the numerous potential uses include the detection of SO_2 with sulfates, NO_2 with nitrates, CO and CO_2 with carbonates. Whilst we will focus on the detection of gases, note that these conductors are widely used to create micro-sensors for chemical spaces in solutions, specifically those using selective membranes [FAB 92]. Primary examples of this include the detection of Na+ ions using a material called Nasicon and, more generally, a series of glass conductors to dose cations (H^+, Na^+, K^+, Li^+, NH_4^+, etc.) or inorganic insoluble Ag, Cu, Cd, etc., based salts to dose anions (Cl^-, Br^-, I^-, S^-, Se^-, etc.).

The "Lambda" oxygen sensor is the best example of a gas sensor which has various industrial versions both in Europe, the United States and Japan with a certain amount of success since these sensors are installed into vehicle motors by a number of manufacturers. Control motors using oxygen sensors save energy and reduce toxic emissions. The effectiveness of three way catalysts (CO, NO_x and unburnt hydrocarbons) highly depends on the composition of the "stoichiometric" mixture for a relationship of λ = fuel/air = 1 (Figure 5.4a). The oxygen sensors at the solid electrolyte are particularly well adapted to this relationship. During the change from

$\lambda = 1$, we progress from an oxygen rich mixture to one which is low in oxygen, hence the name "Lambda" sensor. In terms of measurements, these sensors can be placed directly in the exhaust gases. One side of the solid electrolyte is in contact with the exhaust gases. The other is in contact with the outside air which will serve as a reference point for oxygen pressure (Figure 5.4b). The solid electrolyte is a mixture of zirconium dioxide and yttrium oxide. The trivalent yttrium ions are substituted with tetravalent zirconium ions in the crystal lattice and, due to the electrical neutrality of these materials, have holes of oxygen which ensure ionic conduction of the oxygen ions.

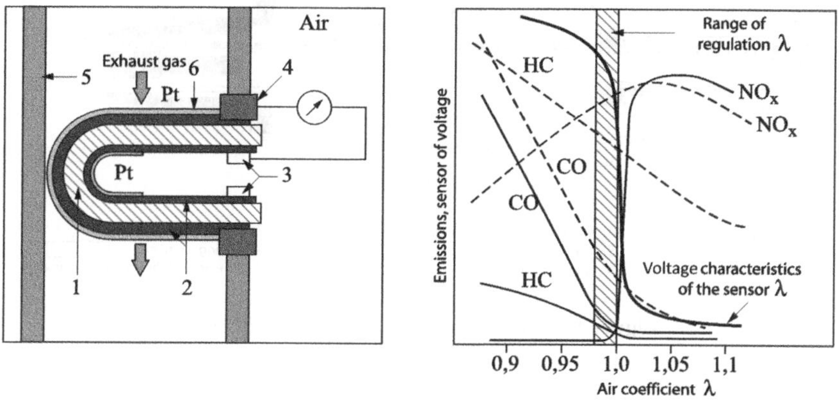

(a) (b)

Figure 5.4. *a) Principle of how a "Lambda" sensor functions in a car exhaust; b) regulation window in a 3 path catalyst system for CO, HC and NO_x*

The example here focuses on the "Lambda" sensor marketed by Bosch (Figure 5.5). The ceramic part (ZrO_2 (Y)) of the sensor is a tube closed on one side. The surface of the ceramic is covered in a microporous platinum layer which serves as an electrode and a catalyst. An extra-porous ceramic layer is superposed on the platinum layer next to the exhaust gases to prevent the erosion of the catalyst's platinum layer by combustion emissions. There are two versions, a non-heated sensor and a sensor with a heating system.

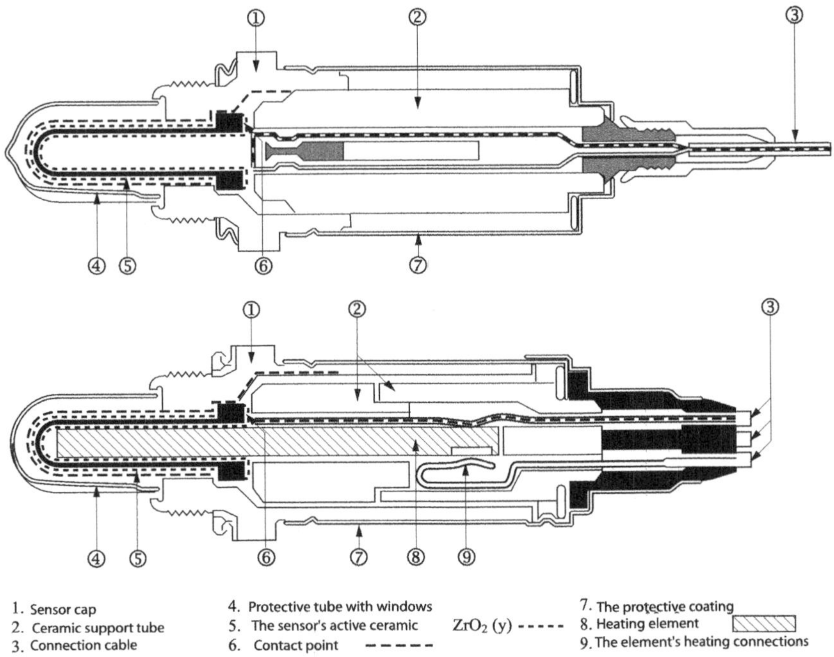

Figure 5.5. *"Lambda" oxygen sensor by BOSCH*

The use of sensors in exhaust gases creates temperatures of several hundred degrees which are sufficient for them to work properly (minimum 350°C). When the application requires heating of the sensor element, there is generally an exterior heating element with an adapted sensor which is not technologically easy to manage. In terms of oxygen reference points, oxygen sensors have been developed with a Pd-PdO mixture as an internal reference. To combat the two limitations (heating and reference) mentioned previously, there have been a number of developments relating to thin and thick layer technologies [TAK 86]. As a prime example, the sensor created by Thomson [VEL 82] in 1982 was developed using a sapphire substrate. Micro-heating consisting of a platinum resistive layer is placed on one side (technology normally used in a number of gas sensors, particularly semi-conductors). On the other side, there is a series of thick layers made from the solid electrolyte (2 µm of CSZ), active electrodes (15 ppm of porous platinum) and references (20 µm of

Pd/PdO_x). The whole is surrounded by a protective glass which allows the lateral separation of the two electrodes as well as access by the gas to the active electrode by micro-channels. Current progress has focused on the development of "proportional" oxygen sensors, which work in amperometric mode (Figure 5.6a) and NO_X sensors (Figure 5.6b), both for diesel motors.

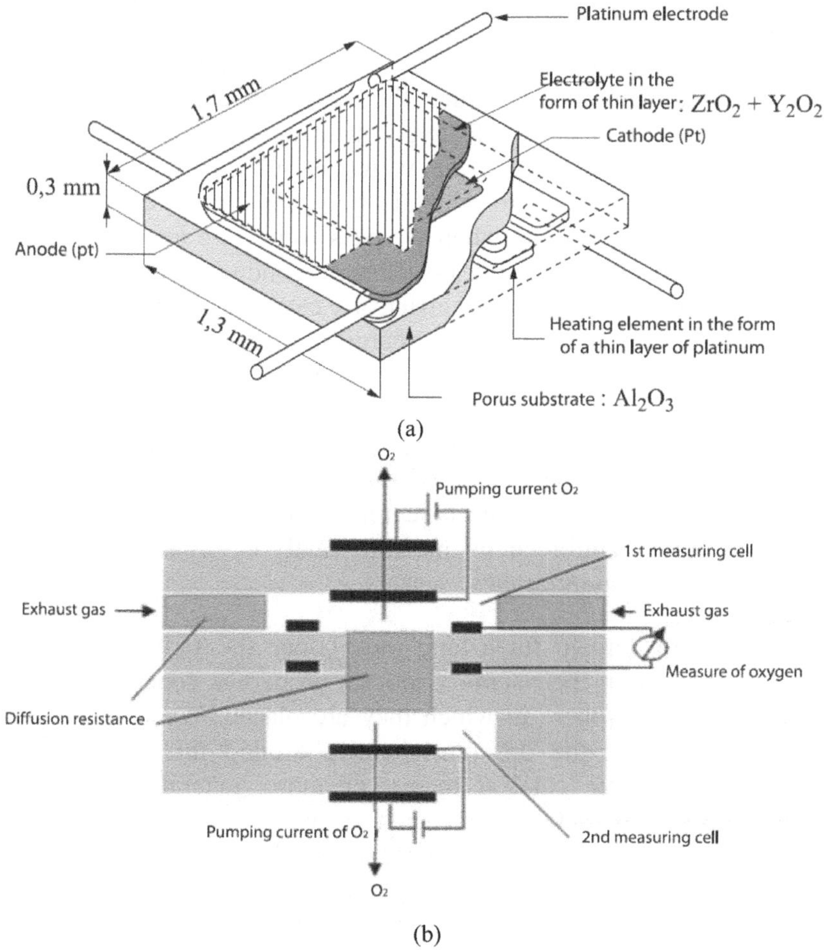

Figure 5.6. *a) Amperometric sensor for measuring proportional oxygen; b) NOx sensor by NGK using several O_2 sensors*

5.4. Semi-conductor sensors

5.4.1. *Introduction [MOR 82, HEI 82, GOP 91, SEI 92]*

Here we are using materials which present variations in electrical resistance depending on the type of gas atmosphere into which they are placed. The choice of material depends in particular on the type of gas being detected. The most common use is the detection of reductor gases (CO, H_2, CH_4, hydrocarbons, etc.) in air, therefore in oxidizing environments. Some semi-conductive metallic oxides are therefore well adapted materials. As such they have an electrical resistance which is not too great to effectively measure electrical conductivity with traditional instruments, but which is sufficiently high at high temperatures in fresh air to maintain good sensitivity, depending on the resistance discrepancy when we pass from pure air into polluted air. In this case we are using n-type semi-conductor metallic oxides (see SnO_2) in which the major carriers are electrons. The adsorption of a reductor gas providing electrons therefore leads to an increase in the electrical conductivity (reduction in resistance). In contrast, a p-type semi-conductor is used to detect an oxidizing gas. In both cases, this transfer of charge requires ionization of the gas in contact with the material, justifying the choice of material with a large amount of catalytic activity. This is the case for certain metal oxides but this aspect is often reinforced by metallic catalysts, which allow us to increase the selectivity of these sensors.

It has been known for a long time (since the 1950s) that metal oxides can vary in their semi-conductor properties depending on the type of gas atmosphere in which they are placed. We will therefore examine some of the different theories around this question, for example, electronic catalyst theory or more generally, adsorption laws. The use of metal oxides such as gas sensitive materials was initiated in the 1960s by Japanese researchers. Around the same period, Seiyama demonstrated that variations in electrical resistance in a ZnO film allowed the detection of reductor gases, and Taguchi proposed the first sensor made from porous SnO_2 ceramic.

A number of semi-conductor materials have subsequently been studied and used in the production of gas detectors. Metal oxides certainly constitute the vast majority of studies carried out over the past 20 years. This ranges from simple oxides, such as NiO, ZnO, MgO, SnO_2, TiO_2, Fe_2O_3, V_2O_5, etc., to complex structured compounds such as some ternary oxides, perovskites, piezoelectric materials (titanates, etc.). It is obviously impossible to provide an exhaustive list of all the metal oxides studied regarding gas sensors in this article. However, in the literature, there are studies focusing on all types of oxides, whether pure or doped as well as combinations of several oxides. In this context however, we will provide the basic principle and notions required for the use of these materials. Given that the most commonly used oxide is tin oxide, there are a number of specificities and examples focusing specifically on this material.

5.4.2. *Basic principles [KOH 89]*

Generally a gas sensor must have at least two basic functions: one being to recognize a specific gas amongst a mixture of other gases (recognition function) and the other being to transform the recognition of a gas via an electrical or optical signal (transduction function). Evidently, resistive sensors are the simplest systems for developing gas sensors. Any interaction between a gas and the material causing an electronic transfer towards the material could be used for detection.

In the system described here, metal oxides have different advantages. For example, they can often provide two functions of recognition and transduction: a pure tin oxide sensor without the addition of a catalyst presents significant variations in electrical conductivity depending on specific gas atmospheres, which are directly measurable without a complex electrical system. This is the second main advantage of metal oxides: depending on the example in question (gap and densities of charge states), these materials can have higher electrical conductivities in temperature ranges affected by the phenomena involved. For example, if we use interactions resulting from the adsorption and reactivity of certain gases on the surface of the metal oxide, we will focus on variations in electrical conductivity between 350 and 500°C for problems of reversibility and interferents (see water action). The usable materials are therefore semi-conductors which have adequate conductivity values, including those which are high enough to be easily measured (10^{-3} to 10^{-6} $ohms^{-1}cm^{-1}$), and which are not too big so as to maintain high gas sensitivity (relative resistance variations $\Delta R/R$) and which do not have too great a dependency on temperature. A number of metal oxides can, for example, be a thousand times less conductive than traditional semi-conductors such as silicon and germanium and still satisfy these criteria.

The general detection process can be summarized as follows: the gas being detected is adsorbed at the oxide's surface or reacts with preadsorbed species (essentially oxygen species) on the oxides, thus causing a variation in the space charge in the surface layer environment, which is converted into variations in the material's electrical resistance (see Figure 5.8a). Gas recognition is ensured by the series of surface chemical processes and more specifically by adsorption sites and reactions which play the role of gas receptors. However, the conversion of this kind of interaction into an electrical signal variation is enabled by a physical process linked to electron transport. Nonetheless, this is not only dependent on the semi-conductor properties of these oxides but also the microstructure of polycrystalline materials by the intermediary of grain size and boundaries. We will describe in further detail the main phenomena involved in metal oxides.

5.4.3. *Semi-conductive properties of metal oxides and redox reactions*

Metal oxides with a general MO formula can provide stoichiometry deviation. The semi-conductive properties of these oxides, originates from this aptitude for sub- or super-stoichiometry. These deviations are due to the presence of more or less ionized point defects in the oxides' crystal lattice. These can consist of cationic vacancies or interstitial anions which release electrons, making the material a p-type semi-conductor.

Depending on the type of predominant faults in the solid, there are four types of oxides: MO_{1+X}, $M_{1+X}O$, MO_{1-X}, $M_{1-X}O$ (M = metal, O = oxygen, x = deviation to stoichiometry). The four types of solids are represented by Figure 5.7. Oxides a and b are n-type and the two others (c and d) are p-type. For example, SnO_2 has a tendency to be sub-stoichiometric in oxygen (written as SnO_{2-X}). Subsequently to, laws governing point defects, oxygen vacancies causing free electrons in the oxide's crystalline lattice and therefore an increase in the concentration of free carriers, confirm the semi-conductor's n-character.

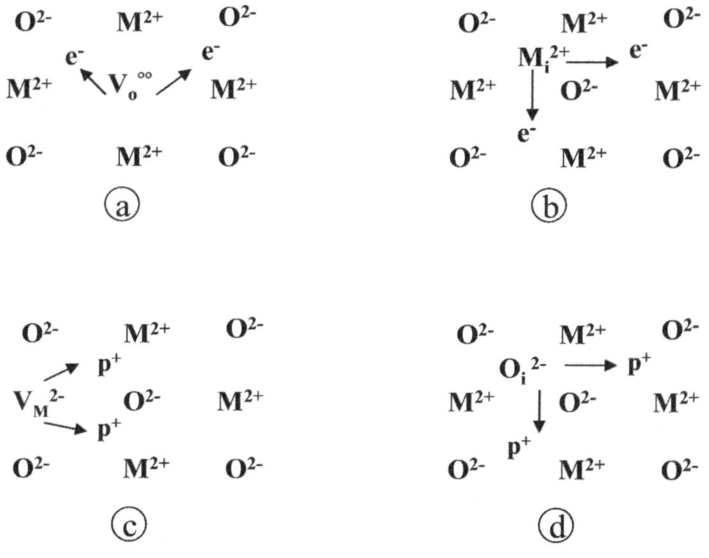

Figure 5.7. *The four types of semi-conductive metal oxides: a) oxygen vacancies (n-type); b) metal in interstitial position (n-type); c) metal vacancies (p-type); and d) oxygen in interstitial position (p-type)*

By reacting with an oxidizing gas or a reductor gas, the metal oxide can evolve by increasing or decreasing the value x representing the deviation to stoichiometry. These interactions are the cause of the variation in the materials' electrical properties. The most significant reaction is that of oxygen. All these metal oxides are in equilibrium with oxygen gas via the mass law, which causes the occurrence of point defects. In the case of n-type oxides with oxygen vacancy (SnO_2, ZnO, etc.) the oxygen atoms in the network (expressed as O^x_o) are in equilibrium with the gas molecules of O_2 and the oxygen vacancies in the lattice, which are simply or doubly ionized (expressed as V°_{o} and $V^{\circ\circ}_{o}$):

$$\frac{1}{2} O_2 \text{ (gas)} + V^{\circ}_{o} + e- = O^{x}_{o} \quad \text{therefore: } [V^{\circ}_{o}]\,[e^{-}]\,P^{1/2}_{O2} = K_p$$

and by applying the condition of electroneutrality, we obtain:

$$[e-] = k\,P^{-1/4}_{O2}$$

i.e. the material's conductivity is directly dependent on the partial pressure of oxygen. In the case of double ionized oxygen vacancies, we therefore obtain:

$$\frac{1}{2} O_2 \text{ (gas)} + V^{\circ\circ}_{o} + 2e- = Oxo \quad \text{therefore: } [V^{\circ\circ}_{o}]\,[e^{-}]^2\,P^{1/2}_{O2} = K_p$$

and by applying the condition of electroneutrality, we obtain:

$$[e-] = k\,P^{-1/6}_{O2}$$

This same calculation can be applied to the four types of metal oxides introduced previously (point defects and different oxygen dependency laws).

5.4.4. *Electronic surface states, space charge and gas adsorption*

The energy levels of an n-type semi-conductor can be represented by the band diagram in Figure 5.8b. The last permitted energy bands (BV and BC) are separated by the gap where localized states are found (whether donor or acceptor), which could be occupied by electrons. For SnO_2, the gap is 3.5 eV and the donor levels are simply or doubly ionized oxygen vacancies. As with any semi-conductor, an increase in temperature leads to the movement of electrons from these traps towards the conduction band. On the surface, the band diagram is modified by the electronic surface states, therefore creating a specific zone called the space charge zone. This can result from intrinsic defects in the material's surface but also particularly by the adsorption of atoms and foreign molecules. Variations in the layer's space charge are directly related to the extraction or injection of electrons by acceptor or donor species. When a surface is charged, a double electrical layer appears with an equal charge but an opposite value. This leads to a curve in the bands in the energy diagram. The concentration of electrons near the surface therefore depends on the type and concentration of gases adsorbed on the surface. This is the basis of gas detection properties by metal oxides.

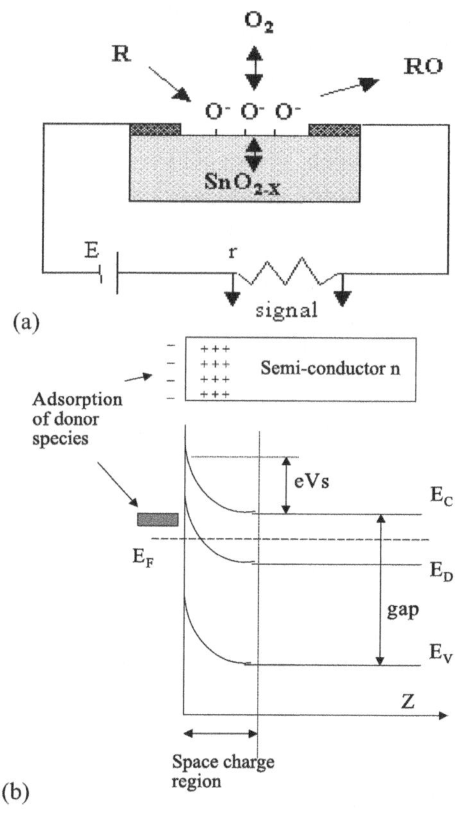

Figure 5.8. *Diagram showing how an SnO_2 sensor functions: a) influence of oxygen gas on the stoichiometry deviation x, and redox reactions with oxygenized surface spaces: b) band diagram for an n-type semi-conductor and the influence of the adsorption of electron accepting molecules*

The adsorption of oxygen on an n-type semi-conductor, such as SnO_2, leads to an electronic transfer towards the adsorbed molecules and thus a decrease in the electrical conductivity. This can be expressed as the following equation:

$$\frac{1}{2} O_2 + 2 e^- = O^{2-}_{adsorbed}$$

The influence of other gases can have two major effects. It can cause the direct transfer of charge following adsorption as previously described, as well as for chemical reactions with pre-adsorbed species, specifically oxygenized species. The reactions involved are essentially redox reactions. Taking the example of an n-type semi-conductor such as SnO_2, the gas' reaction can be illustrated as follows:

– with a reductor gas such as CO: $CO + O^{2-}_{adsorbed} = CO_{2\ adsorbed} + 2 e^-$ leading to an increase in electric conductance and, in terms of space charge zone, a decrease in the size of the electron depletion region.

– with an oxidizing gas such as NO_2: $NO_2 + 2 e^- = O^{2-}_{adsorbed} + NO_{adsorbed}$ leading to a decrease in electrical conductivity and, in terms of space charge zone, an increase in the size of the electron depletion region.

The response of a metal oxide based gas sensor such as SnO_2 can be illustrated as shown in Figure 5.8a: a measure of the variations in the material's electrical resistance, depends on the stoichiometry deviation x, therefore the partial pressure in adsorbed oxygen and on the quantity of adsorbed oxygen, with the latter depending on the range of redox reactions of the gases present.

5.4.5. *Effect of grain size and the complexity of polycrystalline materials [YAM 91, BAR 99]*

The phenomena described previously, concern what we call surface conductivity in contrast to bulk conductivity. A monocrystal therefore acts as two resistances (surface and bulk) in parallel. In reality, this is only rarely the case. The influence of the space charge zone can become highly complex with polycrystalline materials with very small grains, closely related to the size of charge zone, also known as L_D, Debye length. If the grain's diameter is two times less than L_D, the grain is therefore completely affected. In the previous case with SnO_2 in the presence of oxygen, the grain becomes almost isolating with a reducing gas it becomes much more conductive (see Figure 5.9). The quality of grain boundaries therefore becomes the determining element of such a material. If the size of the grain boundaries in this material is less than 2 L_D, then the action of gases (oxidizing to the reductor) induces a significant modulation in the material's conductivity, which could be approached as an almost "all or nothing" system. This in part explains the significance of semi-conductor materials with very fine granulometry (nano-powder) for producing gas detectors which are highly sensitive. However, these materials therefore become the site for a number of phenomena, which are particularly difficult to manage, as we will now see.

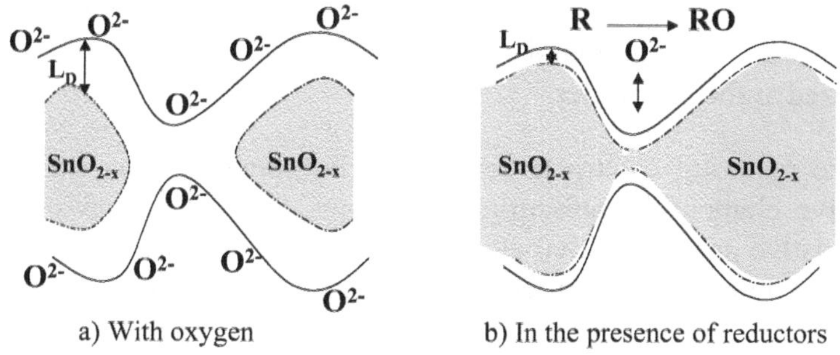

Figure 5.9. *Grain cross-section: blocking with only oxygen a); and a conductor with reductor gases b) affecting the size of the charge space zone*

The interaction between these materials and a gas environment contributing to variations in their electrical conductivity is both part of their value for their use as a gas sensor as well as the difficulty encountered for managing their technological development. In a number of cases, the surface states of these oxides, particularly oxygen and adsorbed hydroxyl surface groups, have to be taken into account. Since these species are elements constituting the atmosphere being analyzed, which are susceptible to changing with experimental conditions of pressure and temperature, these different spaces are interferents to the gas being detected. They often affect the material's temporal stability and can also act as a promoter or inhibitor of adsorptions of the gas being detected. Depending on the nature of the reactional mechanisms involved, these semi-conductor materials must function at temperatures ranging between 300 and 500°C. An increase in temperature increases the kinetic of the reaction and with it the sensor's response times and reversibility. However, the choice of working temperature becomes delicate if we consider that this parameter also determines the type and concentration of surface states and, by extension, the sensor's reactivity with regard to various gases.

To measure conductivity in these kinds of polycrystalline materials, we have to form the material (ceramic produced using a sintered powder or thin layer placed on a substrate) and produce metal electrodes for electrical contacts. We can see (Figure 5.10) that this global electrical measure includes different physicochemical phenomena that we can schematize as three main contributions: the grain itself, the grain boundary and the semi-conductor metal interface localized at the electrodes.

The resulting electrical device can be considered as a series of resistive elements, representing the grains' intrinsic behavior linked by resistive and capacitive elements representing potential barriers to different interfaces (semi-conductor metal contacts and grain cross boundaries). As such, gas sensors based on the principle of varying resistance according to the adsorption of gas molecules generally function at a high temperature in order to guarantee good reversibility of adsorption phenomena (desorption of the species detected when its concentration decreases in the gas environment). Typically, these temperatures are in the order of 300 to 500°C in the case of tin dioxide. As such, the majority of the previously described processes are temperature dependent. The crossing of potential barriers at the interfaces is activated by temperature. The variation in the material's intrinsic conductivity with temperature can take different forms since it is the product of mobility and the concentration in major carriers (electrons in our case, with a conductivity which can be expressed as: $\sigma = n.q.\mu$).

Furthermore, the majority of the phenomena involved in the chain examined previously also depend on the presence of gas species in the surrounding atmosphere. This is the case for adsorption on the grain itself, a phenomenon which is primarily predominant for gas sensors. However, adsorption also occurs at the electrode level and particularly at grain boundaries. These have a concentration of important faults and as such are adsorption sites favored by certain gases. It is also necessary to account for intrinsic defects at the surface such as hydroxyl groupings, which are largely present on the surface of the tin oxide, as well as catalytic effects which could take place either on the material itself or on the electrodes. It is therefore easy to understand the complexity which arises from the combination of the phenomena presented here and their respective changes with temperature or the surrounding gas environment. This complexity is manifested particularly with conduction curves obtained depending on the temperature (G = f (T)), which could present maxima according to the type and concentration of gas [LAL 84, LAL 81].

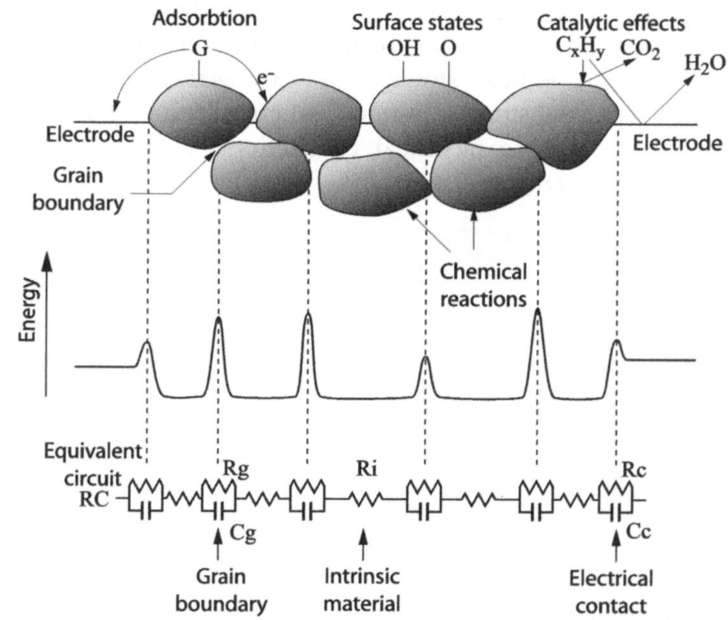

Figure 5.10. *Range of phenomena affecting electrical conductivity in a poly-crystalline material: adsorption, chemical reactions, catalyst (material and electrodes), surface states, interfaces (grain boundaries and metal-semiconductors) and an equivalent electrical circuit*

5.4.6. *Examples of industrial developments and applications [TAK 88]*

The first patents concerning these materials were submitted in 1962 in Japan by Seiyama (ZnO) and Tagushi (SnO2). Their use led to the appearance of the first commercial semi-conductor sensor in 1968 (the Figaro-Capteur TGS made of SnO2), which was designed to detect domestic gas leaks. This Japanese company produced different versions of SnO_2 sensors for micro-wave cookers (1979), domestic gas (1980), hydrogen sulphite (1981), carbon monoxide (1983) or even freons (1989). In total, there is an estimation of more than 80 million SnO_2 sensors currently in use in Japan.

Metal oxide based sensors have created an immense amount of interest globally due to the fact that they are designed as simple systems and offer a number of advantages depending on the sensor materials used and possibilities for miniaturization and integration in these systems. A typical structure for a metal oxide gas sensor contains a heating substrate (ceramic or micro-electronic) in which the sensor element and the electrical connections, even a temperature sensor, are applied (Figure 5.11a).

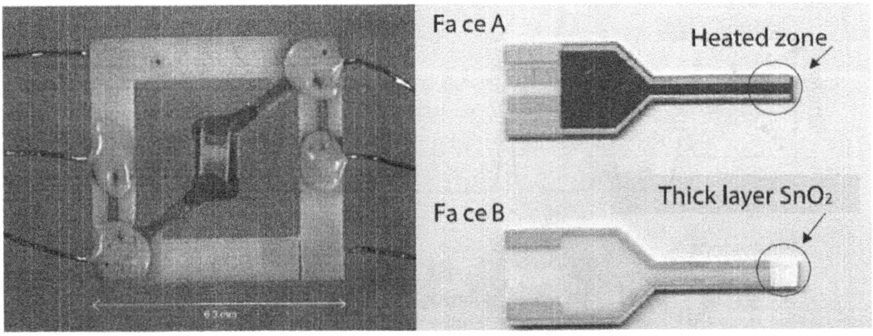

(a) (b)

Figure 5.11. *Metal oxide gas sensor: a sensor material place on a heated or sintered substrate (a), or in thick layers (b)*

Despite the large range of materials (ZnO, Fe_2O_3, NiO, CeO_2, TiO_2, etc., in fact all binary or ternary metal oxides) presenting variations in electrical conductivity for specific gases, only SnO_2 sensors have been industrially developed.

We have already highlighted the wide gap between the research carried out over the past 10 years in over 50 university and industrial laboratories into this kind of semi-conductive detector, and the extremely small number of industrial projects resulting in commercial products. As a result, we will concentrate on SnO_2 sensors.

Historically, the first developments were undertaken in Japan by Figaro (TGS sensors). This explains the fact that most companies producing SnO_2 sensors are Asian (five or six in Japan, China, South Korea, Hong Kong, etc.). In Europe, research in this field occurred more recently with the most significant developments taking place in Germany (UST), the UK (EEV) and Switzerland (Microsens and MICS). Surprisingly, only a few technological transfers have been carried out in the US (only a couple of small companies).

The first Japanese sensors (TGS) to appear on the market used a ceramic technique with the sensor element made from a sintered material. The sensor powder is placed on a porous support in the ceramic in the form of a tube (Ø=1 mm and l = 3 mm) (Figure 5.12).

The heated part of this sensor element is made from a metal wire coiled inside the tube. The deposition of the sensor layer has been the subject of various studies. Firstly, the palladium doped tin powder is produced using tin salts (ammonia attack, precipitate calcinations, palladium chlorine impregnation followed by heating). Secondly, γ aluminum powder (which is also doped with palladium) is used as a bond. Lastly, the third step consists of mixing the powders so as to coat the ceramic tube with the sensor material (SnO_2 + Pd + $A12O_3$). Electrical contact is ensured by metal wires fixed to the two ends of the tube with a conductive paste. The last process consists of mounting the sensor element on its base (element supported only by connecting wires) and placing it in gas permeable casing (metal mesh) with a diameter of approximately 1 cm.

Different versions of this type of sensor have since appeared on the market. Figaro in particular have developed versions using thick and thin layers, which enable important miniaturizations of the sensor element (increase in consumed power) in a similar way to developments in microelectronics. A recent version of the TGS sensor incorporates a thin layer of SnO_2 created using alkoxide tin solutions in isobutanol (deposits, drying and annealing at 500°C, repeated four times). This layer of SnO_2 is then similarly doped with palladium (metal alkoxide in isobutanol: deposits, drying and annealing).

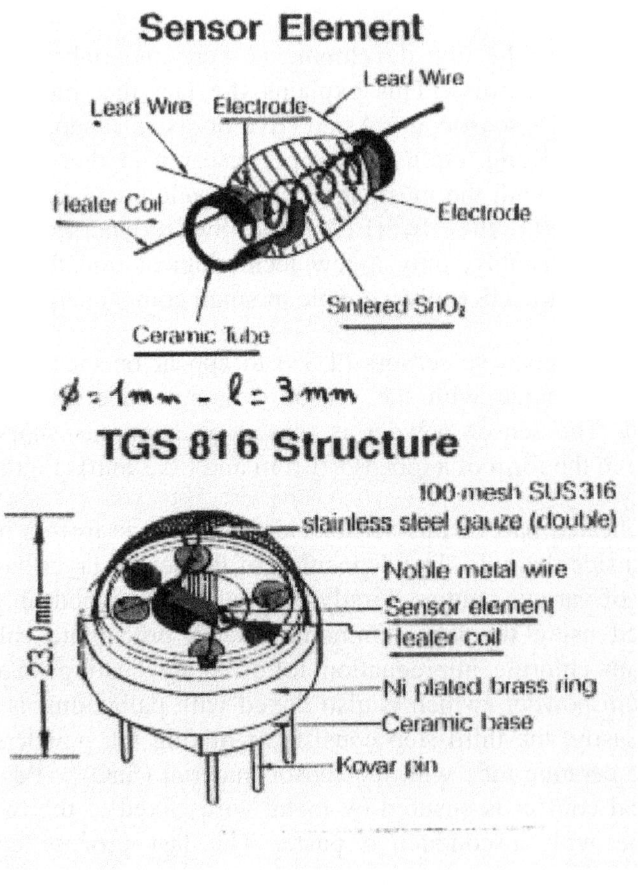

Figure 5.12. *Illustration of a sintered TGS sensor: paste placed on a ceramic tube containing a heating filament*

Whilst sintered sensors have a number of disadvantages in terms of production (costs, dimensions, reproducibility, etc.) they do perform well in terms of gas detection. The main areas of use include monitoring atmospheric pollution, domestic and industrial security and hygiene, controlling industrial processes, or even applications linked to the agro-food industry.

Figure 5.13a provides an example relating to atmospheric urban pollution due primarily to vehicle emissions. Sintered sensors (the Saint-Etienne ENSM prototype) were placed on a busy main route in Lyon, France. The sensors' signals (taken every quarter of an hour) were directly compared with measurements obtained using an infra-red analyzer focusing on carbon monoxide. There was a strong correlation between the two signals, demonstrating the significance of these sensors in monitoring urban pollution, particularly for developing measuring networks integrating a large number of sensors, and therefore enabling high resolution mapping.

Regarding different "sintered" techniques, various industrial developments have occurred in micro-electric technology. These technologies enable mass production, reduction in costs and further miniaturizations, therefore reducing the amount of power consumed. An example of this kind of sensor will be examined in further detail at the end of this chapter.

For a number of years, these devices have made up the vast majority of gas sensors used. They are perfectly adapted for mass markets, as is the requirement for millions of low cost sensors. However, in terms of instrumentation in particular used for managing industrial processes, the specifications are somewhat different. For these applications, we do not necessarily want to reduce either cost or miniaturizations. Volumes are generally low. However, they are often focused around very specific detection performances with calibrations and standardizations specific to each use. An intermediary solution consists of developing sensors using hybrid thick layer technology. This allows sensors to be developed using techniques similar to those used in micro-electronics whilst preserving the basic qualities of powder based sensor elements. The resulting sensors are sufficiently robust (ceramic substrates, non-miniaturization, metal packaging etc.) to be well adapted to instrumentation in industrial contexts.

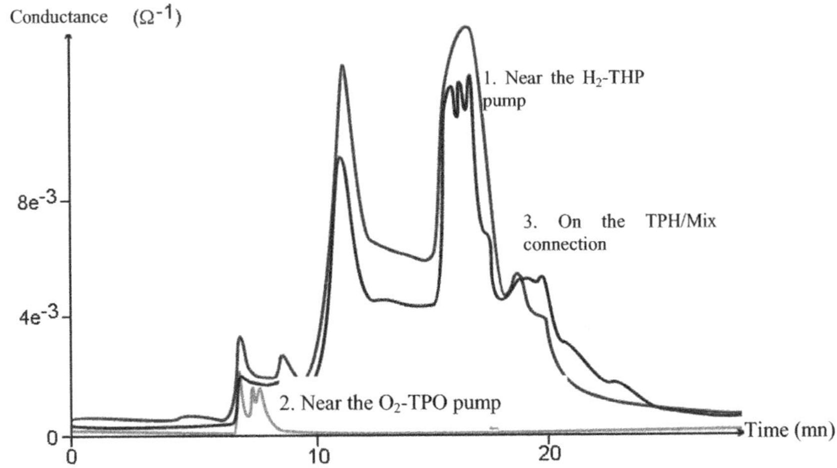

(a)

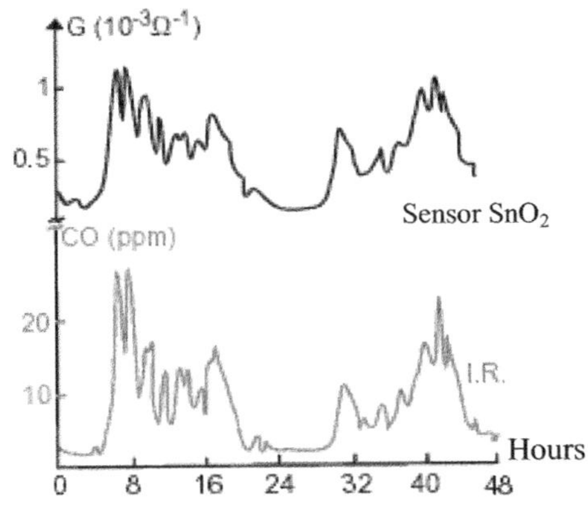

(b)

Figure 5.13. *a) Comparison of responses from a sintered SnO_2 sensor and an IR specific CO analyzer during the urban experiment. b) Results of experimentation with a network of sensors on an Ariane V motor: followed by emissions of H_2 on a Ariane V motor*

An example of this kind of sensor is shown in Figure 5.11b. This prototype, developed by the Ecole Nationale Supérieure des Mines in Saint-Etienne, France, is composed of an aluminum substrate of around 2.5 cm in length, with the heated zone covering 2 × 2 mm. A platinum resistance on one side enables the sensor element to be heated to between 500 and 600°C. The sensor layer is placed on the other side via serigraphy techniques using a SnO_2 paste with controlled granulometry (and the addition of organic and mineral bonds). The gold electrical connections and protective dielectric layers are produced using thick layer techniques (from 10 to 100 μm). The components are placed in a metal box, in isolation, using cement.

A network of around a dozen sensors has been developed in collaboration with SNECMA to control hydrogen emissions in cryogenic motors in the Arian V launcher [PIJ 02]. The aim is to produce maps of hydrogen concentrations as the engine operates. This type of detector therefore seems to have an obvious benefit in comparison to complex analysis techniques requiring pipes for gas sampling (particularly with long response times).

The three curves in Figure 5.13b show the responses from the three sensors in a motor test placed at three different locations: one near the H_2 turbo pump (HTP), one near the O_2 turbo pump (OTP) and one on an intermediary point.

In contrast to applications in industrial instrumentation, "wider public" mass markets impose specifications which can only be satisfied using micro-electronic techniques. The sensor still contains a sensor element placed on a heated substrate but thin film and micro-machining techniques are used to produce miniaturized systems on silicon.

An example of a micro-sensor on a silicon platform is shown in Figure 5.14a. The predicted markets for these types of sensors include CO (air quality, safety, etc.), CH_4 (household safety, etc.), applications in cars or even portable individualized detection systems.

The major benefit of miniaturization is that it produces a micro-sensor with low electricity consumption (only a few 100 mWatts) and uses technologies enabling the significant production of low cost sensors. The silicon substrates are thinned until a membrane of a few microns is produced on which a thin metal layer is placed allowing the SnO_2 sensor layer to be heated to around 500°C.

Different techniques can be used to implant SnO_2 in a thin layer. Standard micro-electronics techniques are then used to form electrical connections and for casing. The system's low thermal inertia enables working modes in thermal cycles and as such the control of a number of problems linked to selectivity and interferents (influence of humidity in particular). The difficult points needing examination are, as with all sensors, validations and tests, long-term stability, selectivity and reproducibility as batches of sensors.

5.5. Future developments

The essential parameters affecting the performance of gas sensors are time stability of sensor materials (in combination with processing electronics) and device selectivity, an essential quality for satisfying a number of applications. It should be recognized that this performance is fairly difficult to satisfy with metal oxides, particularly in connection to their working conditions at high temperatures. A number of basic studies have been conducted in recent years to model and manage phenomena more easily.

In particular, there have been a number of efforts in research focusing on the addition of doping agents and/or catalysts modifying the interactions of metal oxides with gases and as such improving selectivity and stability. However, it should be said that progress is slow due to the complexity of the physicochemical mechanisms involved. As a result, there is a current tendency to use sensor elements without modification but to design new devices or systems in order to improve the performances.

The main direction focuses on the developments of multi-sensor systems integrating different sensor elements with powerful signal processes (multi-variable analyses, neural networks, shape recognition etc.). These advanced systems can only be produced using micro-electronic or similar techniques with the integration of several sensing layers on a single substrate with eventual temperature cycles, as well as micro-processors allowing information to be processed directly on the chip.

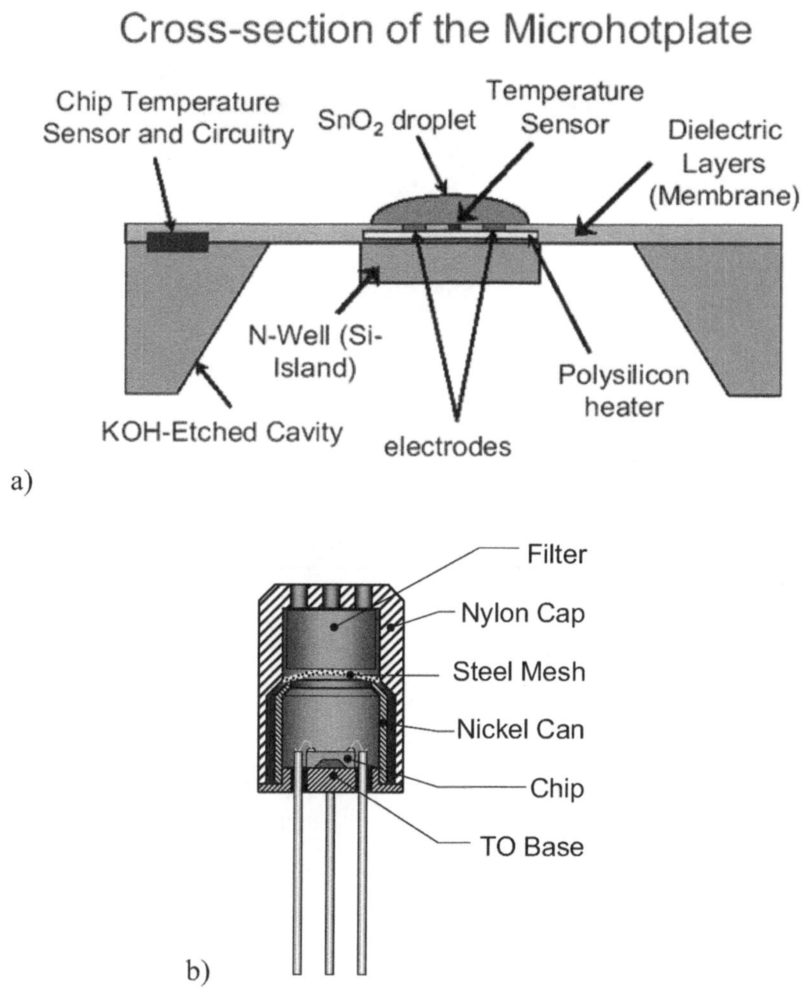

Figure 5.14. *a) Illustration of a sensor on a silicon membrane with integrated heating. b) Use of an active carbon filter to improve performance (MiCS sensor)*

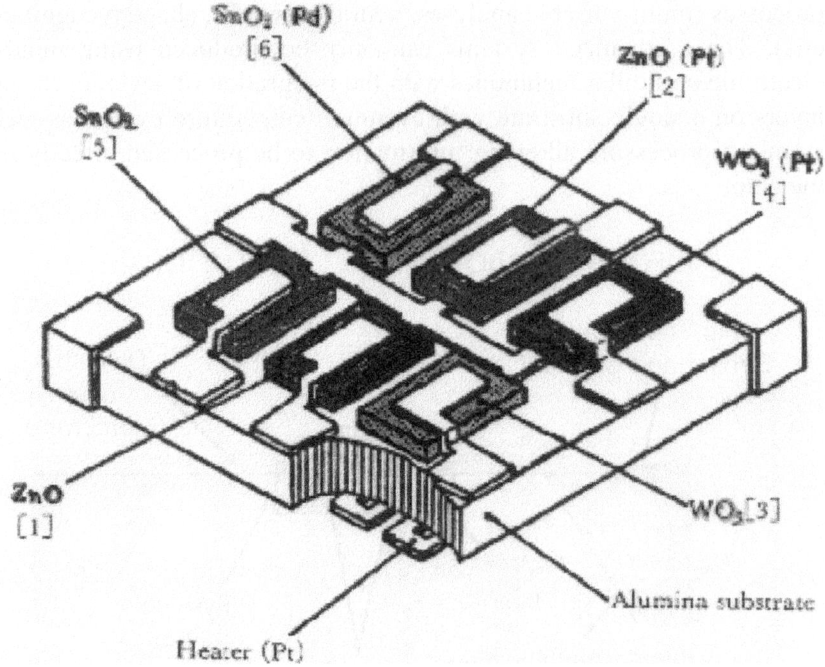

(a)

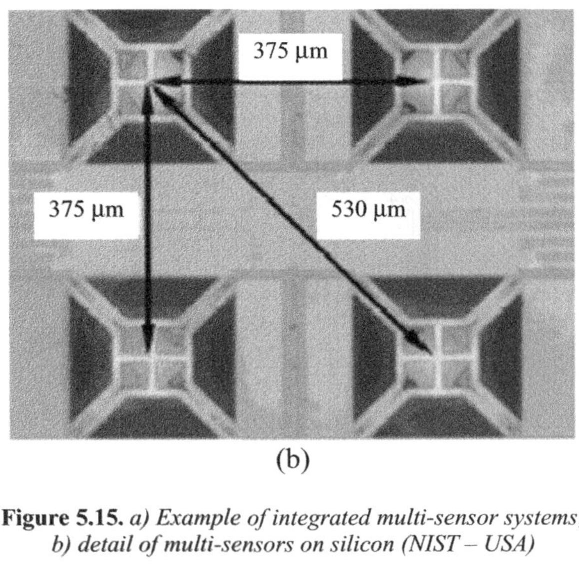

(b)

Figure 5.15. *a) Example of integrated multi-sensor systems; b) detail of multi-sensors on silicon (NIST – USA)*

For such future developments, if microelectronic technology seems to be unavoidable (Figure 5.14a), then we also need to have different elements which are high performing or which are sufficiently different from each other to be used (Figure 5.15a). As such, research into solutions allowing the separation of gases constitutes one of the main areas of interest. Simple solutions include filtering gases from the sensor element.

Figure 5.14b provides an illustration of a device developed by MiCS to filter gas with an active carbon charge included in the casing. Future sensors will obviously have to integrate such filtering layers placed directly on the sensor elements. Several laboratories are currently focusing on such solutions in both thin and thick layers. The challenge of this research lies in creating ceramic membranes on metal oxides, controlling their porosity with the eventual incorporation of active elements in these superficial layers and, more generally, controlling the range of phenomena of diffusions and interfaces in these multi-layer systems.

Once managed, these sensor elements with filter layers could be integrated into silicon platforms with a range of micro-heaters. For example, NIST (in the US) has developed a platform with 32 elements. The four heater elements featured are shown in Figure 5.15b.

5.6. Bibliography

[BAK 59] A. BAKER, Improvements in electrically heated filaments used in connection with the detection of combustible gases in air, UK Patent 892530 A, 1959.

[BAR 99] N. BARSAN, M. SCHEIWER, W. GOPEL, "Fundamental and pratical aspects in the design of nanoscaled SnO_2 gas sensors: a status report", *Fresenius J. Anal. Chem.*, vol. 365, pp. 287-304, 1999.

[FAB 92] P. FABRY, E. SIEBERT, in *Chemical Sensor Technology*, vol. 4, p. l-11, Kodansha, Tokyo, 1992.

[FOU 82] J. FOULETIER, *Sensors and Actuators*, vol. 3, p.295, 1982/1983.

[GOP 91] W. GOPEL, T.A. JONES, M. KLEITZ, I. LUNDSTRÖM, T. SEIYAMA, *Sensors: a Comprehensive Survey. Chemical and Biochemical Sensors*, vol. 2 and 3, V.C.H. Weinheim, 1991.

[HEI 82] G. HEILAND, "Homogeneous semiconducting gas sensors", *Sensors and Actuators*, vol. 2, pp. 343-361, 1982.

[KOH 89] D. KOHL, "Surface processes in the detection of reducing gases with SnO_2-based devices", *Sensors and Actuators*, vol. 18, no. 1, pp. 71-113, 1989.

[KRE 93] P. KREBS, A. GRISEL, *Sensors and Actuators*, vol. B13-14, p. 155, 1993.

[LAL 81] R. LALAUZE, C. PIJOLAT, J.P. COUPUT, Procédé, capteur et dispositive de détection de traces de gaz dans un milieu gazeux, French Patent no. 8119536, 1981.

[LAL 84] R. LALAUZE, C. PIJOLAT, "A new approach to selective detection of gas by a SnO_2 solid-state sensor", *Sensors and Actuators*, vol. 5, pp. 55-63, 1984.

[MOR 82] S.R. MORISSON, "Semi-conductor gas sensors", *Sensors and Actuators*, vol. 2, pp. 329-341, 1982.

[MOS 87] P. MOSELEY, B.C. TOFIELD, in B.E. JONES (ed.), *Solid State Gas Sensors*, Adam Hilger-Bristol, 1987.

[PIJ 99] C. PIJOLAT, C. PUPIER, M. SAUVAN, G. TOURNIER, R. LALAUZE, "Gas detection for automotive pollution control", *Sensors and Actuators*, vol. B 59, pp. 195-202, 1999.

[PIJ 02] C. PIJOLAT, G. TOURNIER, P. BREUIL, D. MATARIN, P. NIVET, "Hydrogen detection on a cryogenic motor with a SnO_2 sensors network", *Sensors and Actuators*, vol. B82, pp. 166-175, 2002.

[SEI 92] T. SEIYAMA, *Chemical Sensor Technology*, vol. 1, Kodanska, Elsevier, Tokyo, 1988; vol. 2, 1989; N. YAMAZOÉ, vol. 3, 1991; S. YAMAUCHI, vol. 4, 1992.

[TAK 86] T. TAKEUCHI, "Oxygen sensors", *Proceedings of the 2nd Int. Meeting on Chemical Sensors*, University Bordeaux 1, Bordeaux, 1986.

[TAK 88] K. TAKAHATA, *Chemical Sensor Technology*, vol. 1, p.39, Seiyama, Kodanska, Elsevier, Tokyo, 1988.

[VAU 91] C. VAUCHIER, D. CHARIOT, G. DELAPIERRE, A. ACCORSI, *Sensors and Actuators*, vol. B5, p. 33, 1991.

[VEL 82] G. VELASCO, J.P. SCHNELL, M. CROSET, *Sensors and Actuators*, vol. 2, p. f371, 1982.

[YAM 91] N. YAMAZOÉ, "New approaches for improving semiconductor gas sensors", *Sensors and Actuators*, vol. B5, pp. 7-19, 1991.

Chapter 6

Molecular Material-based Conductimetric Gas Sensors

6.1. Molecular semiconductors

In this chapter we will examine molecular materials involving weak intermolecular interactions. For neutral molecules without labile hydrogen, these include van der Waal, dipole-to-dipole and π–π interactions. In a solid, molecules maintain their own identity and the molecular material properties are directly linked to the properties of the molecules composing it. As such, molecular properties are governed by energy levels in the molecular orbitals, both HOMO (highest occupied molecular orbital) and LUMO (lowest unoccupied molecular orbital), by definition, localized on the molecules [BOU 95, BOU 03]. In contrast, in an inorganic material such as silicon, the material is composed of atoms connected by strong interactions, bonds and electrical properties are described by energy bands, namely valence and conduction bands. The electronic states are collective. This distinction between molecular and inorganic materials is illustrated in Figure 6.1, which shows the diamond-shape structure of silicon and the crystalline structure of the pentacene with isolated molecules.

Chapter written by Marcel BOUVET.

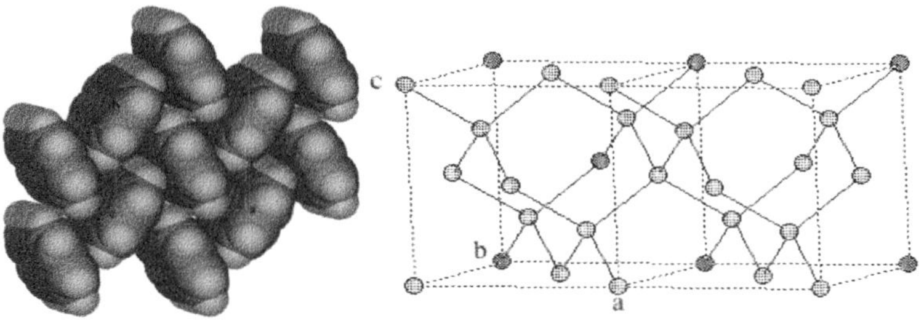

Figure 6.1. *Crystalline structure of pentacene (left), and the diamond structure of silicon (right)*

The creation of charge carriers in molecular materials takes place due to the electronic transfer of the HOMO of one molecule towards the LUMO of another (Figure 6.2). The energy required for this transfer is linked to the ionization potentials and electronic affinity of the molecule and therefore also to the oxidation (E^0_{ox}) and reduction (E^0_{red}) potentials of the molecule. The chemical potential at equilibrium, which is equivalent to the Fermi level for an inorganic material is given by the half-sum $\left(\frac{E^0_{ox}+E^0_{red}}{2}\right)$.

The density n of charge carriers is provided by the following Arrhenius law:

$$n = [A^+] = [A^-] = [A]_0 \exp\left(\frac{E^0_{red} - E^0_{ox}}{2kT}\right)$$

where k is the Botzman constant and T is the temperature in Kelvin. $[A]_0$ is the molecular density, which has a value of around 10^{21} cm^{-3} and is the equivalent of the density of states in the conduction (N_C) and valence (N_V) bands for silicon. This is conveyed by the fact that each molecule can provide an electron from its HOMO or accept one in its LUMO. E_{act} is the activation energy. Taking into account the approximations, it can be directly deduced from the difference between the molecule's redox potentials that:

$$E_{act} = e(E^0_{ox} - E^0_{red})$$

The conductivity σ is the product of this density of charge carriers by the electron charge and their mobility μ: $\sigma = n\ e\ \mu$. Mobility is generally weak in molecular materials since the movement of charges is a process which occurs via successive jumps of electrons from one molecule to another. This is the intrinsic process involved in creating charge carriers.

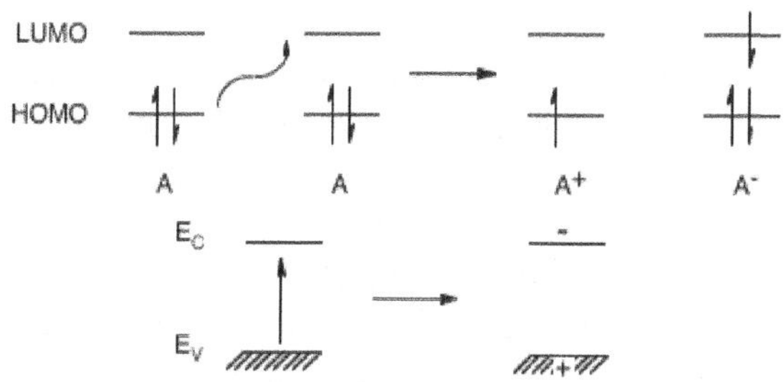

Figure 6.2. *Illustration of the process of creating charge carriers in a molecular material composed of molecules, A (left). As a comparison, the process of creating charge carriers in silicon is illustrated by an electronic transfer between the top of the valence band E_V and the bottom of the conduction band E_C (right)*

A terminological parallel can allow those familiar with inorganic semiconductors to quickly understand the molecular materials' notations (Table 6.1).

Inorganic semiconductor	n	p	N_C, N_V	E_V	E_C
Molecular semiconductor	$[A^-]$	$[A^+]$	[A]	$e\,E^0_{ox}$	$e\,E^0_{red}$

Table 6.1. *Terminological parallel between molecular semiconductors and inorganic semiconductors*

When an impurity, whether voluntary or not, is capable of exchanging an electron with the molecular material, a doping process can take place. If we consider an electron accepting impurity I_A (e.g. an oxidant such as I_2, Br_2 or NO_2), the density of the positive charges created by this process is proportional to the density of impurities and the difference in energy between the acceptor level of I_A (E_a) and the provider level of the molecule A (HOMO).

$$A + I_A \rightleftharpoons A^+ + I_A^-$$

$$p = [A^+] = [A]^{1/2}.[I_A]^{1/2} \exp\left(\frac{E_a - E_{HOMO}}{2kT}\right)$$

Independently of any doping, it is also possible to talk about a p- or n-type material. If we consider the exchange of charge between a molecular material and electrodes, in the case of a resistor for example, some molecules which are highly electron accepting do allow only the circulation of negative charges and are n-type materials (e.g. perfluorinated phthalocyanines $F_{16}PcM$), whilst others transfer electrons towards the electrode and are p-type materials (e.g. pentacene, non-substituted phthalocyanines, PcM). This is illustrated in the following diagram which compares the energy levels in fluorinated and non-substituted phthalocyanines at the Fermi level of an electrode (Figure 6.3) [MUR 06].

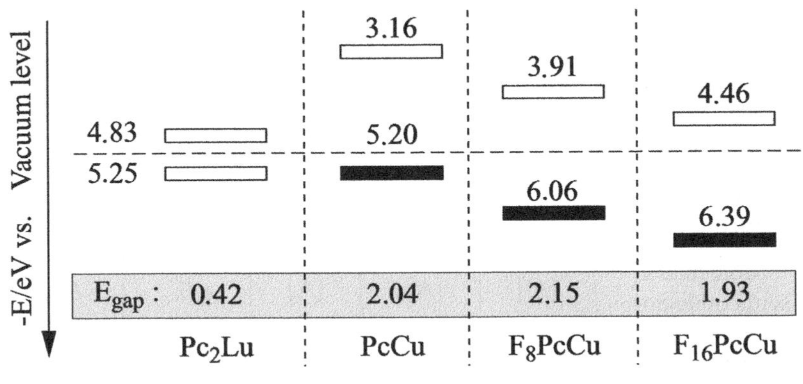

Figure 6.3. *Energy levels, HOMO and LUMO, of fluorinated and non-fluorinated monophthalocyanines compared with those of lutetium bisphthalocyanine. The dotted line indicates the electrode's Fermi level*

The activation energy is of the order of two electron-volts (eV) or more in molecular materials, which leads to a very weak charge carrier density of the order of 10^4 carriers cm^{-3}. Mobility often ranges between 10^{-4} and 10^{-2} $cm^2\ V^{-1}\ s^{-1}$, and more rarely is equal to 1, which leads to an intrinsic conductivity being lower than $10^{-10}\ \Omega^{-1}\ cm^{-1}$. Experimentally, the measured current often corresponds to an extrinsic conduction. It is for this reason that what is often called a semiconductor should be called a doped insulator. There are several cases of molecules which are both easily reducible and oxidable and therefore have weak activation energy E_{act}. They result in intrinsically semiconductive molecular materials. These include radical phthalocyanines such as lutetium bisphthalocyanine (Pc_2Lu), where E_{act} has a value of 0.5 eV, and is the intrinsic conductivity $10^{-5} - 10^{-6}\ \Omega^{-1}\ cm^{-1}$.

The main molecular materials used for their electrical properties are phthalocyanines, pentacene, oligothiophenes, fullerenes and perylene derivatives (Figure 6.4). They are used in photovoltaic cells, field effect transistors and conductimetric chemical sensors [BOU 06].

Figure 6.4. *Examples of molecular materials (from left to right and from top to bottom): phthalocyanine PcM, perylene diimide PTCDI, fullerene C_{60}, pentacene and α-sexithiophene*

6.2. Molecular material-based conductimetric devices

In the field of chemical sensors, conductometric transducers made from molecular semiconductors are principally resistors but can also be p-n contacts and field effect transistors, standard electrical devices. More recently, a new structure has been proposed which combines a doped isolation heterojunction and an intrinsic semiconductor and has been termed a molecular semiconductor-doped insulator (MSDI) (Figure 6.5) [PAR 09a, PAR 09b].

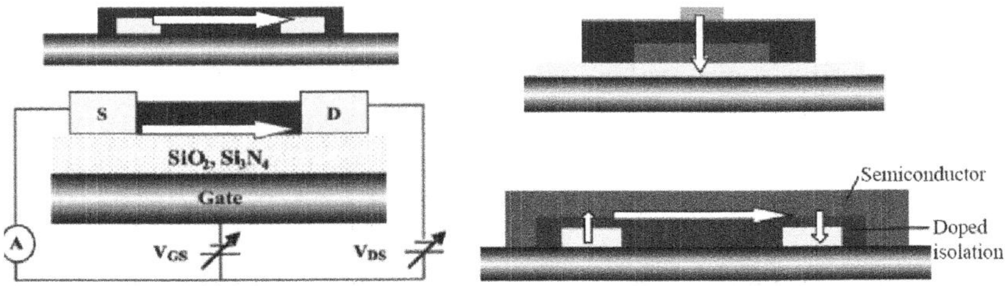

Figure 6.5. *Conductimetric transducers using molecular semiconductors as sensing materials: resistor, p-n junction, field-effect transistor and MSDI. The arrows indicate the circulation of current in the devices*

An important point concerns the production of these devices. The molecular materials enable various deposition processes, particularly when the molecules are soluble. Unsoluble molecules are evaporated under vacuum, but the soluble molecules enable the preparation of films by spin-coating, by Langmuir-Blodgett or Langmuir Schäfer techniques, by solvent casting but also by covalent grafting onto surfaces: for example, via silanol functions (SiOH) on the surface of silicon substrates. Incorporation into polymers, either by simple mixing in a solution or by covalent grafting allows us to modulate the properties of the films obtained. As such, crystallinity, morphology, rugosity and the hydrophilic/hydrophobic character can be modulated as well as the affinity for a target species. These parameters must be managed in order to improve the sensors' selectivity, reversibility and also stability in time [CHE 10].

Chemistry allows the introduction of moieties on molecular semiconductors which ensure a high selectivity of chemical sensors. As such, the introduction of ether-crown moieties renders it sensitive to cations, whilst the introduction of calixarenes or cyclodextrines enables selectivity regarding aromatic volatile organic components [ROS 98, HAS 08]. The type of bond can be covalent but

π–π interactions enable selective substituents to be fixed, for example on carbon nanontubes via pyrene moieties.

For a resistor, the current-voltage curve, I(V), is a straight line when behavior is ohmic and is superlinear if a process of charge injection from electrodes occurs. For a p-n junction, the I(V) characteristic is that of a diode (Figure 6.6). For a field-effect transistor, there is a modulation of both the source to drain current I_{DS} according to the gate voltage (V_G), and a saturation of source-drain-current I_{DS} at a source-to-drain voltage V_{DS} higher than a threshold voltage.

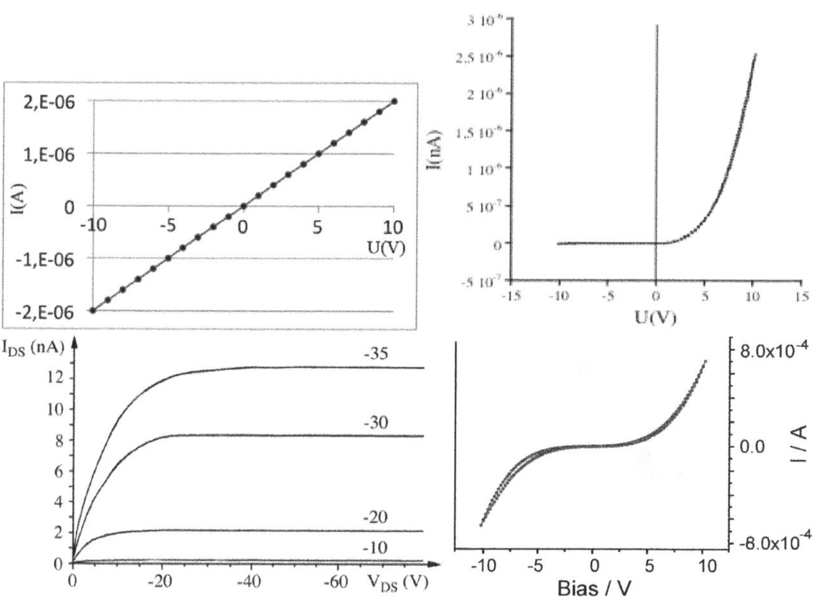

Figure 6.6. *I(V) characteristics of a resistor (Pc_2Lu), a junction p-n ($PcNi/F_{16}PcNi$), a field-effect transistor (PcNi) and a MSDI ($Pc_2Lu/F_{16}PcCu$) (from left to right and top to bottom)*

In the case of MSDIs, there is an energy barrier which induces a curve I(V) resembling that of a forward polarized diode but, inversely, the device is symmetrical and the positive and negative polarizations result in the same type of curve.

6.3. Oxidizing or electron providing compound sensors

As we saw previously, an oxidizing species absorbed on the surface of a p-type molecular semiconductor leads to an increase in the density of the majority charge carriers and therefore an increase in the current. This has been used to develop ozone (O_3) sensors to monitor atmospheric pollution, as well as for resistors and transistors (Figure 6.7) [BRU 03, BOU 01].

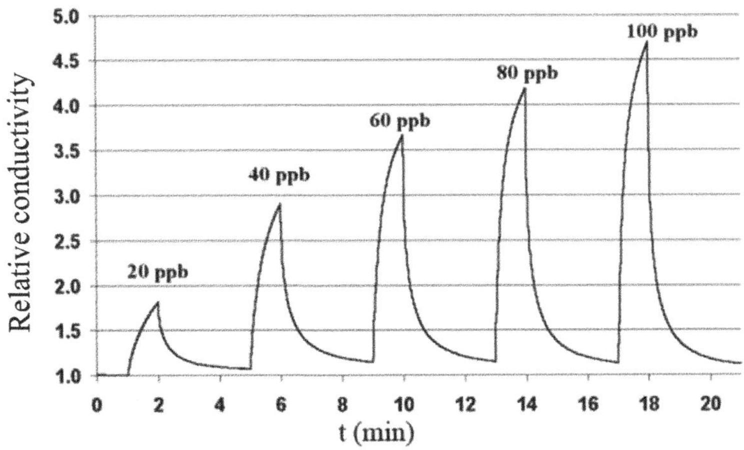

(a)

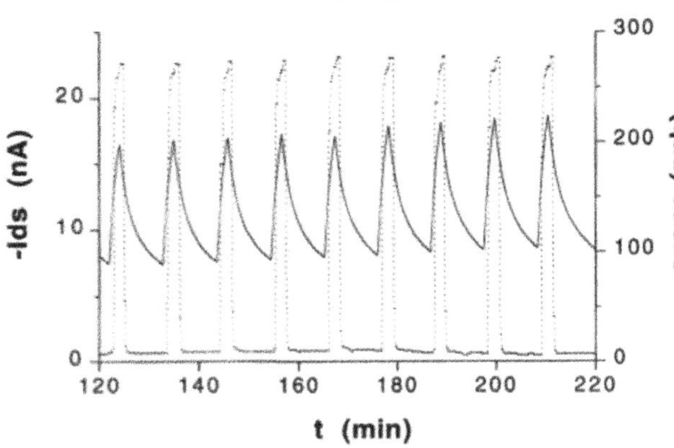

(b)

Figure 6.7. *Responses of ozone sensors: a) a resistor composed of a PcCu film (300 nm in thickness) held at 80°C, $t_{exposition}$ = 1 h, $t_{recovery\ in\ clean\ air}$ = 3 h; b) a transistor (solid line) composed of a PcNi film (300 Å in thickness), $t_{exposition}$ = 2 min, $t_{rstatic\ recovery}$ = 8 min, all at ambient temperature [BRU 03, BOU 01]*

The main interferent to O_3 in this application is nitrogen dioxide NO_2. This is an oxidizing gas which occurs in the same concentrations as O_3 in the range of 10-100 ppb, with one having a maximum concentration when the other is minimal and vice versa depending on time. The main difference between the two gases lies in O_3's significant instability. It has been shown that in short amounts of time, O_3 induces a greater response than NO_2, whilst for longer time periods the opposite is true. This can be explained by the ability of NO_2 to diffuse inside the sensor layer, whilst O_3 is destroyed when it crosses the material (Figure 6.8).

NOTE.– It is for this reason that O_3 can easily be selectively eliminated, without affecting the concentration of NO_2: via a filter composed of a molecular material such as indigo or carbon nanotubes placed in front of the sensor or directly in the form of a film on the resistor [BRU 05, BRU 10].

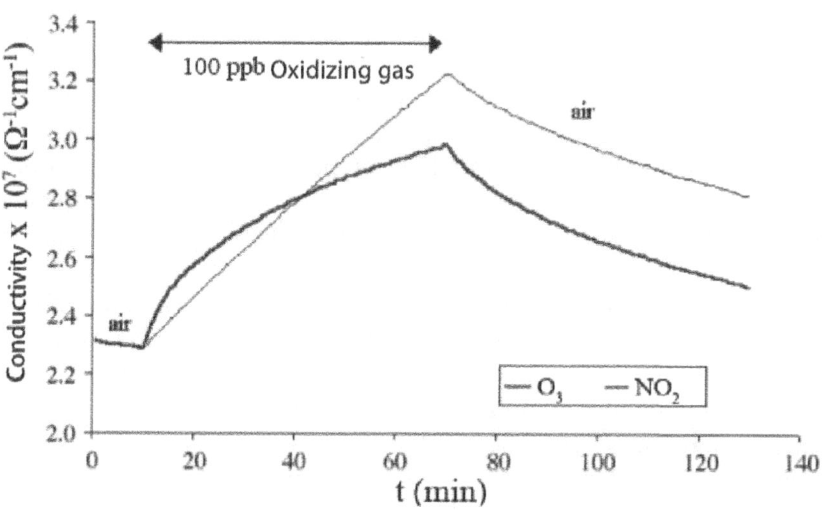

Figure 6.8. *Response of a PcCu resistor as a function of time (300 nm) held at 50°C and exposed to 100 ppb of oxidizing gas (O_3 and NO_2) [BRU 03]*

Inversely to oxidizing gases, ammonia (NH_3), which is an electron donating species, induces a decrease in the p-type conductivity and an increase of conductivity in n-type materials as illustrated in Figure 6.9 for a perfluorinated phthalocyanine resistor.

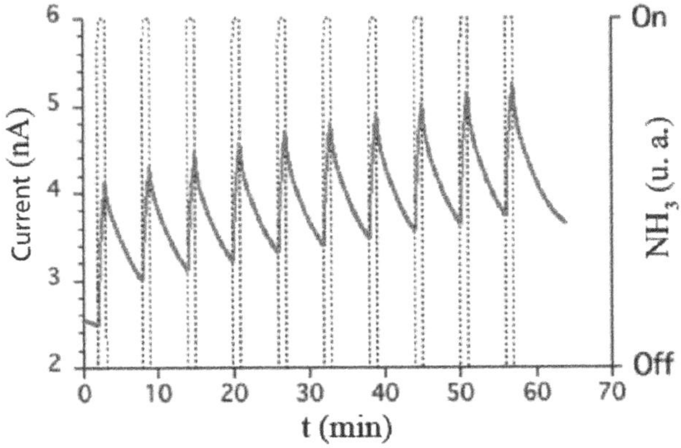

Figure 6.9. *Response to NH_3 of a resistor composed of $F_{16}PcCu$ during rest-exposure cycles*

n-type materials see their resistance increase strongly in the presence of ozone due to the neutralization of major negative carriers. However, when covered by p-type materials in p-n junctions, they demonstrate their advantage. The sandwich structure Au/F_{16}PcNi/PcNi/Al in which the p-type material covers the n-type material sees its I(V) characteristic strongly affected by the presence of NH_3. This leads to a decrease in p-carriers in PcNi and therefore a decrease in the energy barrier at the junction between the two materials resulting in an increase in the current, irrespective of the junction's direction of polarization, whether direct or inverse. The rectification factor ($I_{direct}/I_{inverse}$) is strongly modified. The diode's significance is that it allows us to access several experimental parameters, I_{direct}, $I_{inverse}$ and the rectification factor and therefore a multimodal detection [MUZ 07].

Intrinsic semiconductors such as Pc_2Lu, which are easily oxidized and reduced, are particularly sensitive to oxidizing gases or electron providers. Their sensitivity is such that when exposed to NO_2, their curve I=f(t) reaches a maximum (Figure 6.10). This is explained by the density of oxidized species, Pc_2Lu^+, which does not allow the transfer of charge with a neutral neighboring molecule. Mobility therefore declines. The limited case is where all the molecules are oxidized. We therefore obtain the equilibrium:

$$2Pc_2Lu^+ \rightleftharpoons Pc_2Lu + Pc_2Lu^{2+}$$

which requires more energy than the transfer of electrons between two neutral Pc_2Lu since it involves the second oxidization potential.

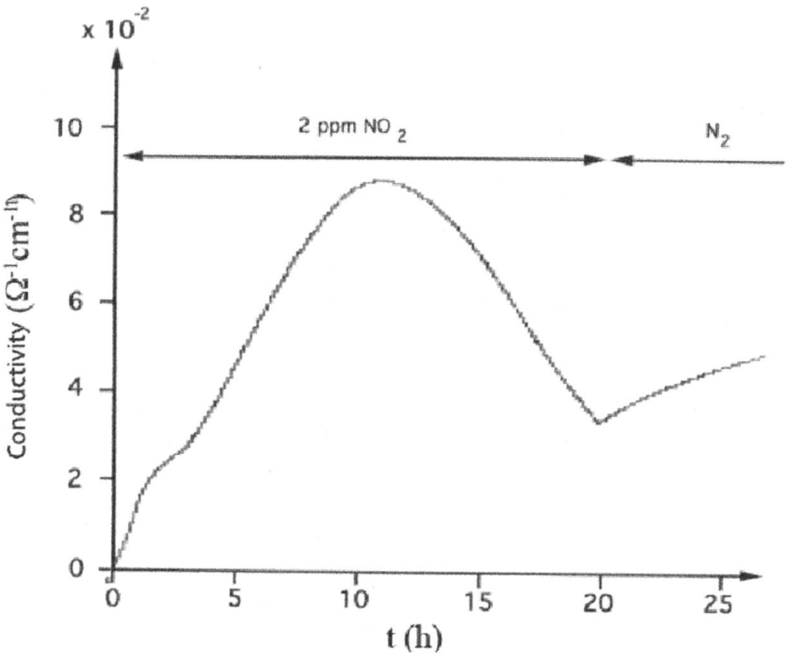

Figure 6.10. *Response of a Pc_2Lu sensor in contact with NO_2 (300 nm in thickness, ambient temperature) [BOU 06]*

To avoid this saturation phenomenon and benefit from the high sensitivity of radical phthalocyanines, two distinct approaches have been proposed. The development of hybrid phthalocyanine-polymer materials in which the molecular semiconductor is dispersed in a non-conductive matrix, such as poly(methyl methacrylate) (PMMA), has enabled the production of an O_3 sensitive resistor [PAR 09a].

On the other hand, MSDIs, where the semiconductor is the only material in contact with the atmosphere being analyzed, enable the preservation of Pc_2Lu's high sensitivity without presenting saturation, because the current inside the device is not directly proportional to the semiconductor's conductivity but to the energy of the barrier between the semiconductor and the sub-layer (doped insulator). This energy barrier depends on the charge density in the two materials, but not on the charge mobility as is the case in a resistor made of just Pc_2Lu [PAR 09b].

Another characteristic of MSDIs is that we can still have a configuration where the response to electron accepting or providing species is positive, thanks to the choice of sub-layer (Figure 6.11). As such, when the sub-layer is n-type ($Pc_2Lu/F_{16}PcCu$), the current increases in the presence of NH_3, following the decrease in charge carriers in the semiconductor. To detect ozone, we will choose a p-type sub-layer ($Pc_2Lu/PcCu$). In this case, the increase in density of the majority charge carriers in the semiconductor results in an increase in current. MSDIs have also been developed using other doped insulators as sub-layers, such as oligothiophene (p-type) and perylene diimide derivatives (n-type) [BOU 10]. Similarly the use of lanthanide trisphthalocyanine has resulted in MSDIs with a high sensitivity to O_3 and NH_3 [CHE 11].

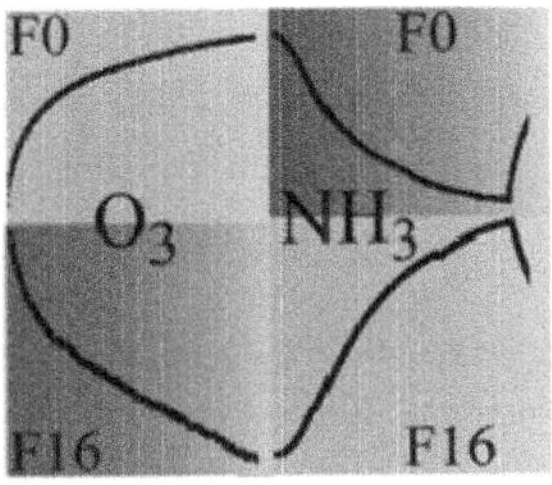

Figure 6.11. *Qualitative response of the MSDIs to ozone (100 ppb) and ammonia (50 ppm). F0 = $Pc_2Lu/PcCu$ and F16 = $Pc_2Lu/F_{16}PcCu$ [PAR 09b]*

6.4. Volatile organic compound (VOC) sensors

Whilst the redox active molecules lead to a variation in density of charge carriers, other molecules can induce a variation in material conductivity by modifying charge mobility. This can be the result of a variation in the dielectric constant of the sensing layer following the adsorption of gas or, more generally, a trapping of charges. Whilst selectivity is difficult to obtain for volatile organic compounds, categorical selectivity can however be obtained. Differentiation can originate from the hydrophilic or hydrophobic character of VOCs, their aromatic nature or simply a selection by size, through the use of sensor materials with adapted characteristics. However, what is most common is the use of multisensors produced using distinct sensing materials which, after adapted data processing, enables a system which can be used to detect COVs.

Prime examples of this include the research undertaken by I. Lundström in Sweden [WIN 97] applied to discrimination between different fruit juices, and that undertaken by R. Paolesse in Italy [DIN 96] focusing on studying the freshness of fish. These systems, whilst providing good results in the laboratory, have the major disadvantage of learning difficulty and a lack of temporal stability in individual sensors. We can also cite the use of a series of organic field affect transistors used for detecting volatile organic compounds by L. Torsi in Italy [CRO 01], but even categorical selectivity was not very satisfying. As such, several transduction modes have been combined in a single conductometric, electrochemical and optical microsensor by a team headed by M.-L. Rodriguez-Mendez in Spain and applied to analyzing wine [ROD 04]. Again research and development has provided a new analytical tool, but one which is much closer to a classic analyzer than a sensor microsystem with regard to its dimensions as well as its cost.

6.5. Bibliography

[BOU 95] M. BOUVET, E.A. SILINSH, J. SIMON, "Determination of energy gap values in molecular crystals II. Intrinsic dark conductivity and electrochemical methods", *Mol. Mat.*, vol. 5, pp. 255-277, 1995.

[BOU 01] M. BOUVET, A. LEROY, J. SIMON, F. TOURNILHAC, G. GUILLAUD, P. LESSNICK, A. MAILLARD, S. SPIRKOVITCH, M. DEBLIQUY, A. DE HAAN, A. DECROLY, "Detection and titration of ozone using metallophthalocyanine-based field-effect transistor", *Sensors and Actuators B*, vol. 72, pp. 86-93, 2001.

[BOU 03] M. BOUVET, Chapter 118 "Radical Phthalocyanines and Intrinsic Semiconduction", Part II, vol. 19 "Applications of Porphyrins and Phthalocyanines", in K.M. KADISH, K.M. SMITH, R. GUILARD (eds), *The Porphyrin Handbook*, Academic Press, pp. 37-103, 2003.

[BOU 06] M. BOUVET, A. PAULY, "Molecular semiconductor – based gas sensors" in C.A. GRIMES, E.C. DICKEY, M. V. PISHKO (eds), *The Encyclopedia of Sensors*, American Scientific Publishers, vol. 6, pp. 227-270, 2006.

[BOU 10] M. BOUVET, H. XIONG, V. PARRA, "Molecular semiconductor – doped insulator (MSDI) heterojunctions: Oligothiophene/ bisphthalocyanine ($LuPc_2$) and perylene/bisphthalo-cyanine as new structures for gas sensing", *Sensors and Actuators B*, vol. 145, pp. 501-506, 2010.

[BRU 03] J. BRUNET, J.-P. GERMAIN, A. PAULY, B. LAURON, C. VARENNE, "Methodology of measurement for phthalocyanines gas sensors dedicated to the detection of ozone and nitrogen dioxide. Results in urban air pollution context" in P. SICILIANO (ed.) *Sensors for Environmental Control*, World Scientific Publishing, Singapore, pp. 188-192, 2003.

[BRU 05] J. BRUNET, A. PAULY, L. MAZET, J.-P. GERMAIN, M. BOUVET, B. MALÉZIEUX, "Improvements in real time detection and selectivity of phthalocyanine-based gas sensors dedicated to oxidizing pollutants evaluation", *Thin Solid Films*, vol. 490, pp. 28-35, 2005.

[BRU 10] J. BRUNET, L. SPINELLE, A. PAULY, M. DUBOIS, K. GUERIN, M. BOUVET, C. VARENNE, B. LAURON, A. HAMWI, "All-organic device with integrated chemical filter for the selective measurement of NO_2 in air", *Organic Electronics*, vol. 11, no. 7, pp. 1223-1229, 2010.

[CHE 10] Y. CHEN, M. BOUVET, T. SIZUN, Y. GAO, C. PLASSARD, E. LESNIEWSKA, J. JIANG, "Facile approaches to build ordered amphiphilic tris(phthalocyaninatol) europium triple-decker complex thin films and their comparative performances in ozone sensing", *Phys. Chem. Chem. Phys.*, vol. 12, pp. 12851-12861, 2010.

[CHE 11] Y. CHEN, M. BOUVET, T. SIZUN, G. BAROCHI, J. ROSSIGNOL, E. LESNIEWSKA, "Enhanced chemosensing of ammonia based on the novel molecular semiconductor-doped insulator (MSDI) heterojunctions", *Sensors and Actuators B*, vol. 155, no. 1, pp. 165-173, 2011.

[CRO 01] B. CRONE, A. DODABALAPUR, A. GELPERIN, L. TORSI, H. E. KATZ, A. J. LOVINGER, Z. BAO, "Electronic sensing of vapors with organic transistors", *Appl. Phys. Lett.*, vol. 78, pp. 2229-2231, 2001.

[DIN 96] C. DI NATALE, J. A. J. BRUNINK, F. BUNGARO, F. DAVIDE, A. D'AMICO, R. PAOLESSE, T. BOSCHI, M. FACCIO, G. FERRI, "Recognition of fish storage time by a metalloporphyrins-coated QMB sensor array", *Meas. Sci. Technol.*, vol. 7, pp. 1103-1114, 1996.

[HAS 08] A.K. HASSAN, C. GOY, A.V. NABOK, "Interaction of volatile organic vapours with azo-calix[4]-resorcinarene and poly(9-vinylcarbazole) thin films using SPR measurements", *Thin Solid Films*, vol. 516, pp. 9006-9011, 2008.

[MUR 06] R. MURDEY, N. SATO, M. BOUVET, "Frontier electronic structures in fluorinated copper phthalocyanine thin films studied using ultraviolet and inverse photoemission spectroscopies", *Mol. Cryst. and Liq. Cryst.*, vol. 455, pp. 211-218, 2006.

[MUZ 07] I. MUZIKANTE, V. PARRA, R. DOBULANS, E. FONAVS, J. LATVELS, M. BOUVET, "A novel gas sensor transducer based on phthalocyanine heterojunction devices", *Sensors*, vol. 7, pp. 2984-2996, 2007.

[PAR 09a] V. PARRA, C. LOCATELLI, H. XIONG, M. BOUVET, "Electrical transduction in phthalocyanine-based gas sensors: from classical resistors to new functional structures", *J. Porphyrins and Phthalocyanines*, vol. 13, no. 1, pp. 84-86, 2009.

[PAR 09b] V. PARRA, J. BRUNET, A. PAULY, M. BOUVET, "Molecular semiconductor – doped insulator (MSDI) heterojunctions, an alternative transducer for gas chemosensing", *Analyst*, vol. 134, pp. 1776-1778, 2009.

[ROD 04] M.L. RODRIGUEZ-MENDEZ, A.A. ARRIETA, V. PARRA, A. BERNAL, A. VEGAS, S. VILLANUEVA, R. GUTIERREZ-OSUNA, J.A. DE SAJA, "Fusion of three sensory modalities for the multimodal characterization of red wines", pp. 348-354, 2004.

[ROS 98] S. ROSLER, R. LUCKLUM, R. BORNGRÄBER, J. HARTMANN, P. HAUPTMANN, "Sensor system for the detection of organic pollutants in water by thickness shear mode resonators", *Sensors and Actuators B*, vol. 48, pp. 415-424, 1998.

[SPI 10] L. SPINELLE, M. DUBOIS, J. BRUNET, K. GUERIN, V. PARRA, C. VARENNE, B. LAURON, M. BOUVET, A. HAMWI, A. PAULY, "The use of nanocarbons as chemical filters for selective detection of nitrogendioxide and ozone", *J. Nanosc. Nanotech.*, vol. 10, pp. 5653-5661, 2010.

[VIR 06] J.-P. VIRICELLE, A. PAULY, L. MAZET, J. BRUNET, M. BOUVET, C. VARENNE, C. PIJOLAT, "Selectivity improvement of semiconducting gas sensors by selective filter for atmospheric pollutants detection", *Materials Science and Engineering C*, vol. 26, pp. 186-195, 2006.

[WIN 97] F. WINQUIST, P. WIDE, I. LUNDSTRÖM, "An electronic tongue based on voltammetry", *Analytica Chimica Acta*, vol. 357, pp. 21-31, 1997.

Chapter 7

Responses and Electrical Properties of Gas Microsensors

7.1. Introduction

Gas microsensors are miniaturized devices which deliver a signal when gas or specific vapors are present. The term signal should be interpreted here in its wider sense. For example, reaction to the presence of a gas or a mixture of gases can result in the variation of a sensor's given property. This can be optical, thermal or electric. This chapter will focus on sensors emitting an electrical signal and, more specifically, those with a sensitive layer made from a semi-conductive oxide such as SnO_2, ZnO, TiO_2 and WO_3. They have a number of advantages such as high sensitivity, rapid response and recovery time, a long lifespan, good compatibility with microelectronic technologies, small size and low production costs. In contrast, these sensors often require a higher working temperature (200–500°C) and low gas selectivity. In this chapter we will use the term gas but note that this can also refer to vapors as well as gases.

The main advantage of developing gas detection devices is the possibility over recent years of producing low cost and miniaturized

---

Chapter written by Khalifa AGUIR.

components for a number of applications such as environment surveillance (pollution mapping, for example), health and industrial security, military security (explosive and toxic gas detection, etc.), domestic security (gas alarms, controlling ventilation, etc.), diagnosing illness, detecting food quality, monitoring procedures, etc.

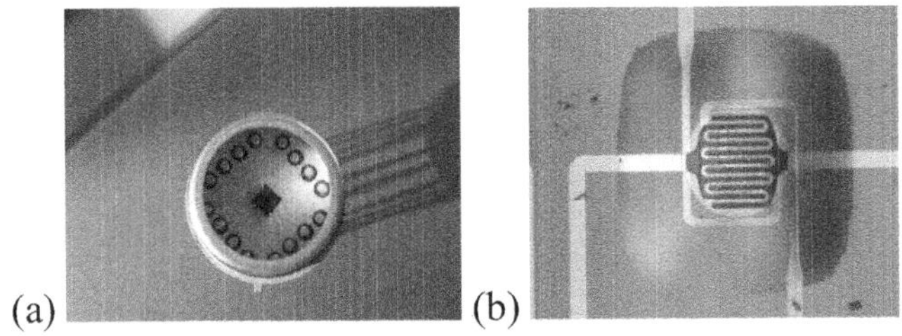

Figure 7.1. *(a) Multisensor including four microsensors integrated onto the same chip. Each microsensor is connected to four pins, two for heating and one for the sensor's signal. (b) Close up of a microsensor's membrane*

Gas sensors are part of the wider family of sensors. They are classified in relation to the signal emitted:

Chemical: gas, vapors, humidity, volatile organic pollutants, pH, etc.

Electrical: stress, current, dielectric permittivity, charge, resistance, capacity, inductance, polarization, frequency, etc.

Radiation: γ, X, UV, visible and IR radiation, micro-waves, radio waves, etc.

Thermal: temperature, heat, thermal flux, entropy, calorific capacity, etc.

Mechanic: displacement, constraints, speed, acceleration, force, pressure, masse, flux, acoustic waves, etc.

Magnetic: magnetic field, flux, time period, permeability, etc.

Biological: cells, proteins, sugar, antigens, hormones, etc.

7.2. Response of a gas sensor

Ideally, the input signal will develop according to the output signal which is shown in Figure 7.2.

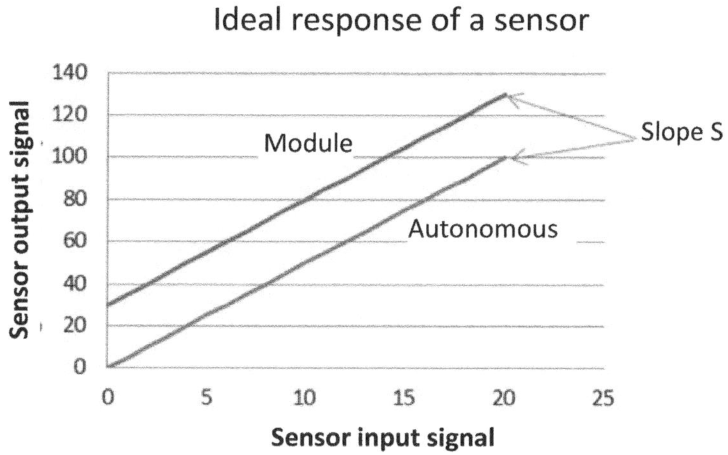

Figure 7.2. *Output signal versus input signal for an ideal sensor (y(t) = S.x(t))*

In the case of a linear response, the slope S is constant. It is often referred to as the sensitivity.

In reality, a sensor's response is not always instantaneous. It requires a certain amount of time to reach equilibrium after modifying the input signal. If this response is linear over time, the equation *y(t)=F(x(t))* can be expressed as:

$$a_n \frac{d^n y}{dt^n} + a_{n-1} \frac{d^{n-1} y}{dt^{n-1}} + \ldots + a_1 \frac{dy}{dt} + a_0 y(t) = x(t) \qquad [7.1]$$

a_i is the i^{nd} slope, which can be physically interpreted.

NOTE.– In many cases, the sensor can be modeled using a first order system:

$$a_1 \frac{dy}{dt} + a_0 y(t) = x(t) \qquad [7.2]$$

a_0 represents the system gain and a_1 indicates the system's response time.

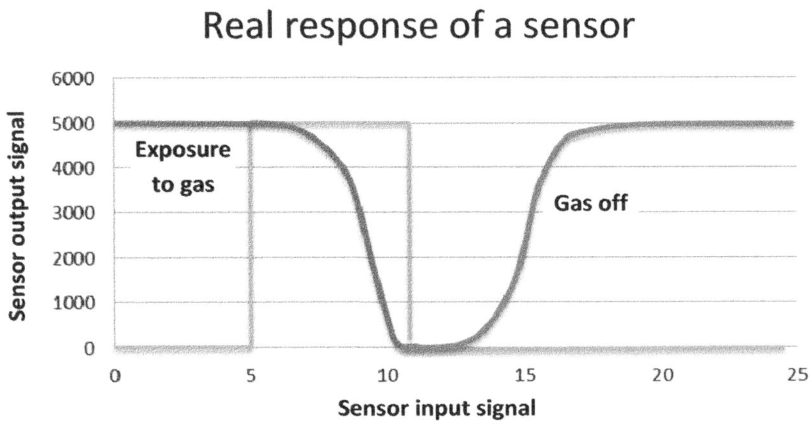

Figure 7.3. *Gas sensor response under gas exposure*

Figure 7.3 provides an example of a gas sensor's response. In this case, the sensor's output signal decreases when a gas is present and increases again when the gas disperses.

Gas sensors (or chemical sensors) are devices which can convert a chemical quantity into an electrical quantity. The chemical being detected may be hydrogen, for example, and we talk about a sensor for gas or even water vapor, a humidity sensor or even for detecting ions in water (e.g. potassium) using an ISFET (ion-sensitive field-effect transistor).

It is therefore useful to classify gas sensors according to the detection principle (see Table 7.1). This classification allows us to make a connection between the detection principle, the measurand and the type of sensor being used.

To conclude this classification and gain a better understanding of gas sensors, we should also examine the materials used for the sensitive element (see Table 7.2).

Principle	Mesurand	Type of sensor
Conductometric	Resistance/conductivity	Tungsten oxide sensor
Potentiometric	Tension/f.e.m	ISFET for pH
Capacitive	Capacity/charge	Polymer humidity sensor
Amperometric	Current	Electrochemical cell
Calorimetric	Heat/temperature	Pellistor, thermistor, etc.
Gravimetric	Mass	Piezoelectric or SAW sensor
Optical	Frequency/absorption	IR methane detector
Resonant	Frequency	Surface plasmon
Fluorescent	Intensity	Fiber optics

Table 7.1. *Classification of gas sensors according to the detection principle*

Sensitive material	Examples	Detection principle	Mesurands
Thin layer (<90 nm)	SnO_2, WO_3, ZnO	Surface conductivity	Combustible gas, CO, NH_3, NO_x
Thick layer	SnO_2, ZnO, TiO_2	Volume conductivity	Combustible gas
Organic layer	Phtalocyanines Substituted	Volume conductivity	NO_x
Conductive polymers	Poly (pyrrole)	Volume/mass conductivity	Polar compounds, NH_3
Catalytic metals	Pd.TiO_2, Pd.MOS, Pd.MOSFET	Surface potential, Stress threshold	H_2, CO, H_2, NH_3, CO
Iono-selective materials	AgCl, AgBr	Electro-chemical potential	Cl^-, Br^-
Langmuir-Blodgett layer	Stearic acid	Piezoelectric/SAW	Polar molecules

Table 7.2. *Examples of materials used to detect gases in the environment and industry*

7.3. Chemical microsensors

This is probably the simplest way of detecting a given chemical space. The sensitive layers used are commonly semi-conductor oxides. Other examples are provided in Chapter 5 of this book.

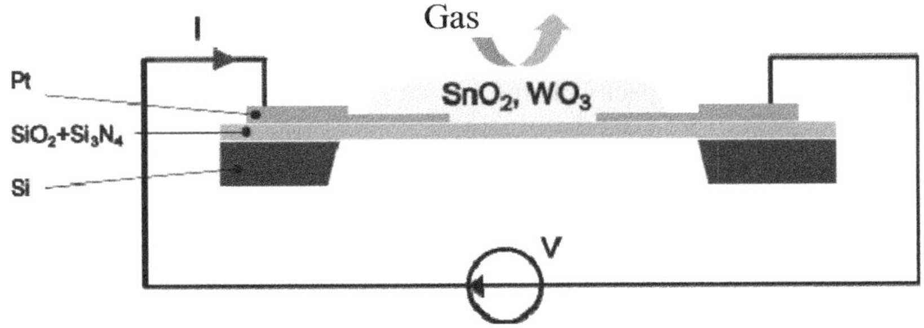

Figure 7.4. *Diagram of the principle of a semi-conductor gas oxide microsensor*

In 1953, Brattain and Bardeen discovered that contact between a gas and the surface of a semi-conductor (oxide or polymer), leads to an important change in its conductivity.

Let us take the case of a semi-conductor oxide, for example SnO_2 or WO_3 and add oxygen vacancies to this oxide. This produces an exchange of free charges and an excess of electrons occurs in the sensitive material, significantly modifying the space charge zone. We therefore obtain an n-type semi-conductor whose conductivity varies according to the number of carriers, which in turn depends on the surrounding gas. These modifications are possible because the act of adding vacancies corresponds to the movement of oxygen atoms from their equilibrium and material positions.

Vacancies can occur on the oxide's surface or volume. In the case of gas sensors, the most active vacancies are obviously those created by the gas environment, which generally contains oxygen, e.g. air.

The presence of oxygen plays a major role in the density equilibrium and the density of spaces present on the surface of the sensitive layer (vacancies, molecules or atoms of the gas being detected, etc.).

It is this phenomenon which is the basis of the detection principle including a reversible reaction between the oxygen in the atmosphere and vacancies in the sensitive layer. The reduction of the number of electrons in the semi-conductor leads to the creation of oxygen ions: O_2^-, O^-, and $O2^-$. Note that these ions can react irreversibly with certain gases (X).

$$\frac{m}{2}O_2 + [\text{site vacant}] + e^- \xrightarrow{k_1} \left(O^-{}_m\right)_{site} \qquad [7.3]$$

$$X + \left(O^-{}_m\right)_{site} \xrightarrow{k_2} \left(XO_m\right)_{site} + e^- \qquad [7.4]$$

where *m* represents the various oxygen species.

EXAMPLE.– The measurand X can be the result of CH_4, which reacts with the oxygen atoms in the network and produces CO_2 and water vapor. This leads to an increase in the density of carriers Δn, which could induce an increase in the conductivity σ.

Note that conductivity is written as:

$$\Delta\sigma = \mu_n.e.\Delta n; \quad \alpha \quad [X]^r \quad \text{with} \quad 0.5 \le r \le 1 \qquad [7.5]$$

where e is the electron's charge.

To be more rigorous, we have to take into account the dependency between the concentration and mobility of carriers μ_n.

These electrical conduction phenomena are also important in crystalline oxides. Electron mobility is governed by the barrier potential between grains. This is the "hopping conduction" model.

Conductivity, G, between grains can be expressed using a simple model:

$$G = G_0.\exp(-\frac{\varphi_B}{kT}) \qquad [7.6]$$

Conductivity G_0 depends slightly on temperature.

The height of the barrier Φ_B, or to be more precise its variation $\Delta\Phi_B$ due to the presence of gas can be linked to the variation in the concentration of carriers before, during and after the presence of the gas has been detected.

$$\Delta\varphi_B \approx \ln\left(1+\frac{\Delta n}{n'}\right) \qquad [7.7]$$

The volume's conductivity is supposedly dominated by adsorption sites. Note that Δn also depends on the temperature.

Lastly, the concentration of carriers can be linked to reaction rates by the following relationship:

$$\Delta\varphi_B \approx \ln\left(1+\frac{\Delta n}{n'}\right) \qquad [7.8]$$

As such, sensitivity and selectivity are governed by the reaction rate.

These phenomena are illustrated by the diagrams in Figure 7.5 which show the model of the barrier potential on a grain boundary in the case of a n-type semi-conductor oxide, such as WO_3. In air the barrier is linked, to a certain extent, to oxygen vacancies and reactivity in the sensor's surface with oxygen in the ambient air. With reductor gases such as CO or NH_3, the barrier's height decreases proportionally with the density of vacancies, which themselves depend on the number of atoms absorbed in the gas being detected.

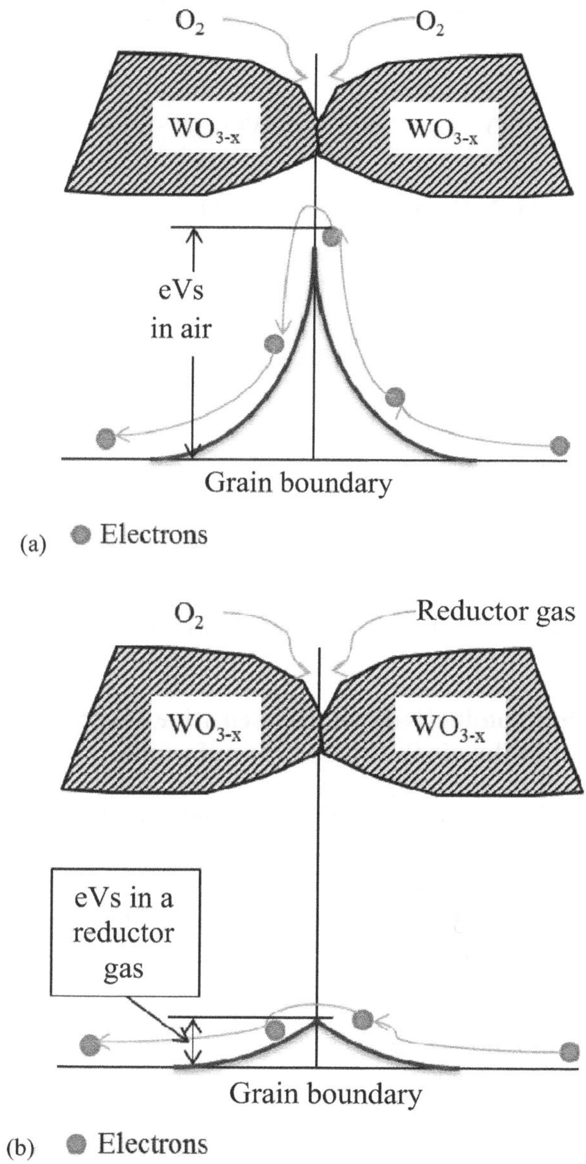

Figure 7.5. *Illustration of the barrier potential model at the grain boundary in the case of the semi-conductor oxide WO_3: a) in air and without gas; b) with a reductor gas, such as CO or NH_3*

The reaction between CO and the oxygen absorbed on WO_3 can be expressed as:

$$1/2O_2 + (WO_{3-x})^V \rightarrow O^-ad\ (WO_{3-x})$$

$$CO + O^-ad\ (WO_{3-x}) \rightarrow CO_2 + (WO_{3-x})^V$$

It should be noted here that this can become more complex if we consider reactivity dynamically because several parameters can intervene including other pollutants or OHs.

Furthermore, it should be noted that the adjunction of metal catalysts on the sensitive material's surface can strongly influence the variation in the barrier's height $\Delta\Phi_B$.

These elements can be easily found in the rich and abundant literature in these fields. For now, however, let us return to aspects of conduction and the electrical response of gas microsensors.

7.4. Modeling conductivity of the sensitive layer with WO_3 gas microsensors in the air in the presence of ozone

In this section we will describe modeling conductivity in gas WO_3 microsensors, by examining the "drift-diffusion" model for the sensitive layer, which is an n-type semi-conductor and according to Wolkenstein's adsorption-desorption theory is the best adapted for microsensors exposed to air in the presence of ozone.

7.4.1. *Model of a microsensor*

The sensitive layer is completely porous and composed of a regular stack of identical and spherical semi-conductive nanograins. Conduction from grain-to-grain is ensured by a small spherical cap at the boundary between the touching grains. Electrodes are parallel to one another to free transversal currents.

Each grain lies in the surrounding atmosphere and gives its surface a homogenous N* density from the adsorption sites S to the surface. For diatomic gases such as oxygen O_2, there are two adsorption mechanisms for non-dissociative (ND) or dissociative (D) kinetic (D):

$$\text{ND kinetic: } S + O_2 \rightarrow (S\text{-}O_2);\ (S\text{-}O_2) + e^- \rightarrow (S\text{-}O_2)^- \qquad [7.9a]$$

$$\text{D kinetic: } 2S + O_2 \rightarrow 2(S\text{-}O);\ (S\text{-}O) + e^- \rightarrow (S\text{-}O)^- \qquad [7.9b]$$

For triatomic gases such as ozone O_3, there is only one possible mechanism:

$$S + O_3 \rightarrow (S\text{-}O) + O_2;\ (S\text{-}O) + e^- \rightarrow (S\text{-}O)^- \qquad [7.10]$$

7.4.2. *Wolkenstein model*

This model enables us to express the adsorption-desorption dynamic of chemical gas spaces on the sensitive material in gas sensors.

Adsorption generally takes place on adsorption sites whose density depends on several parameters such as ambient atmosphere, temperature, etc. In the case of semi-conductive oxides, these sites are oxygen vacancies. Depending on the type of gas adsorbed (i.e. oxidizing or reductor), an increase or decrease in the microsensor's electrical conductivity can be observed. Similarly, this change in conductivity can be seen by modifying the type of semi-conductor (n to p) for a single adsorbed gas. Note that these variations increase in significance with the sample's thinness.

Two phases characterize the adsorption process

Phase 1: Approaching the particle (either an atom or molecule) which has just been adsorbed onto a site on the sensitive layer. Bonding is achieved by weak chemisorption: van der Waal forces, e.g. a deformation of orbitals but no transfer of charge. This results in:

a loss of free energy in the system = weak chemisorption energy = E_w

Consequences: The creation of an acceptor surface state E_{ss} in the energy gap (EG) but no effect on conduction.

Phase 2: An electron in the energy conduction gap (CG) E_c is captured by the adsorbed particle. The bond energy increases from $E_s = E_c - E_{ss}$. This results in an increase in the superficial charge and the barrier potential V_s ($V_s<0$). The CG level on the surface is $E_{cs} = E_c - qV_s$ ➔ $E_s = E_c - E_{ss} = E_{cs} - E_{ss} + qV_s$. The difference $E_{cs} - E_{ss}$ also corresponds to the difference $\chi_{ads} - \chi_{sc}$ between the electrical relationships between the neutral adsorbent and the semi-conductive oxide. Therefore the energy of strong chemisorption is:

$$E_w + E_s = E_w + \chi_{ads} - \chi_{sc} + qV_s$$

Consequences: The decrease in conductivity between grains by the space charge zone (SCZ). The bond energy of strongly chemisorbed particles decreases when they increase in number, therefore facilitating desorption.

NOTE.– It is important to take into account the potential degradation of the surface by irreversible chemical reactions, which make controlling and maintaining chemical properties in chemical semi-conductive sensors more difficult. This difficulty can often be overcome if there is a good choice of the sensitive layer and its production process.

7.4.3. *Modeling*

One of the most important parameters in the adsorption-desorption process is the level of gap (number of adsorbed particles versus the number of sites) $\theta = N_{ads}/N^*$. Its development is determined by the equilibrium between the adsorption and desorption of neutral and ionized particles, considering both dissociative and non-dissociative kinetics.

With the non-dissociative kinetic, the variation in gap rate depending on time is expressed as:

$$\frac{d\theta}{dt} = k_p\, p(1-\theta) - \nu \exp(\frac{-E_w}{kT})\theta_0 - \nu \exp(\frac{-E_w - E_s}{kT})\theta_- \qquad [7.11a]$$

where k_p is a proportionality constant which can be obtained using the kinetic gas theory, p is the partial pressure of the gas in question, v is a constant related to oscillation frequency in the crystal's phonons, T is the temperature in Kelvin, k is Boltzmann's constant which is $1.3806504 \times 10^{-23}$ m^2 kg s^{-2} K^{-1} (or J K^{-1}).

In the case of a dissociative kinetic, the variation in the temporally dependent gap rate is expressed as:

$$\frac{d\theta}{dt} = k_p\, p(1-\theta)^2 - \nu \exp(\frac{-E_w}{kT})\theta_0^2 - \nu \exp(\frac{-E_w - E_s}{kT})\theta_-^2 \qquad [7.11b]$$

The distribution of the gap rate between the ionized θ_- and neutral θ_0 species can be calculated using the Fermi-Dirac statistic:

$$\eta_- = \frac{\theta_-}{\theta} = \frac{1}{1 + 2\exp\frac{E_{SS} - E_F}{kT}} \qquad [7.12]$$

$$\eta_0 = \frac{\theta_0}{\theta} = 1 - \eta_- \qquad [7.13]$$

Equations [7.11]–[7.13] are relative to superficial quantities but depend on the band curvature qV_s, which must be calculated using Poisson's equation:

$$\Delta V = \frac{q}{\varepsilon}\left(n - p - N_d^+\right) \qquad [7.14]$$

where n is the density of electrons, p represents the holes and N_d^+ is the ionized oxygen vacancies which behave as donors.

$$N_d^+ = \frac{N_d}{1 + 2\exp\frac{E_f - E_d}{kT}} \quad [7.15]$$

$$n = N_c \exp\left(\frac{E_f - E_c}{kT}\right) \quad [7.16a]$$

$$p = N_v \exp\left(\frac{E_v - E_f}{kT}\right) \quad [7.16b]$$

N_c and N_v are the density of effective states in the conduction and valence bands. E_f and E_d are the Fermi and donor level energies, E_c is the conduction band's base energy below which the density of states is equal to zero. E_v is the energy at the top of the valence band.

The interaction between the atmosphere and the material can be described by the (Gaussian) relation:

$$\boldsymbol{E}_n = \frac{\sigma}{\varepsilon} = \frac{-qN^*\theta_-}{\varepsilon} \quad [7.17]$$

$\boldsymbol{E}_n$ is the normal electrical field on the surface and σ is the superficial charge density.

The equation above, which proves the existence of a Fermi level, models the material at thermodynamic equilibrium, without a current. This enables us to identify the evolution of the gap rate θ and its two components, both θ_0 neutral and θ_- ionized, but it does not allow us to calculate the conductivity.

If two (or more) gas environments are present in the atmosphere and if we consider the non-dissociative kinetic, equation [7.11a] is replaced by two coupled equations:

$$\frac{d\theta_1}{dt} = k_1 p_1 (1-\theta_1-\theta_2) - v_1 \exp(\frac{-E_{w1}}{kT}) \left[\theta_{10} + \theta_{1-} \exp(\frac{-E_{s1}}{kT}) \right] \quad [7.18a]$$

$$\frac{d\theta_2}{dt} = k_2 p_2 (1-\theta_1-\theta_2) - v_2 \exp(\frac{-E_{w2}}{kT}) \left[\theta_{20} + \theta_{2-} \exp(\frac{-E_{s2}}{kT}) \right] \quad [7.18b]$$

In the case of a dissociative kinetic, equivalent expressions will be used.

In the presence of polarization, the Fermi level is only defined in the equipotential zone (central grain zones) and the cycle of equations should be completed.

NOTE.– The adsorption-desorption mechanisms on the grain surface are much slower than the effects of recombination generation in the semi-conductor. In addition, given this, the complimentary equations retained for conductivity are as follows.

The continuity equations for the electrons and holes are:

$$Div\,\boldsymbol{J}_n = qU \quad . \quad [7.19a]$$

$$Div\,\boldsymbol{J}_p = -qU \quad [7.19b]$$

with U being the recombination-generation rate derived from the Shockley-Read-Hall model:

$$U = \frac{np - n_i^2}{\tau_n(p + n_i) + \tau_p(n + n_i)} \quad [7.20]$$

where τ_n and τ_p are the lifespans of the electrons and holes and n_i is the intrinsic density.

The transport equations ("drift-diffusion" model) for electrons and holes:

$$\boldsymbol{J}_n = nq\mu_n \boldsymbol{E} + qD_n \boldsymbol{Grad}\, n \quad [7.21a]$$

$$\boldsymbol{J}_p = pq\mu_p \boldsymbol{E} - qD_p \boldsymbol{Grad}\, p \quad [7.21b]$$

where $\boldsymbol{E}$ is the electrical field, μ_n and μ_p are mobilities, D_n and D_p are the diffusion coefficients for electrons and holes and q is the electron's charge.

This should be accompanied by the mobility model:

$$\mu(T) = \mu_0 \frac{300}{T} \quad [7.22]$$

With each phase in time, the calculation procedure consists of two stages.

Stage 1

The values θ_1 and θ_2 are calculated using equations [7.18a] and [7.18b] in accordance with Poisson's equation [7.14] to satisfy equation [7.17]. Equations [7.18a] and [7.18b] are strongly nonlinear due to the exponential terms E_{s1} and E_{s2}, which themselves depend on θ_1 and θ_2. Supposing the variation of these terms is low in the interval of one time phase, we can analytically integrate these equations and readjust E_{s1} and E_{s2} at each phase. Poisson's equation is solved using the fourth order Runge–Kutta method. This first calculation stage is also carried out without applied polarization, the layer being in thermodynamic equilibrium with its environment.

Stage 2

In the second stage, a weak current $\boldsymbol{J}$ transiting between two adjacent grains and injected into the induced electrical field is calculated by equations [7.14] and [7.19]–[7.22]. To do so, weak perturbation is added to equation's solution:

$$n = n_0 + \delta n \quad p = p_0 + \delta p$$

$$\boldsymbol{E} = \boldsymbol{E_0} + \boldsymbol{\delta E} \quad U = 0 + \delta U$$

and the equations are linearized around this solution.

Poisson's equation therefore becomes:

$$\frac{d}{dx}\delta \boldsymbol{E} = \frac{q}{\varepsilon}(\delta p - \delta n) \tag{7.23}$$

Equations [7.19a] and [7.19b] show that $\boldsymbol{J} = \boldsymbol{J_n} + \boldsymbol{J_p}$ = const.

Equations [7.21a] and [7.21b] therefore provide, by addition:

$$\delta \boldsymbol{E}(n_0\,\mu_n + p_0\,\mu_p) + \boldsymbol{E_0}(\delta n\,\mu_n + \delta p\,\mu_p) + \frac{kT}{q}\left(\mu_n\frac{d}{dx}\delta n - \mu_p\frac{d}{dx}\delta p\right) = \frac{\boldsymbol{J}}{q} \tag{7.24}$$

Equations [7.19b] and [7.20] provide:

$$p_0\frac{d}{dx}\delta \boldsymbol{E} + \delta \boldsymbol{E}\frac{dp_0}{dx} + \boldsymbol{E_0}\frac{d}{dx}\delta p + \delta p\frac{d\boldsymbol{E_0}}{dx} - \frac{kT}{q}\frac{d^2}{dx^2}\delta p + \frac{C}{\mu_p}(n_0\delta p + p_0\delta n) = 0 \tag{7.25}$$

with:

$$C = \frac{1}{\tau_n(p_0 + n_i) + \tau_p(n_0 + n_i)} \qquad [7.26]$$

Second order discretization in the series of equations leads to a system of $5 \times 3 \times N$ linear equations, with N being the number of discretization points i between the center of two adjacent grains.

Solutions $\delta n(i)$, $\delta p(i)$ and $\boldsymbol{\delta E(i)}$ ($0<i<N$) are obtained by inverting this system (Gaussian elimination). We therefore deduce the potential difference (ddp) between grains, followed by the sensitive layer and finally the sensor's response.

7.4.4. *Some results from the calculations developed above [GUE 08]*

7.4.4.1. *Influence of grain size*

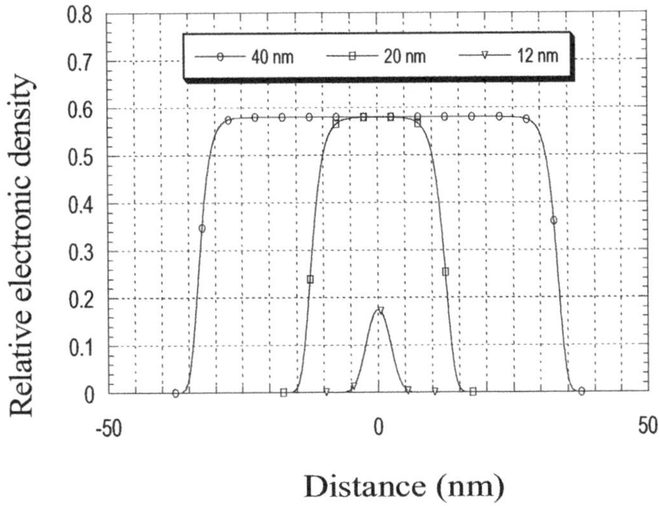

Figure 7.6. *Relative electronic density (n/Nc) for a sensor in air without ozone at 523 K. The nominal value is reached in the central part for grains with a radius greater than 20 nm, but is not for grains of 12 nm*

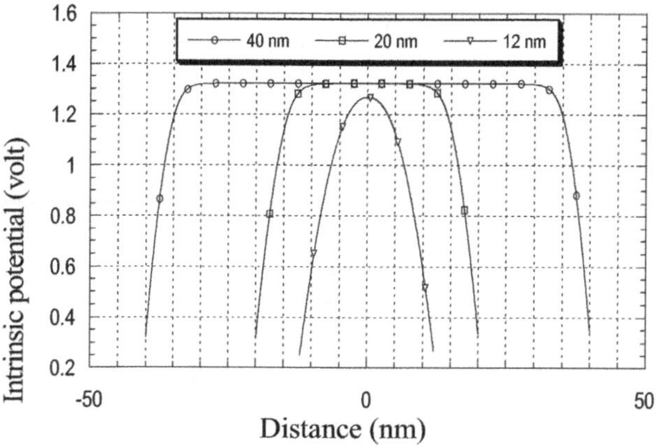

Figure 7.7. *Intrinsic potential for a sensor in air, without ozone at 523 K. The same applies to relative density*

7.4.4.2. *Static responses for WO_3 sensors in gas*

The results shown here have been obtained using three gases: dissociative oxygen (θ_1), non-dissociative oxygen (θ_2) and ozone (θ_3). The result is the relationship between conductivity in an air–gas mixture and in air.

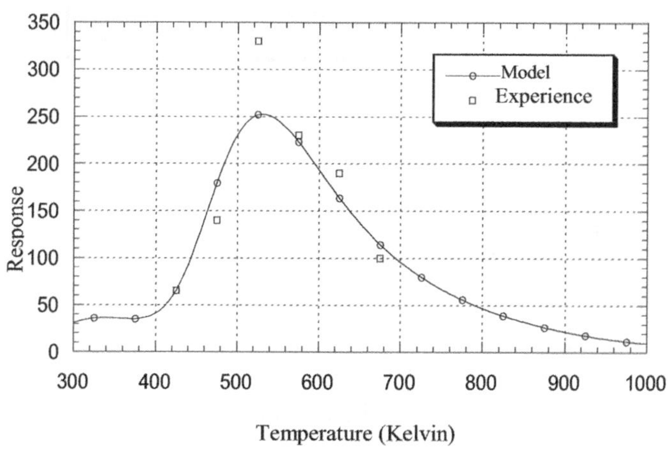

Figure 7.8. *Response according to temperature with 0.8 ppm of ozone. Comparison with the experimental model*

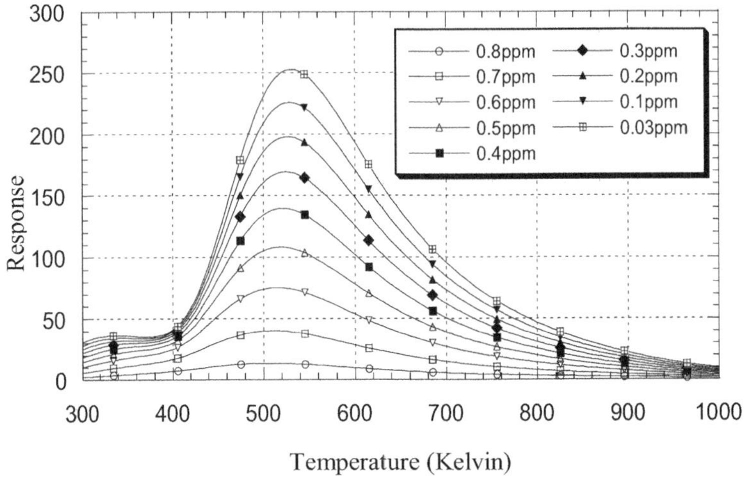

Figure 7.9. *Response according to temperature with different concentrations of ozone*

Note the slight decline of the sensitivity maximum with temperature as concentration increases.

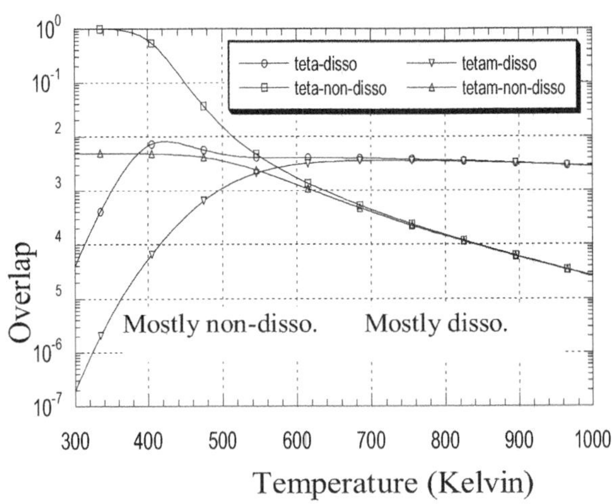

Figure 7.10. *Variation in recovery rates due to oxygen with 0.8 ppm of ozone with temperature*

Largely non-dissociative adsorption at low temperatures becomes dissociative below 550 K.

7.4.4.3. *Dynamic responses of the WO_3 sensor in gas*

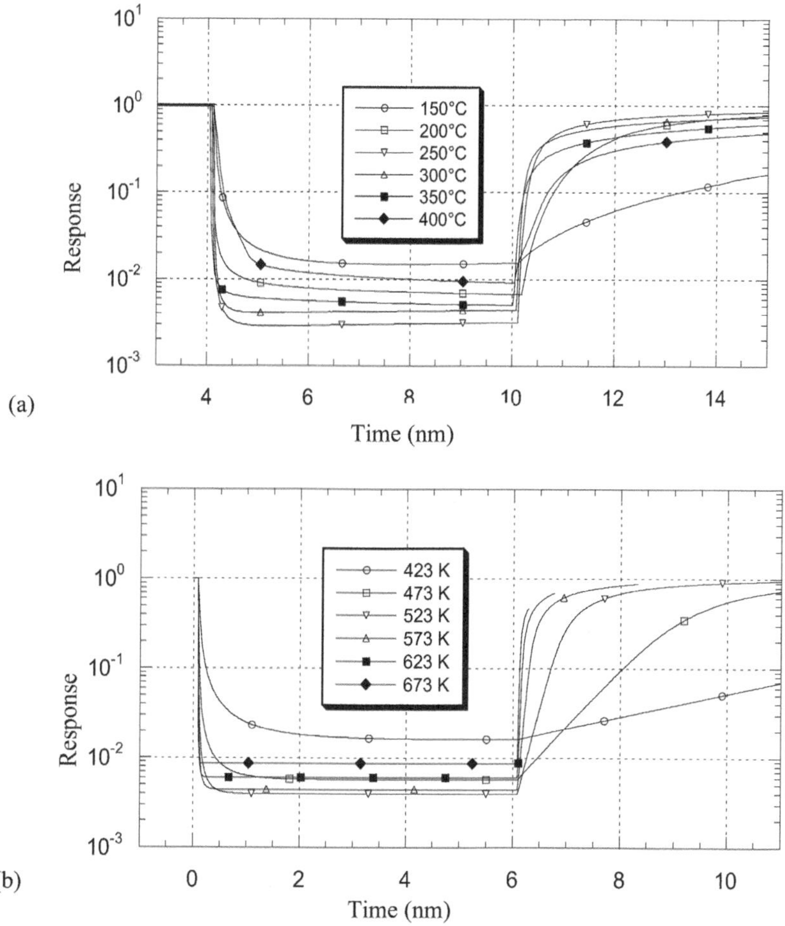

Figure 7.11. *Responses recorded (a) and modeled (b) for a sensor with a concentration of ozone of 0.8 ppm over six minutes, according to time, with the operating temperature as a parameter: range of temperatures 423 to 673 K*

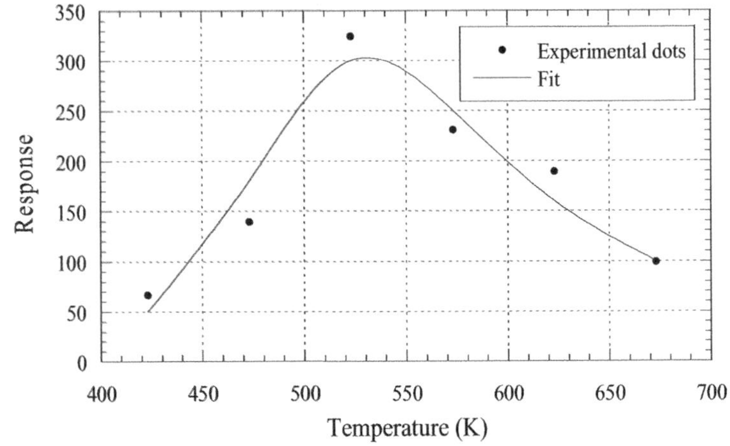

Figure 7.12. *Variation in the sensor's response according to temperature. Experimental conditions (static conditions): temperature range = 150 to 400°C; gas mixture = dry air + 0.8 ppm of ozone*

It is evident in the case of the WO_3 sensor that the kinetic is non-dissociative at low temperatures and dissociative at high temperatures (>523 K).

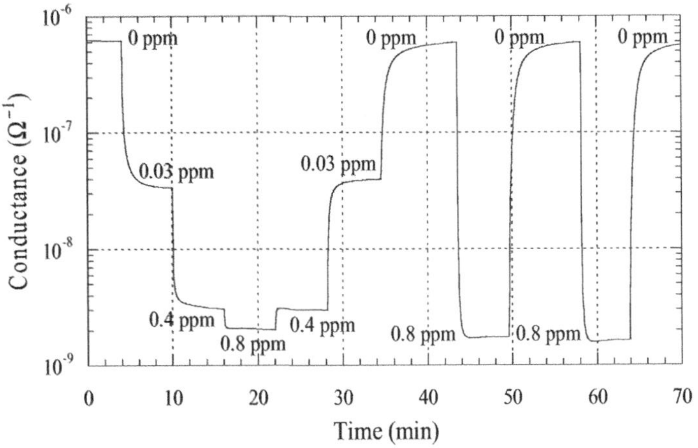

Figure 7.13. *(a) Conductivity of the WO_3 sensor measured at 250°C. Experimental conditions: gas mixture = dry air + ozone in concentrations of 0.03, 0.4 and 0.8 ppm*

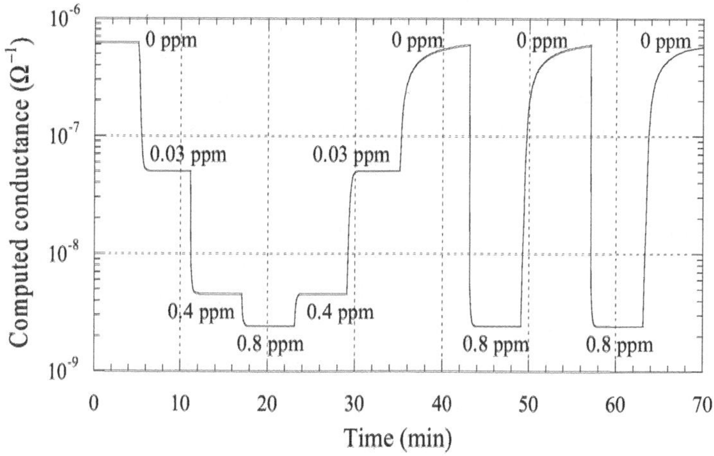

Figure 7.13. (continued) *(b) Conductivity in the WO_3 sensor simulated at 250°C. Simulation conditions: non-dissociative adsorption, gas mixture = dry air + ozone (0.03, 0.4 and 0.8 ppm)*

7.5. Impedance spectroscopy on gas sensors

In direct current (DC), measures allow us to obtain global information about the sensor's response. Nevertheless, they allow us to monitor the sensor's state in real-time and therefore its surrounding gas environment.

In alternating current (AC), measures provide a powerful tool for examining the nature of conduction processes and the gas/solid interaction mechanism. The processes with different types of constants can be distinguished by varying the frequency [LAB 05]. In contrast, these measures require a stable gas environment, the time required to recover the impedance spectrum. This is a powerful electrical characterization tool even if this method cannot normally be used for real-time gas detection.

The sensor's conductivity (specifically on its surface) is modified by adsorbing the target gas. In the case of an oxidizing atmosphere, the oxide's surface is covered by negatively charged adsorbed oxygen. This results in a reduction in electrons in the adjacent space charge zone. The sensor's impedance increases. In reducing conditions, oxygen adsorbates are eliminated by reacting with the reducing gas, and electrons are reinjected in the space charge zone. As a result, the sensor's impedance decreases. A good understanding of the transport properties within the grains is therefore necessary to optimize the development of this type of sensor.

Having discussed impedance, the main significance of these measures in AC is that they highlight both the real and imaginary parts of the signal. In appropriate conditions, the contributions of grains, grain boundaries and electrodes can be solved by exploiting the sensor's response for an applied AC signal with variable frequency [ABR 00]. Quantitative data on each of these areas can be extracted by considering an equivalent circuit, which is generally considered as taking a series of R-C parallel cells.

Figure 7.14 shows an example of an impedance spectrum for a WO_3 sensor in air with 0.1 ppm of ozone, with temperatures ranging from 150 to 375°C.

These spectra are obtained after the period of time required to stabilize the parameters considered (several minutes). Impedance increases as the operating temperature rises to 300°C and then decreases below this temperature.

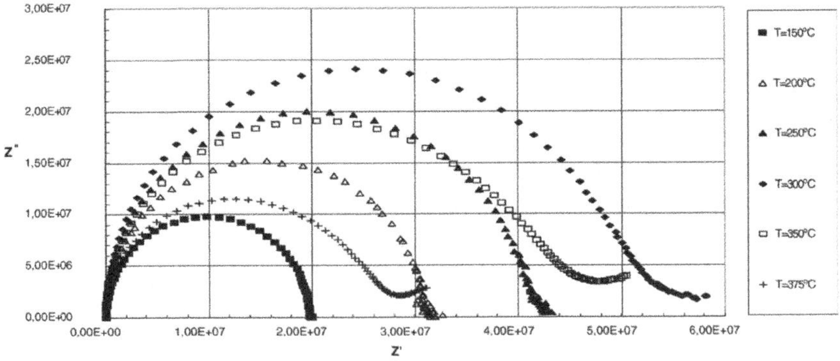

Figure 7.14. *Variation in impedance of a WO_3 sensor according to temperature in dry air and ozone with a concentration of 0.1 ppm*

This phenomenon could be interpreted as a competition between adsorption and desorption rates as the Lennard-Jones equation shows [MAD 89]:

$$\frac{d\theta}{dt} = k_{ads} e^{-\frac{\Delta E_A}{kT}} - k_{des} e^{-(\frac{\Delta E_A + \Delta H_{chem}}{kT})} \qquad [7.27]$$

where θ is the fraction of available adsorption sites, k_{ads} and k_{des} are the adsorption and desorption reaction speed constants respectively, E_A is the chemisorption's activation energy and H_{chem} is the chemisorption heat.

In order to interpret the sensor's impedance spectrum according to frequency, the response of Nyquist in dry air with 0.1 ppm of ozone has been divided into three terms corresponding to parallel R-C cells, see Figure 7.14. Three semi-circles are therefore obtained: the first relating to the volume of grains, the second for grain boundaries and the third for diffusion to electrodes. The values R-C are shown in Table 7.3.

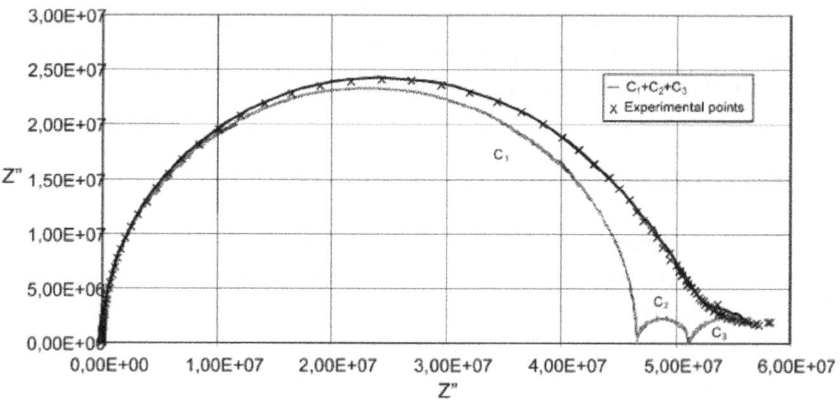

Figure 7.15. *Nyquist diagram of a WO_3 sensor operating at 300°C, in an atmosphere of dry air and ozone of 0.1 ppm. The semi-circles C_1, C_2 and C_3 are the results of modeling the impedance spectrum using three R-C cells*

	R_1 (Ω)	C_1 (nF)	R_2 (Ω)	C_2 (nF)	R_3 (Ω)	C_3 (nF)	n
Air	6.1 10^5	0.17	4.1 10^4	3430	4.3 10^4	1.5	1
Air+O_3	4.7 10^7	0.16	4.5 10^6	9.7	6.7 10^6	122	0.77

Table 7.3. *Model of the R and C values of the three R-C cells*

When the mixture of air + ozone comes into contact with the sensor, we can see that the resistances attributed to grain volume, grain boundaries and electrodes (R_1, R_2 and R_3, respectively) increases, the volume capacity of C_1 does not vary, whilst for grain boundaries C_2 decreases and for the electrodes C_3 increases.

The adsorption of negatively charged ions leads to the increase in resistances R_1 and R_2, by replacing the vacant surface sites. This results in a decrease in the concentration of electrons. The size of the space charge zone at the grain boundaries increases and the space charge capacity C_2 decreases.

The fact that C_1 does not vary and the slight increase of C_3 should be interpreted with caution, particularly given the parasitical abilities linked to the measuring system itself.

7.6. Selectivity in gas sensors

There are number of definitions for selectivity. The simplest is to consider selectivity as a sensor's ability to recognize and identify a gas from a mixture or more generally to separate two gases. It depends on the nature of chemical reactions involved. Selectivity can be obtained by choosing the sensitive material M, which reacts preferentially to the measurand X. The material can be used in a thin or thick layer.

A good means of obtaining satisfactory selectivity is to use molecular recognition ⇒ biological sensors (the key lock system principle). It can also be improved by using enzymes aiding a specific biological reaction.

Semi-conductive oxide sensors should also be selective. The majority of oxides employed often react to several gases. These materials are used for their various qualities: temporal stability, reproducibility, sensitivity, etc. As a result, there is a need to work on improving their selectivity. This can be achieved by "doping" the sensitive material using catalytic metals such as: palladium (Pd), platinum (Pt), rhodium (Rh) and gold (Au), etc. Measuring doping should be done with caution, however, because it is often a question of placing the catalytic metal in the form of nanograins on the sensor's surface. This results in an important modification in the reactivity regarding the targeted gases and therefore the sensor's conductivity. Selectivity can also be obtained using filters and sensor networks, etc.

Whilst desired, selectivity is difficult to obtain without appropriate treatment of the sensors' signals (PCA, FFY, neuronal networks, etc.). This can consist of analyzing data as its principle component (ACP method) or other similar methods such as electronic noises, as well as an analysis of the fluctuations in conductivity generated by the adsorption-desorption dynamic on the sensors' surface. This last method is called electronic noise spectroscopy. This technique will be examined in further detail in the following section.

7.7. Electronic noise spectroscopy in gas microsensors

This technique involves extracting electronic noise information produced by the sensor with a semi-conductive oxide sensitive layer.

The noise has a random magnitude. It is expressed via the density of noise power in V^2/Hz or A^2/Hz.

Note that there are numerous sources of noise in this type of device. Table 7.4 shows those encountered in semi-conductive oxides.

Source of noise	Noise power spectral density
Thermal noise	$S_v^{Th} = 4kTR$
Shot noise	$S_I^{Shot} = 2qI$
Excess noise	$S_v^{Ex} = \frac{A}{f^{\alpha}}$
Recombination-generation noise	$S_v^{GR} = \sum_{i=1}^{n} B_i \frac{2\tau_i}{1+(2\pi f\tau_i)^2}$
Adsorption–desorption noise	Lorentzian (see below)

Table 7.4. *Noise power spectral densities for the main sources of noise*

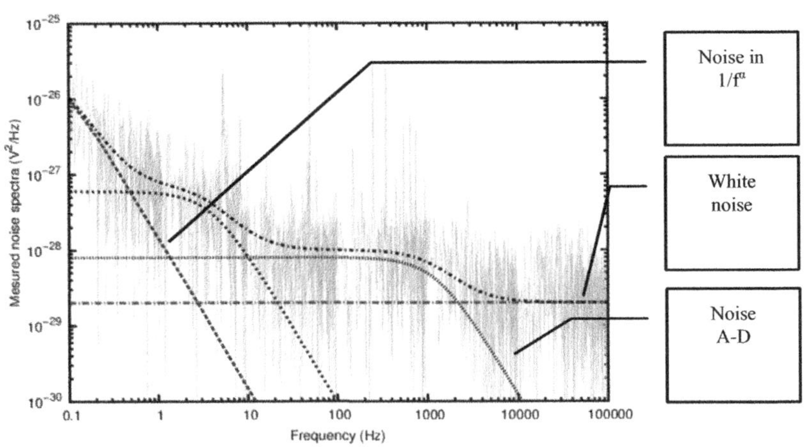

Figure 7.16. *Global measure of a WO_3 gas sensor's noise [CON 11]*

The measure illustrated by Figure 7.16 shows that it is possible, in certain conditions, to highlight adsorption–desorption noise.

Adsorption–desorption noise

The random arrival of atoms, molecules and ions on the sensor's surface generates fluctuations in the number of free electrons in the material. These fluctuations generate variations in the sensor's conductivity and can be measured using a random signal, noise.

Modeling adsorption–desorption noise is carried out in relation to the following path:

$$g = e\mu_n \frac{S}{L} n \Rightarrow \delta g(t) \propto \delta n(t) \Rightarrow S_{\delta g}(f) \propto S_{\delta n}(f)$$

where g is the conductivity, S is the section, L is the length and S_i is the spectral noise density.

The spectral noise density $S_{\delta n}$ is linked to the fluctuations in the number of gas molecules adsorbed per surface unit δN.

Adsorption-desorption theory should also be taken into account. The use of Wolkenstein theory allows us to take a more realistic approach.

These calculations will not be examined further in this chapter but further reading is available in the references [GOM 06, WOL 91].

Equation [7.28] is one of the important results of these calculations. It conveys the noise power spectral density which results from fluctuations in conductivity in Lorentzian's form:

$$S_{\delta g}(f) \propto \frac{H_0}{1+\frac{f^2}{f_C^2}} \qquad [7.28]$$

The origin of these Lorentzians can be associated with grains and grain boundaries composing the sensitive layer, as well as other elements which can affect the sensors' conductivity such as, for example, catalytic metals and filters applied to their surfaces.

The amplitude H_0 and the cut-off frequency f_C depend on the parameters related to the sensor and gases being considered.

Regarding measurements, the more significant adsorption–desorption noise is in relation to other sources, the easier it is to measure it. A simple measure involves taking four identical sensors. Their exposure to the target gas should generate identical responses in each sensor. The deterministic output signal is therefore zero. Only the random (noise) signal can be measured because each sensor behaves as an independent noise source. The bridge is permanently out of equilibrium and the bridge's output voltage is due solely to a noise voltage. This bridge configuration eliminates the effect of variations in impedance between the sensor and the reference.

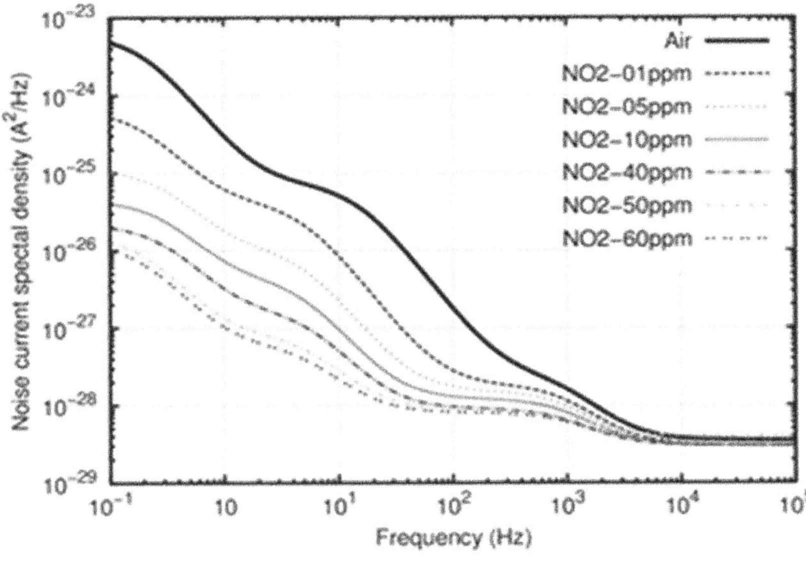

Figure 7.17. *Noise power spectral density according to the frequency of the WO_3 sensor with different concentrations of NO_2*

Figure 7.17 provides an example of a WO_3 sensor's response to different concentrations of NO_2. Note that spectral density in the noise current is essentially dominated by the adsorption–desorption noise. White noise is only visible after a frequency of 10 kHz and the Flicker noise $1/f^{\alpha}$ is invisible in the frequency range considered.

7.8. Conclusion

In this chapter we have introduced several concepts which are essential for a good understanding of the electrical conduction phenomena in gas microsensors. Readers will have discovered that the notions described here are classical concepts taken from physics, microelectronics and chemistry, and as such "sensors" is a multidisciplinary field. There is an abundance of literature, focusing on rich and active applications aimed at the environment, security and biology, etc.

The field of microsensors has opened new, highly innovative perspectives, particularly with the development of new technologies and the possibilities for miniaturizing components and microsystems. This chapter is an introduction to problems relating to measuring the electrical properties of microsensors, but should be considered as an initial approach rather than an in-depth examination of this field.

Readers should nevertheless bear in mind that measuring a microsensor's electric characteristic should be carried out more thoroughly, in both measuring and interpreting results, because the signal sent by the sensor to the user is often electric.

Having said this, an in-depth understanding of the concepts developed here should enable readers to understand the numerous studies in the literature, carry out effective electric measures and develop and/or design high performing devices and gas detection systems.

Note finally that research in the field is still rich with the potential for new ideas to be discovered and developed.

7.9. Bibliography

[ABR 00] J.C.C. ABRANTES, J.A. LABRINCHA, J.R. FRADE, "An alternative representation of impedance spectra of ceramics", *Mater. Res. Bull.*, vol. 35, pp. 727-740, 2000.

[CON 11] T. CONTARET, J.L. SEGUIN AND K. AGUIR, "Physical-based characterization of low frequency responses in metal-oxide gas sensors", *IEEE Sensors 2011 Conference*, Limerick, Ireland, 28-31 October, 2011.

[GOM 06] S. GOMRI, J.L. SEGUIN, J. GUERIN, K. AGUIR, "Adsorption–desorption noise in gas sensors: modelling using Langmuir and Wolkenstein models for adsorption", *Sensors and Actuators B*, vol. 114, pp. 451-459, 2006.

[GUE 08] J. GUERIN, M. BENDAHAN, K. AGUIR, "A dynamic response model for the WO_3-based ozone sensors", *Sensors and Actuators B*, vol. 128, pp. 462-467, 2008.

[LAB 05] A. LABIDI, C. JACOLIN, M. BENDAHAN, A. ABDELGHANI, J. GUERIN, K. AGUIR, M. MAAREF, "Impedance spectroscopy on WO_3 gas sensor", *Sensors and Actuators B*, vol. 106, pp. 713-718, 2005.

[MAD 89] J.M. MADOU, S.R. MORRISON, Chapters 3 and 12, *Chemical Sensing with Solid State Devices*, Academic Press, 1989.

[WOL 91] T. WOLKENSTEIN, *Electronic Processes on Semi-conductor Surfaces During Chemisorption*, Consultant Bureau, 1991.

Chapter 8

Gas Microsensor Technology

8.1. Introduction

In both literature and industry, there are various examples of gas detection devices which use electrochemical sensors, resistive metal oxide based sensors, catalytic or even piezoelectric sensors.

The growing interest in these sensors, both in industry and research is due to several reasons. Among others, we can cite low production costs aided by the development of microelectronic technologies. This allows us to reduce the size of components and therefore produce a large number of sensors on one silicon wafer. As such, with the emergence of Microsystems (as of the beginning of the 1980s), we are seeing an ever increasing development of miniature, portable and "intelligent" devices integrating sensors, their supply, processing electronics and other elements. These are therefore termed as integrated electronic noises.

Producing these devices involves standard microelectronic technologies as well as new techniques for designing and integrating new materials, which are often nanostructures or even new micro-production or assembly methods. This enables industry to respond to

---

Chapter written by Philippe MENINI.

the needs of the market such as low production costs, the lowest possible energy consumption, good stability, reproducibility, reliability and increased portability. Even in 2005, 70% of gas detectors were portable with a constant progression of this rate each year (starting at 76% today).

Among the sensors developed so far, semiconductive gas sensors today respond to these needs best with moderate production costs (which decrease as the number being produced increases). They are therefore not only well adapted to microelectronic techniques but also integrate a large number of materials such as metal oxides, semiconductive polymers and other components. The first metal oxide based gas sensors to be marketed were developed by Seiyama and Taguchi in the 1960s using ZnO and SnO_2 as sensitive materials for detecting liquid petroleum gases (LPG). Since then, extensive research has been carried out to improve performance, which still remains to be perfected in terms of stability, reproducibility and particularly selectivity.

In the middle of the 2000s, the "Grand Research Challenges" as set out by the MNT Gas Sensor Forum (December 2006) included the following:

– COV characterizing inside and vehicle cabin air quality.

– Identifying gases from patients in medical analysis (measuring abnormal variations in gaseous markers of disease).

– Improving selectivity and stability for semiconductor gas sensors.

– A combinatorial methodology for optimizing sensing materials.

– Integrated MEMS using sensor networks (combination of sensing optima).

This chapter will be split into two main parts:

– technological development of metal oxide sensors;

– developing a new generation of wireless communicative gas sensors, whilst responding to these current main challenges.

8.2. Metal oxide gas sensor technology

8.2.1. *Overview of how gas sensors function*

In semiconductor gas sensors, chemical information is sent to the material's electronic structure and its surfaces as measurable electrical characteristics, such as changes in conductivity. As measurable interactions, mostly consisting of chemisorption, they are mainly redox reactions, which occur during the exchange of electrons between the gas and sensitive material.

The main factors affecting the detection of gas spaces on a "sensitive" layer are surface temperature, concentrations of O_2 and H_2O (relative to the number of molecules in contact with the surface) and the flux of gas on the surface (relating to the number of gas/surface collisions). The existence of an oxygenized space on the surface of the sensory layer strongly depends on the energy provided, i.e. the sensor's operating temperature and its variations. In other words, sensitivity to gas depends on the surface state and the sensitive layer's temperature.

Finally, the material's morphology plays an important role in global transduction. It defines the specific surface and transport properties inside the material. Sensitivity and selectivity to gas depend on the sensitive material's structure, whether compact, porous or nanostructured.

8.2.2. *Existing structures*

As we have seen, in order to optimize adsorption phenomena and exchanges of electrons between the gas and the sensitive material (metal oxide), it is necessary to heat the surface to high temperatures (between 300 and 500°C). Metal oxide based gas sensors are therefore composed of:

– a sensitive layer composed of the part which interacts with the gas environment;

– electrodes for electric measures of this sensitive layer;

– a heating element which brings the sensitive layer to the required temperature. This part must be well isolated electrically from the measuring electrodes.

A diagram showing how the sensor functions is shown in Figure 8.1. Only four technological elements are required: the three parts described plus the insulating layer.

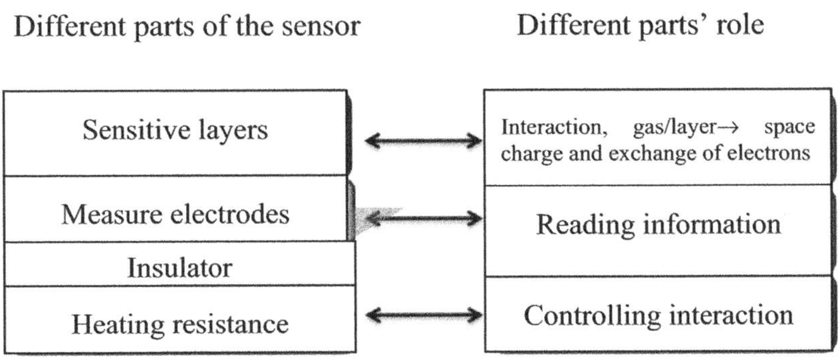

Figure 8.1. *Functional diagram of a semiconductor gas sensor*

This schema demonstrates the simplicity of these systems and therefore their "cost-based" characteristic.

It should be noted that the collection of thermal resistance/measuring electrodes is termed a "microhotplate" in the literature.

8.2.2.1. *Electrodes*

Electrodes create an electrical contact with the sensitive layer to measure its conductivity (or resistance). They enable the material's charges to be conducted towards the circuit receiving the signal. Electrodes are said to be optimal if they establish good ohmic contact with the sensitive layer and enable the maximum transfer of the material's charges towards the circuit. The parameters involved in their design are their geometry and the materials used given that they must withstand high temperatures.

Material for electrodes

The materials sought must be good conductors and remain stable over time, particularly with very high temperatures. The optimization of contact between the electrode and sensitive layer in terms of response (resistance, capacity etc.) requires the use of pure metal contacts. The choice is normally one between metals such as aluminum (AL, simple), gold (Au, noble), platinum (Pt, noble), tungsten (W, refractory), tantalum (Ta, noble) or Chrome (Cr, noble). These electrodes can be a superposition of these materials to obtain the desired characteristics.

It has been shown that platinum electrodes are best adapted for a gas sensor with an SnO_2 layer, which is the most commonly used material today. Generally associated with a "bonding layer" in titanium, platinum has highly stable characteristics in various temperatures and temporally (it only oxidizes at 650°C). This also has a catalyzing effect on certain gases such as CO. However, when detecting oxidizing gases, it has been shown that adding a layer of gold improves performance.

Electrodes' geometry

The electrodes' geometry determines the lines of current (the possible paths for carriers) in the material. These conception parameters are the surface, form, inter-electrode spacing and their positions. Based on the existing studies on numerical simulation into the form and position of electrodes on the response of a gas sensor, it is evident that the electrodes' geometry has an influence on the sensor's sensitivity and selectivity. A poor choice of electrodes can result in a poorly performing sensor even if the sensitive material is well adapted. In addition, the greater the surface area is between the electrode/sensitive layer, the weaker the measured resistance (which is interesting for highly resistive materials).

There are several adapted geometries for two electrode measures (measuring via two parallel, perpendicular points with interdigitized contacts, etc.), four point measures, transmission lines or even micro contact. In the literature there are more complex multi-electrode systems of different sizes, forms and spacings. It is therefore possible to add further information to the previously examined configurations. A multiple electrode sensor is an equivalent system to a multi-sensor.

Heating elements

The heating element is of great importance for sensors. It allows the sensitive layer to reach very high temperatures (500°C, 600°C) which, depending on the type of layer and gas being detected, enables optimal reaction between the gas and the surface. Again, the choice of material is essential in withstanding these temperatures without being damaged.

Materials employed

The main factors affecting resistance to heating are firstly the possibility of increasing to temperatures which are sufficiently high to adsorb molecules as well as enable desorption to render the sensor reversible. The higher the temperature range, the more the number of adsorbed, desorbed spaces is important. The maximum heating temperature depends strongly on the material used.

Commercial sensors mostly have heating resistors made of polysilicon or platinum. Polysilicon is easy to integrate with a resistance value which is adjustable by doping but its properties drift less over time. Its resistance value changes gradually and the temperature provided by the heating element changes over time. In order to enable it to function in identical temperature conditions, it is necessary to compensate for these faults using regular calibration or even power regulation, which accelerates aging of the heating resistor in both cases. As such, polysilicon sensors have an operating limit of 450°C. The use of metals such as platinum (or Mo, Ti, Cr, TiN, etc.) allows us to reach much higher temperatures (600°C or more) and provides better stability of performances (less drift overtime).

Geometry employed

Constance of temperature is very important. It is preferable that the sensitive layer is heated uniformly. Hot points should be avoided in order to standardize surface reactions. Significant discrepancies in temperature can result in very different spaces over the surface and, furthermore, increase the possibility of different reactions. As such, a strong temperature gradient on the sensitive layer can lead to instabilities, even uncontrolled drifts over time (partial oxidizations of the material). Normally the forms used are rectangular tablets of polysilicon which create good uniformity in temperature or a platinum coil. In practice, these heating structures can be highly varied:

– heating resistor covered in a sensitive material (the first Figaro sensors);

– metal resistance on an insulating substrate (aluminum, glass, etc.);

– integrated resistance on a closed membrane;

– integrated resistance on a suspended platform;

– platform on a suspended substrate.

Sensitive layers

In the same way that there are different types of materials for detecting gas (polymers, elementary semiconductor, organic, etc.), metal oxides are currently the subject of attention because they are often an ionic material with a high degree of gas sensitivity. Thanks to new synthesis technologies, it is possible to gain good geometric and structural control (nanostructure) with high porosity, which enables a larger surface area for exchange with the gas.

Metal oxides can be found in the form of powders, thin or thick films, grains and nanoparticles, etc.

Two approaches are used for synthesis methods:

– "top-down" physical methods (mechanical, electrical or thermal erosion etc.) are applied to a large solid to divide it;

– "bottom-up" methods involve condensing the atoms, ions or molecules (in gas or liquid).

The metal oxide can be synthesized and then installed into the sensor or produced during (chemical) bonding. The main synthesis methods are summarized in Figure 8.1.

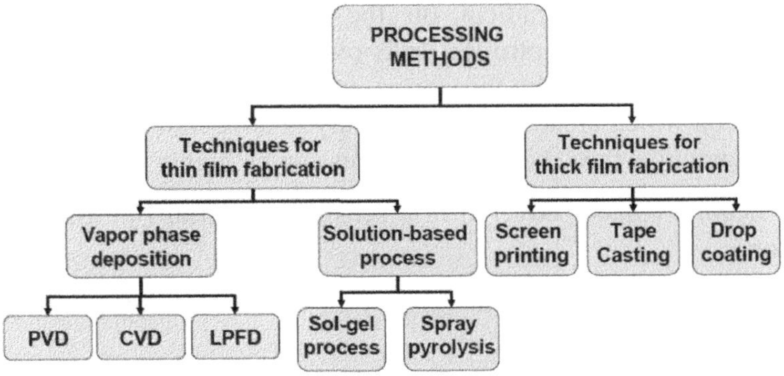

Figure 8.2. *The main techniques used to produce thick or thin layers*

The different deposition techniques (linked to synthesis methods) are as follows:

– collage (slurry deposition);

– screen-printing;

– drop deposition;

– dip coating;

– spray deposition;

– chemical vapor deposition (CVD);

– thermal CVD;

– plasma activated CVD;

– laser induced CVD;

– electroless plating;

– physical vapor deposition (PVD);

– sputtering;

– evaporation.

The choice of synthesis method and deposition technique defines the material's morphology and its detection properties.

For a number of years "bottom-up" methods have been the most commonly used in research, notably for creating nanostructured materials (nanoparticles, nanorods and nanowires) to obtain sensitive layers with an effective strong surface.

8.2.3. *Different technological approaches*

8.2.3.1. *Heating platforms on a solid substrate*

In the literature there are a number of platforms based on a solid substrate made from aluminum, glass, sapphire, gallium arsenide (GaAs) and, more recently, polymers.

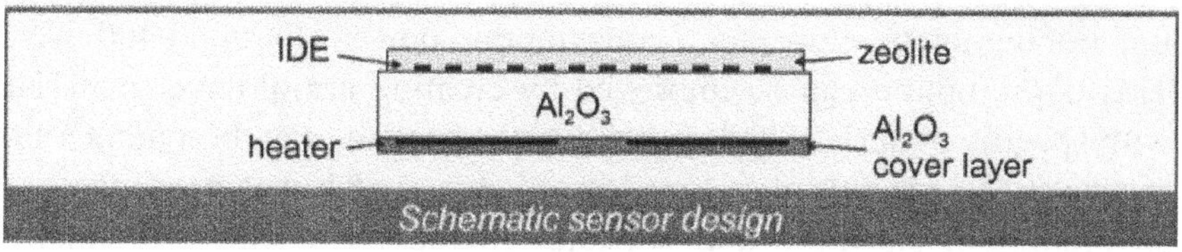

Figure 8.3. *Illustration of a heating platform with an aluminum solid substrate*

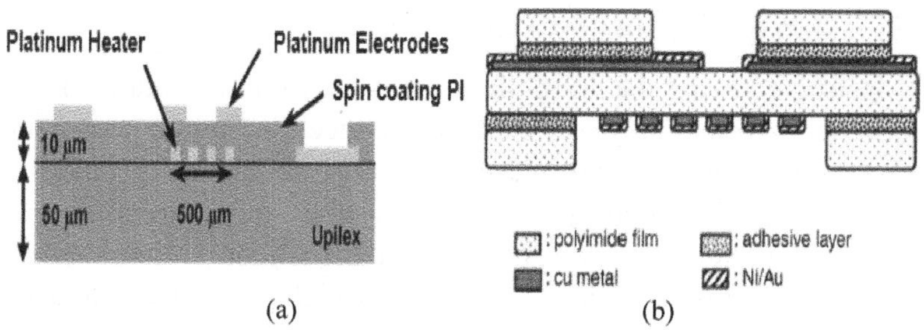

Figure 8.4. *Platform on a substrate: a) plating on the face of a substrate; b) on a thinned substrate*

The heat resistor can either be located on the back of the substrate, placed on the front or "embedded" as shown in Figure 8.4, which represents a polyimide structure with platinum plating. This "subtle" structure means it can reach temperatures of 325°C for a consumption power of 130 mW.

Figure 8.4b is an example of a platform on a thin layer with a heating resistor at the back. However, this system is only designed to function at low temperatures (around 100°C with 80 mW).

The choice of materials is clearly an essential factor when building the heating platform. Even if the maximum temperature reached by previous structures in polymers is decent (350°C), it is still not high enough for applications such as the desorption of some molecules found on the surface of the sensitive layer.

Today there are devices being marketed based on the structure of a solid substrate on ceramic (aluminum). Examples of this include the SP series developed by FIS Inc. These platforms reach 400°C (typical functioning temperature) for a consumption power of around 400 mW. This consumption can be improved by creating a multilayer material with specific materials. This is notably the case for a carbon monoxide sensor (CO), recently developed by Figaro, which was composed of an aluminum substrate and a heater (made from RuO_2) thermally insulated by glass. The power consumed was therefore reduced by half in continuous mode.

8.2.3.2. *Heating platforms on a membrane*

Constructed using the same platform as previously, a platform on a membrane is principally designed to reduce the power consumed whilst minimizing thermal inertia. This is aided by advances in microelectronics, which have enabled the miniaturization and integration of these devices. It is based on silicon technology and generally contains a silicon substrate which is placed under a membrane, heating element and electrical connections. In this section we will examine the different existing structures followed by an overview of the different production techniques as well as the materials used.

There are two main types of architecture which can be distinguished:

– *Suspended structures* are said to be compatible with CMOS technology because the structure is open on the side in front of the substrate. The sensor's active zone is therefore reduced to a small surface suspended on the substrate by suspension arms as shown in Figure 8.5. From a purely thermal perspective, it is better for the structure to be closed because this creates better insulation (less thermal loss by conduction towards the substrate) and therefore minimizes energy consumption. Another advantage of this platform is due to its production (speed and cost). Fabrication procedures take place on the front side with masking alignments taking place on the side opposite the closed platforms, thus requiring a double-side alignment.

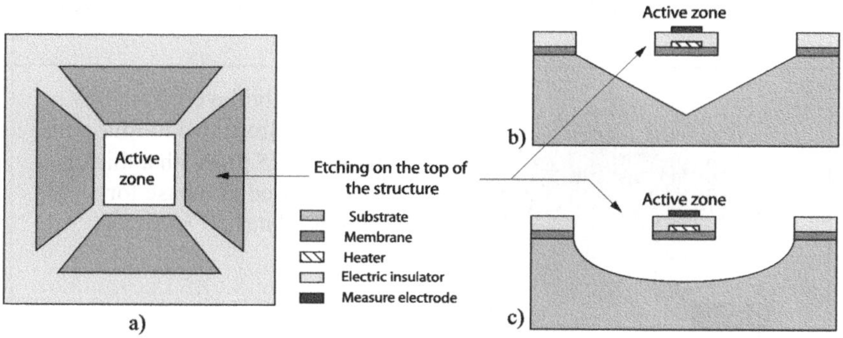

Figure 8.5. *Suspended platform: a) from in front; b) and c) two examples of etchings*

Table 8.1 lists the main devices created in research laboratories (none of these structures have been marketed to date). As we can see, it is possible to reach normal functioning temperatures with relatively low energy consumption depending on the material and geometry used.

Suspended platform	Material and performance
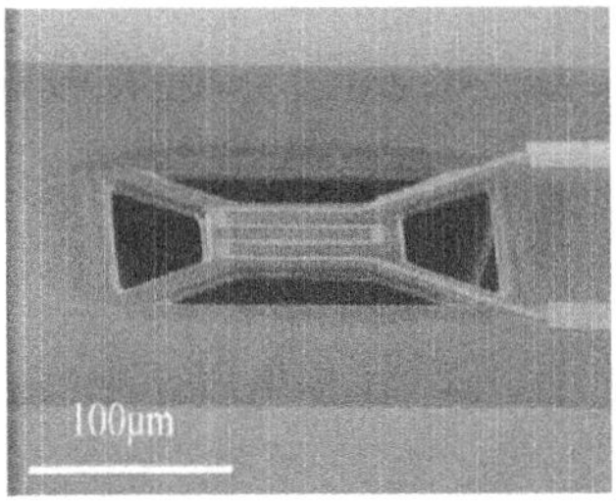	Structure suspended by four suspension arms: – Si substrate; – porous Si membrane 4 μm + SiO_2 TEOS; – Ti/Pt heater or polysilicon. Approx. 15 mW required to reach 300°C.
	Structure suspended by four suspension arms: – Si substrate; – SiO_2 thermal membrane; – polysilicon heater; – double SiO_2 insulation layer; – multilayer measure electrodes (titanium-tungsten-gold). Approx. 12 mW required to reach 300°C.
 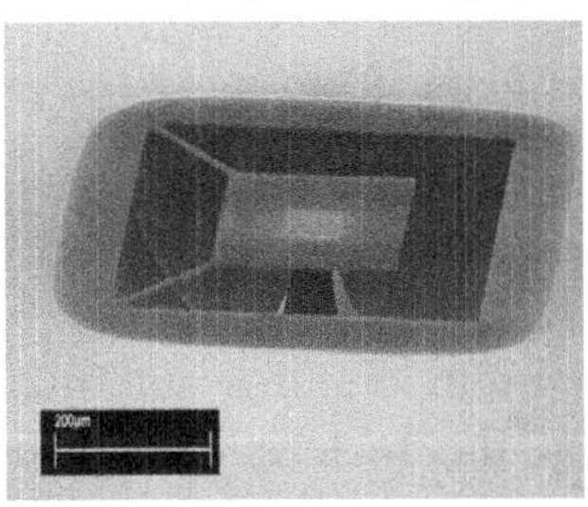	SiO_2/Si_3N_4 biolayer structures suspended from two to four arms 100 μm × 25 μm with a thickness of 1 μm, used as a base for microhotplates.

Table 8.1. *Some examples of suspended platforms*

Suspended platform	Material and performance
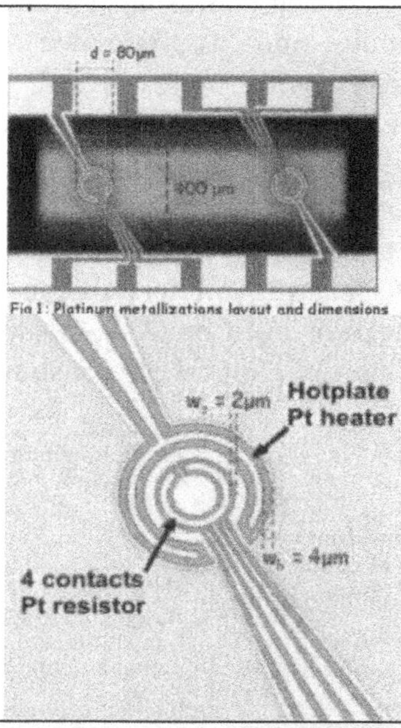	Circular (SiO_2/Si_3N_4) structure, 80 µm in diameter suspended from two arms. Platinum heater 400°C – 10 mW
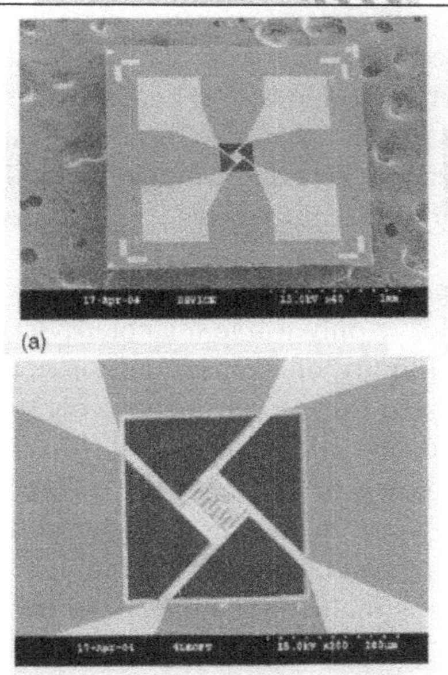	Suspended structure with 85 µm depth and 10 µm thickness. 350°C with 220 mW. The structure at 4 µm thickness causes consumption to fall to 61 mW at 350°C.

Table 8.1. (continued) *Some examples of suspended platforms*

Despite highly interesting thermal performance, the device's thermal stability at high temperatures can be confusing if working at variable temperatures (pulsing mode) and we integrate highly constrained materials (sensitive layer) with more diverse layers integrating the platform (membrane, heater element and electrical connections). The main limitation of all these structures is their mechanical fragility linked to deformations in temperature. These deformations can be significant and, in addition to issues of production and total rupture, generate shearing and torsion constraints on the sensitive layer leading to cracks or even breaks. As such, the sensitive layer must be thin (and not thick) to limit these constraints.

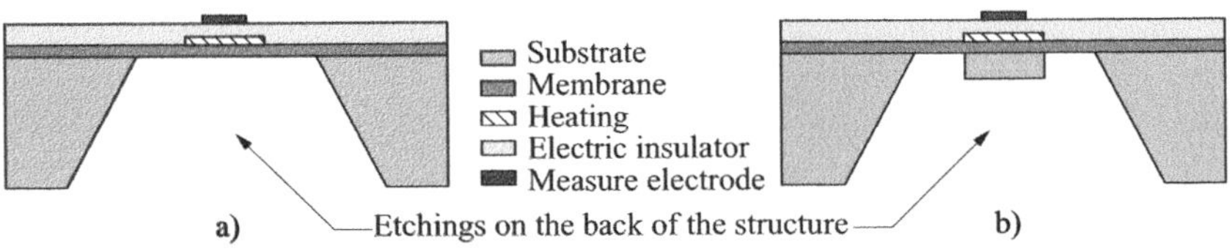

Figure 8.6. *Closed platform: a) without a silicon block; b) with a silicon block*

The majority of these suspended platforms were developed at the beginning of the 2000s and are yet to be adopted by industry. This is because their lack of robustness means that this type of heating platform and therefore the sensor cannot be manufactured, particularly if it has to be able to function at variable temperatures.

The "closed" structures shown in Figure 8.6 consist of an open front on the substrate in order to leave only a fine membrane serving as a mechanical support. The uniformity of temperature distribution on the active surface can be improved (to the detriment of consumed power) by leaving a thermal conductor block (made of silicon, for example) in the active zone as shown in Figure 8.6b). Laboratory research has shown that it is possible to increase the temperature uniformity by a factor of four with a power consumption multiplied by two (for T = 300°C, P = 75 mW for structure a), and P = 140 mW for structure b).

Some commercial sensors have a structure comparable to Figure 8.6a (without a block). Examples of these include the MICS sensors produced by E2V Sensors. These devices, based on Silicon technology have a heating resistor made of polysilicon or platinum. They require an average energy consumption of 50 to 80 mW to reach a temperature of around 350-400°C.

8.2.4. *Technological improvements in the heating layer*

So that these heating platforms can reach much higher temperatures (to facilitate and accelerate desorption) whilst maintaining the necessary similar minimum energy (<100 mW), a minimal response time (<30 ms) and lastly with as few deformations in the membrane as possible, several optimization methods have been explored in research: notably at LAAS-CNRS, focusing on more robust technology for closed membranes.

8.2.4.1. *Membrane: mechanical support and thermal isolation*

The first generation of membranes were square and composed of a SiO_2/SiN_X bilayer deposed on two of the substrates sides (specificity of the deposition technique in a furnace). The SiO_2 layer with a thickness of 1.4 μm is obtained by thermal oxidization in a furnace with a temperature of 1,150°C. Lastly, a layer of 0.6 μm of SiN_X is deposed by LPCVD at a temperature of 750°C. This technique consists of creating a chemical deposit in a vapor environment with low pressure by injecting SiH_4 and NH_3 gases, which react and synthesize the material being deposed with good stoichiometric control (x = 1,2).

To free the membrane, the silicon on the back has to be attacked using deep reactive ion etching (DRIE). Firstly, this type of etching is not uniform over the entire silicon wafer's surface. Generally, etching occurs more rapidly on the edges of the wafer than in the center, resulting in over-etching on the periphery. This is an effect of the etching reactor. Nevertheless, uneven etching (under-etching) of silicon on the back-side can substitute the silicon blocks which could reach 50 μm in angles causing the membranes to curve (Figure 8.7). These phenomena can also be explained by poor thermal transfer or incomplete etching of the silicon.

A third, equally important, point concerns the structures' mechanical aspects. Figure 8.7 shows important deformations (cross-shaped) in the membrane. These deformations are the result of residual constraints due to the superposition of different materials in different technological procedures at higher or lower temperatures (oxidization occurs at 1,150°C, thermal anneals at 550°C, etc.), required for fabricating the device. In addition to these deformations, others appear during the platform's functioning at high temperatures. Previous research has shown that deformations can form a 10 µm arrow during functioning (8 µm when not in use). These deformations can lead to rapid damage in the platform and as such cause problems in the structure's stability. These deformations can also damage the sensitive layer (cracks or even unsticking) therefore affecting the sensor's performance, particularly when the sensor is used in pulsed operating temperature mode.

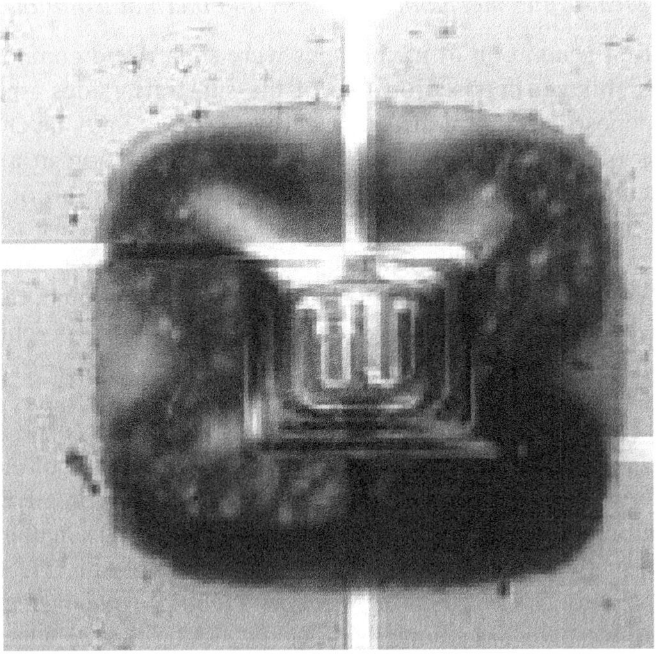

Figure 8.7. *Effect of the heater's geometry on the mechanical behavior of the previous platform*

To compensate for these recurrent problems, making the membrane's geometry circular has been explored to suppress the side effects examined previously. Subsequently there have been studies into optimizing the deposition of the SiO_2/SiN_x bilayer in order to minimize the residual constraints in the whole device (of the order of around 80 MPa). The decrease in the membrane's thickness minimizes thermal loss which is required to reduce consumption and thermal inertia.

8.2.4.2. *Heating resistance*

The architecture studied is based on the platform structure on the closed membrane described previously. The first generation of platforms, like current commercial sensors, had a polysilicon heater strongly doped with phosphorous. These devices, even if they have very good uniformity on the active surface ($|\Delta T|$=10°C to 15°C), have reliability problems over time when functioning at around 450°C with inevitable degradation in the polysilicon layer and its interface with metalizations which reduce the sensor's performance (and reduces its lifespan).

In order to solve these derivative problems, studies have focused on replacing polysilicon with platinum, which has better thermal properties and is more temporally stable. Its deposition via evaporation or cathodic pulverization requires a bonding layer on the silicon oxide. Titanium is one of the most commonly used materials in the literature. This titanium/platinum bilayer has been optimized using the following thicknesses 15 nm of Ti and 150 nm of Pt following annealing at 550°C.

The rounded spiral design of the Figure 8.8 has been obtained after multiple geometries by multiphysical simulation, such as COMSOL® in a doctoral thesis by P. Yoboue [YOB 09]. This geometry offers a good compromise between thermal uniformity and power consumption, contrary to the other geometries used until now (coiled or wound).

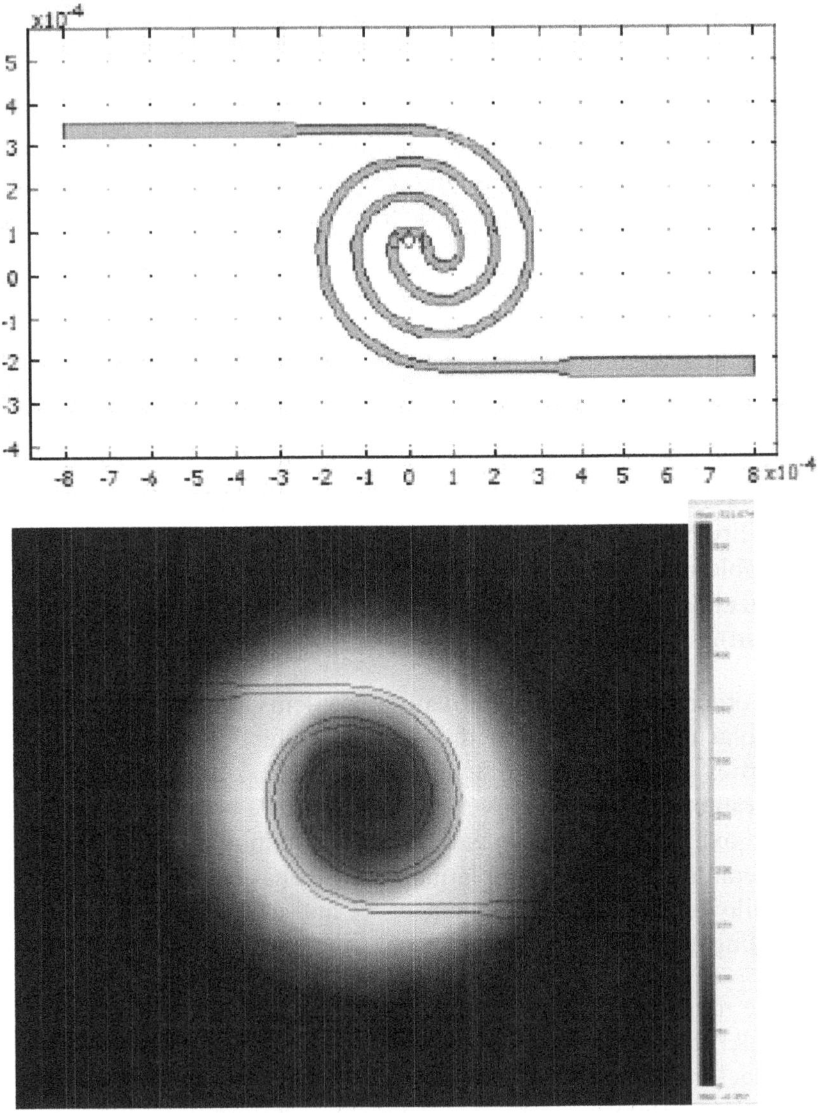

Figure 8.8. *Design of the thermal resistor used in LAAS sensors. An example of thermal simulation*

The spiral's rounded form prevents hot points and enables a more even distribution of temperature across the surface. An example of thermal simulation allows us to see an even distribution of temperature across the sensor's sensitive zone.

Aside from heat resistance, a SiO_2 passivation layer (deposed by PECVD at low temperatures, 350°C maximum) with a 0.5 μm thickness is required to electrically insulate the measure electrodes on the surface.

The heater produced can therefore reach an ambient temperature of 550°C in 25 ms. The cooling time is of the same order, which allows this type of platform to generate very rapid variations in temperature, making it particularly well adapted for functioning in a sensor in pulsed mode.

8.2.4.3. *Electrodes*

Electrodes are mostly produced from titanium/platinum for their electric and thermal properties using the same procedure as for the "heater". Of the different possible geometries studied in Yoboue's thesis [YOB 09], a rounded form has been shown to be particularly interesting for matching a sensitive layer deposed by drops. The use of interdigitized points is generally employed to achieve better contact with the sensitive layer, which generally has strong resistance (Figure 8.9).

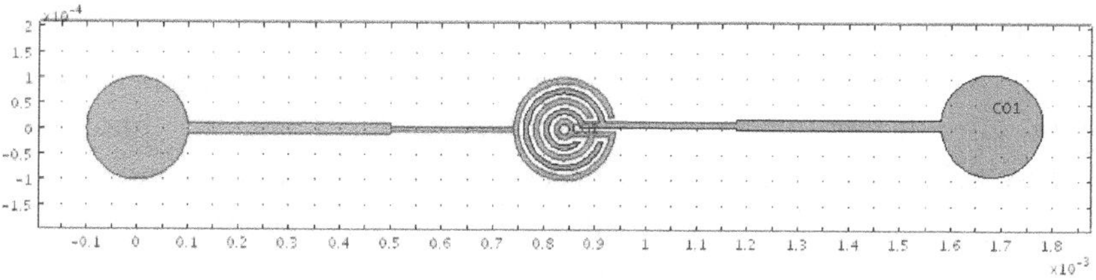

Figure 8.9. *Geometry of measure electrodes for LAAS sensors*

A study has also been carried out into the effect of geometry (number of interdigitized points) on the mechanical deformation of the membrane and uniformity of temperature on the sensor's "active" zone. It has been shown that this kind of structure allows the even distribution of temperature (to the detriment of power required). The only possible limitation is the extra weight added to the membrane but this is not very important overall.

8.2.5. *Integrating the sensitive material*

The major disadvantage related to integrating sensitive materials using liquids (currently used to integrate "nanomaterials") remains the non-reproducibility of depositions in terms of uniformity, thickness (variability in the concentration of nanoparticles across depositions, issue of sedimentation), alignment and centering. These layers therefore have highly dispersed resistances ranging from a few kilo ohms to several tens of mega ohms.

All these problems make it difficult to control the effective material on a given sensor's active zone and as such reproducibility from one sensor to another.

"Inkjet printing" is a technique which has been used in the literature (as in industry) to draw metal lines or even deposition polymer materials in a structured way on a small scale (a few µm). The ink jet principle is based on projecting drops of around a picoliter using nozzles of variable diameters (50 µm – 80 µm) and as such is a contactless printing process.

The first research carried out by LAAS consisted of carrying out metal oxide depositions, diluted in ink on an oxidized silicon substrate with measuring electrodes produced from Ti/Pt, as shown in Figure 8.10, to make gas sensors. Instead of writing on a line (normal printing by an inkjet), the jet operates using a predetermined quantity of microdrops, which occupy the active zone. For these initial tests, the material used was commercial nano-particled ZnO (Sigma-Aldrich) with grain sizes of between 50 and 70 nm.

Integrating the sensitive material into the substrate takes place over three main stages.

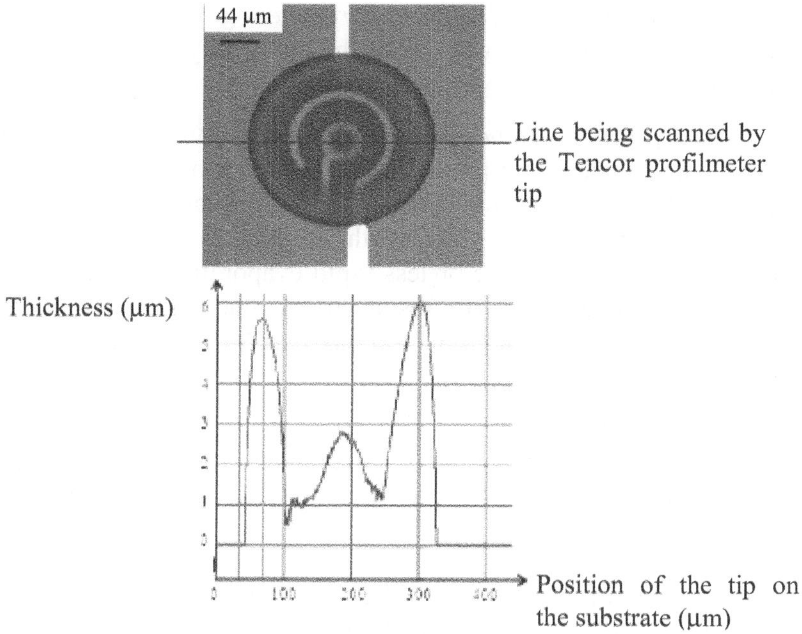

Figure 8.10. *Profile measure using a profilometer: coffee stain effect for a ZnO layer with 80 drops*

Stage 1: Surface treatment

Treating the surface consists of functionalizing the active zone, i.e. the electrode's surface (SiO_2 and Ti/Pt) separately from the rest of the tablet. The active surface remains hydrophile whilst the rest of the tablet is rendered hydrophone by treatment with Octadecyl TrichloroSilane (OTS). To do so, a photolithographic stage is required to protect the area not being treated.

Stage 2: Deposition

Integration of the sensitive layer is carried out using deposit-on-demand (DOD) technology in the inkjet machine. This device is completely automatic with very good control of all the deposition parameters.

There are three main *influencing parameters* governing the quality of the deposition, notably in terms of the uniformity of thickness and localization on the active zone:

– *The deposition temperature* is the temperature at which the "chuck" takes place in the jet to facilitate the evaporation of ethylene glycol (a solvent). This parameter is important because it plays a role in distributing the particles across the active zone. Depending on temperature, there is a more or less rapid evaporation of the solvent. Too high a temperature will result in the uneven distribution of particles (evaporation which is too fast) and degradation in OTS surface treatment.

– *The number of drops* can also affect the quality of deposition, notably on the uniformity of the deposed layer with the appearance of a more or less significant "coffee staining" effect depending on the number of drops ejected.

– *The composition of the ink* is also equally important to the quality of deposition. Tests carried out with ethylene glycol have shown that ink mixed with 2.5% w/w of glycol diethylene in glycol ethylene minimizes the coffee stain effect thus improving uniformity in the ZnO layer.

Stage 3: Annealing

In order to ensure cohesion in the layer, ZnO undergoes annealing in an air oven at 400°C for 20 minutes. Note that this annealing temperature is much lower than that found in the literature, which is greater than 600°C.

Similar studies have shown that this technique provides good reproducibility for this type of device from one sensor or tablet to another for the same parameter, as well as relatively good uniformity of deposition (with a thickness of 0.5 µm ±0.15 µm).

8.2.6. *Summary*

In this section we have introduced a new generation of heating platforms with significant improvements in relation to existing ones (and commercial products). Improvements have meant that a maximum and stable temperature is 550°C for 65 mW, i.e. a gain of 100°C in relation to the minimum temperature reached by previous versions. The advantage from a mechanical point of view is also interesting since low deformations create stability in the sensitive element on the active zone, therefore increasing the sensor's life span.

To complete the sensor's production, once the etching stage (or opening the membrane) on the back side has been completed, the tablet is cut to separate each sensor (around 1,000 sensors on a silicon substrate with four chips). A resin layer is applied to the surface of the tablet to protect the silicon from chipping. Once completed, the platforms are cleaned in acetone and ready for assembly on a TO5 eight pin casing, as shown in Figure 8.11. The technique employed is "wedge-bonding", which attaches the sensor to its casing. An aluminum wire of 25 µm is led by a stylet and then applied to the area being soldered. The bond between the wire and the area being connected is carried out in cold temperatures with a combination of pressure and sonar vibration to ensure good electrical contact.

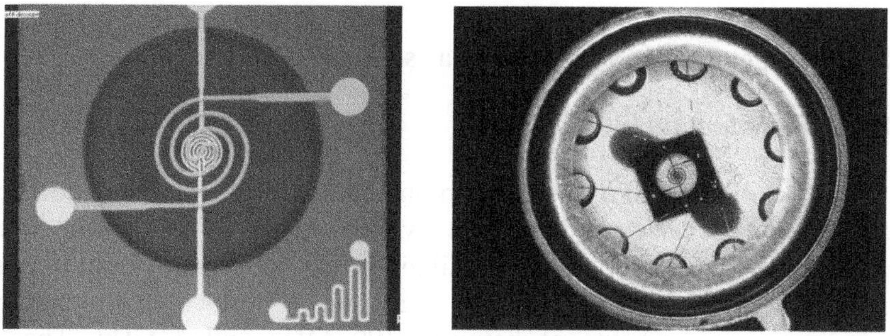

a b

Figure 8.11. *Photos of a) a platform cross-section and b) a cased platform*

Towards the development of multisensors

As shown in Habib Chalabi's thesis [CHA 07], "multisensor" platforms (integrating several heating platforms in a matrix into a single chip) have been developed to move towards integrated electronic noise. These structures can therefore take one or several sensitive layers and function together or individually. Several types of platforms have been produced to show their reliability with two, four and six measure cells. Figure 8.12 is an example of this kind of four celled structure.

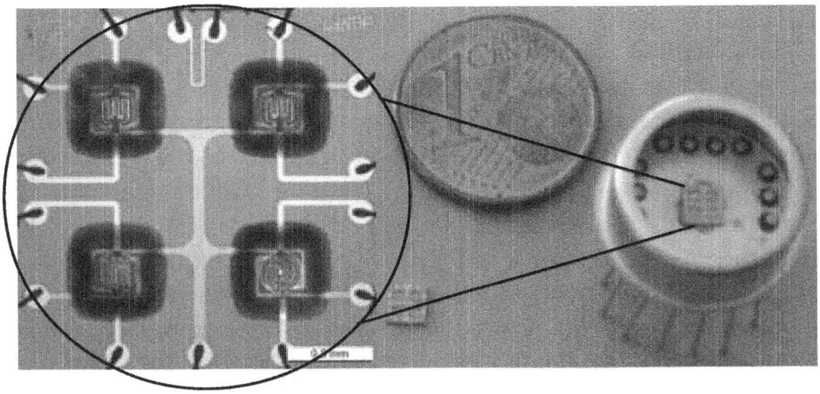

Figure 8.12. *Image of a four celled sensor mounted on a TO-8 16 case*

8.3. New generation of wireless gas sensors

8.3.1. *Introduction*

The explosion of the telecommunications market has led to the appearance of new independent wireless sensors at millimetric frequencies, which rely on the availability of small sensitive and high performing elements as well as new low-cost electronic communication circuits of between 300 MHz and 3 GHz. These components respond to the growing demand for communicative independent sensor networks for applications such as analysis, diagonistics, etc. Of course, gas sensors are part of this and the deployment of a network of autonomous and dispersed gas sensors in a given environment has a number of applications which, as yet, have not been explored.

In terms of research, several microwave emitting sensors and prototypes of nanometric carbon nanotube transistor sensors have been created. For these sensors, the microwave signal reflected by a resonator element in the presence of a gas is compared to the occurring microwave signal. It is at the sensitive element's resonant frequency that the gas' presence becomes evident. As a result, the use of a microwave signal in a wide range of frequencies allows us to obtain much more information and data on the interaction between the gas and the sensitive material employed, thereby allowing, in theory at least, better gas sensitivity.

As such, the scientific community focusing on developing communicative gas sensors should be able to combine techniques taken from different fields such as microelectronics, telecommunications and materials. As we will see, the key points of this technology lie in the design and production of autonomous microsensors (low consumption) in the design of the gas sensitive material (still being researched for the improvement of measuring, research and identification techniques. Their development therefore demands technological advances, even breakthroughs, which require the use of new knowledge from the three fields: materials, microsystems and hyper-frequencies.

To date, there are surface acoustic wave (SAW) sensors which use the variation in the propagation of the electromagnetic wave by environmental conditions. These sensors are highly sensitive and can function on a frequency range of 300 MHz to 3 GHz depending on their design. These sensors suffer however from low interrogation distance (several tens of centimeters in general), which means that their use in, for example, spatially distributed sensor networks is unlikely.

As such, today we are seeing an explosion in the market for RFID sensors. RFID refers to "radio frequency identification". This technology allows us to identify an object or measure a physical size from a distance using a label which emits radio waves which is either attached or incorporated into the object. The radiofrequency label (transponder, RFID label), is composed of a chip linked to an antenna in casing (RFID tag or RFID label). It is read by a reader which captures and transmits information. However, RFID systems also have several limitations:

– *Frequency*: They rely on different standards in terms of transmission frequency or necessary power. They function in wave propagation mode and rely on highly specific and regulated frequencies. For example, in the UHF band, RFID systems function either at 868-950 MHz, or 2.45 GHz.

– *Interrogation distance*: In the first case (868 MHz), communication is limited to a reading distance of 1 m meaning that they have a very low range. In the second case (2.45 GHz), this frequency is that used by Bluetooth and Wi-Fi. As a result, the standard RFID has difficulties such as anti-collision traffic and parasitical reflections.

In addition, in the UHF frequency band allocated to the RFID, there are a number of external elements which are often difficult to identify, but which have to be taken into account: such as adsorption phenomena depending on ambient humidity or even the problem of energy autonomy. Lastly, for this RFID UHF technology, there are two serious challenges; the first concerns several tens or even hundreds of meters of radio ranges, (today ranges are normally lower than 15 m) and the second challenge is that of energy autonomy (the energy required for transmission, which it has to minimize).

To solve these problems it was therefore necessary to design a new generation of wireless gas sensors using new hyper-frequency modes of transduction by removing itself completely from embedded energy needs which allow sensors to be communicated from distances of several tens of meters. This new generation, designed and created at LAAS will be examined in further detail.

8.3.2. *Detection principle*

According to previous studies, the dielectric relaxation phenomenon can be used as a transduction principle in microwave detection systems. This approach uses variations in a material's dielectric properties (permittivity and loss) depending on frequency, notably in terms of hyper-frequencies. This phenomenon has therefore been used for other applications such as integrated MUM (metal-(dielectric)insulator-metal) technologies or other passive functions in integrated HF circuits.

We have therefore proposed a new RF transduction for detecting gas, which is based on changing the permittivity of a metal oxide (used as a resonator) by the phenomenon of dielectric relaxation in the presence of gas, which therefore causes a shift in frequency at the hyper-frequency filter.

The developed principle of transduction is based on modification, by the size being measured, in a resonator's resonant frequency. Several types of resonators can therefore be imagined: planar resonators in coplanar or microstrip technology or even dielectric resonators. For example, the use of a sensor as a variable element in a filter where the characteristic frequency is read with as high a quality factor as possible in order to have the best level of sensitivity. The principle of our gas detector can therefore be illustrated as shown in Figure 8.13.

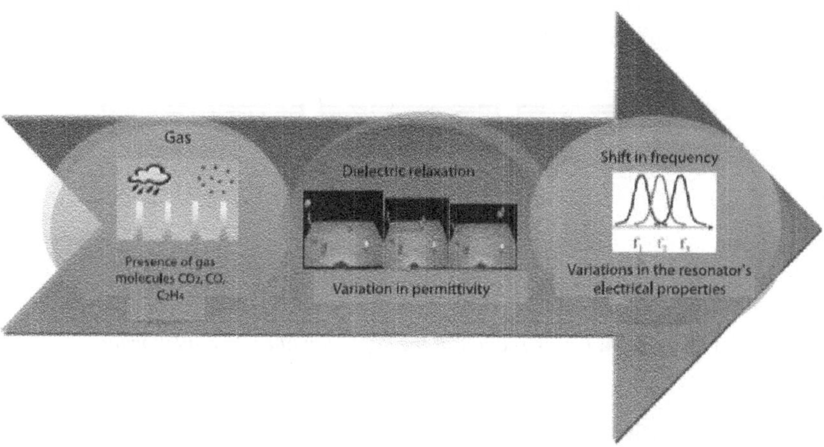

Figure 8.13. *Illustration of the new RF transduction mode for detecting gas*

8.3.3. *Design structure*

As we have seen, the sensor is based on a dielectric resonator (DR) whose dialectic characteristics have been modified by the environment. To obtain the best quality factor possible, we have opted for whispering gallery mode (WGM) by coplanar lines (100 times greater than those of an excited DR on conventional TE or TM modes). This type of excitation for gallery modes is conveyed by a direct coupling between the line and the DR. As a result, the coupling can be used to design a direct filter on the bandwidth, which allows us to measure gas due to the shift in its resonant frequency.

In terms of the filter, the structure is based on coplanar lines on a membrane. These coplanar lines are less dispersive than the microtape which is interesting for use in millimetric frequencies. On the other hand, the choice of structure on a membrane allows us to minimize losses. The air layer placed on this membrane constitutes the dielectric substrate of the created line. The substrate is therefore transparent regarding the guided wave and the effective relative permittivity of the equivalent transmission is close to the unit. The substrate's low thickness therefore results in a strong decrease of dielectric losses. More favorable propagation conditions are therefore found with the TEM mode. A diagram of a cross-section of these coplanar lines on a membrane is provided in Figure 8.14.

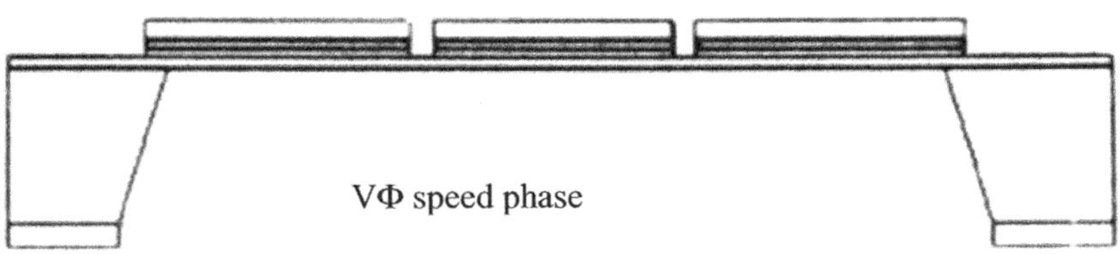

Figure 8.14. *Cross-section of the coplanar lines on a membrane*

The originality of this new communicative gas sensor is based on modifying the resonant frequency of a hyper-frequency resonator (between several GHz and several tens of GHz) in the presence of gas. This can be done using a metal oxide material, whose permittivity varies in the presence of a gas by the effect of dielectric relaxation. The potential advantages of this type of transducer are:

– an important reduction in signal loss due to direct modulation of the hyper-frequency signal by the size being measured;

– high sensitivity of the electromagnetic propagation to the environment used to create the sensor function;

– high spatial and temporal resolution due to increased functioning frequency;

– a more subtle choice of functioning frequency, which can be adapted to different constraints on the sensor's use;

– an integration facility in a measuring chain (radar and antennas).

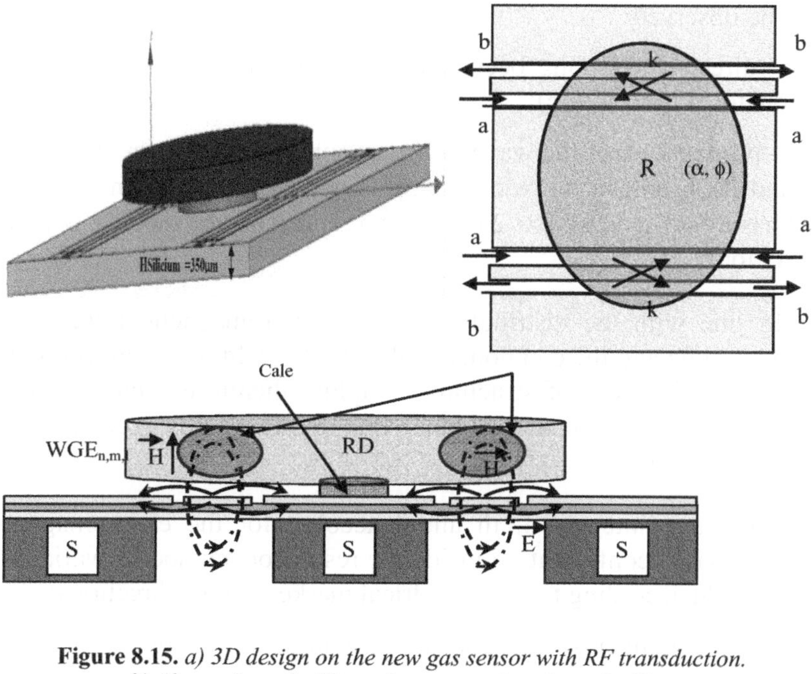

Figure 8.15. *a) 3D design on the new gas sensor with RF transduction. b) Above view of a filter; c) cross-section view of a filter*

The structure of the gas sensor shown in Figure 8.15a is therefore based on a structure which is already used in microwave filters. It is composed of two parts: the micro manufactured coplanar lines and the dielectric resonator maintained by a wedge between and above the two lines. The difference here is that the dielectric resonator is not made from silicon but a metal oxide.

As we have stated, this sensor can be considered as a direct filter (Figure 8.15b). The parameters S of such a system depends on the coupling coefficient between the DR and each of the coplanar lines, the attenuation α and the phase ϕ of the wave, which is propagated in the resonator.

Respectively, a and b, are the incidental and reflected waves normalized in the circuit's coupling. If we consider a resonator coupled with one or two transmission lines, two distinct phenomena can be observed:

– directive coupling occurs between each line and the dielectric resonator (Figure 8.15c);

– propagation of the wave in the resonator produces an attenuation α and dephasing ϕ. In resonance, this wave turns back on itself and dephasing is therefore $\phi = 2n\pi$ where n represents the mode's azimutal variations.

In line with the distribution of the electromagnetic field in the coplanar line and the components of a WGE[1] field, these are primarily excited in this type of structure. Coupling therefore occurs between the magnetic field of the coplanar line and the fields excited during $WGE_{n,m,l}$ mode.

The three integers n, m and l account for the electromagnetic field's spatial configuration inside the resonator (number of variations in the field according to the cylindrical marker's three directions):

– n: the number of variations in the azimutal direction;

– m: the number of variations in the radius;

– l: the number of variations in the axial direction.

---

1 WGE mode: where the electric field is essentially transversal.

During the design stage, it is necessary to determine the optimal geometry and dimensions of lines, their distribution, the electrical resonator (diameter, thickness), the distance between lines and the resonator (the wedge's height) to obtain the best possible performance (minimal losses, resonances in gallery mode in a good bandwidth, with as high a quality factor as possible, etc.).

8.3.4. *Production*

The production process for this new sensor can be split into three main parts:

– creation of coplanar lines on the membrane:

- deposition of a SiO_2/SiN_x (0.8/0.6 μm) bilayer with low residual constraints,

- deposition of metalizations in (Ti/Au – 1μm) by evaporation and then lift-off;

– creation of the electric resonator:

- either from commercial ceramic (BaSmTiOx) such as Temex Ceramics,

- or a metal oxide (TiO_2) by stereolithograpy (at CTTC[2]) or flash sintering (using the PNF2[3] platform at CIRIMAT);

– adding the DR to the lines:

- gluing the DR using the aluminum wedge,

- gluing both DR+Wedge on the coplanar lines by Flip-Chip.

A more detailed description of the process is detailed by Hamida Hallil in his thesis [HAL 10].

The following figures provide an example of these processes.

2 CTTC: Centre de Transfert de Technologies Céramiques de Limoges.

3 PNF2: Plateforme Nationale de Frittage Flash.

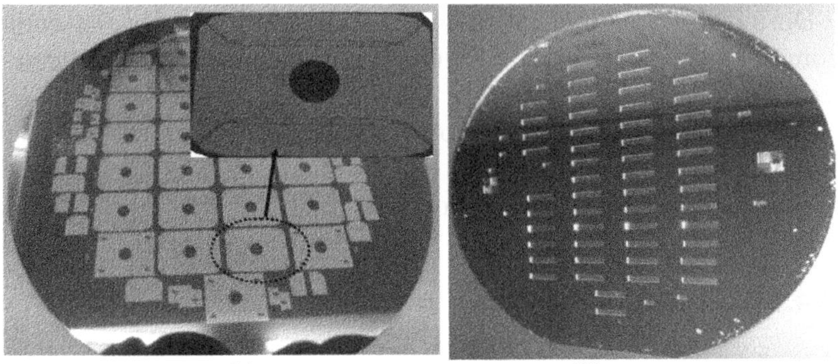

(a) (b)

Figure 8.16. *Photo of a wafer with coplanar lines: a) upper face before opening the membrane; b) the back after etching*

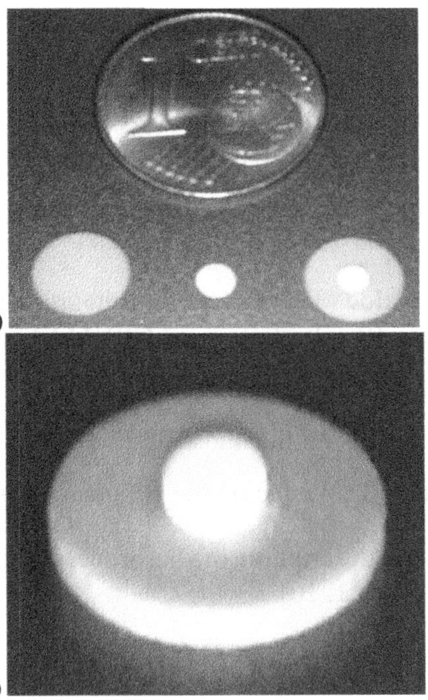

(a)

(b)

Figure 8.17. *Photos showing the assembly of the DR (produced by TEMEX-CERAMICS) with the aluminum wedge: a) (from left to right) the BaSmTiOxide DR and the aluminum wedge; b) the DR and wedge after gluing*

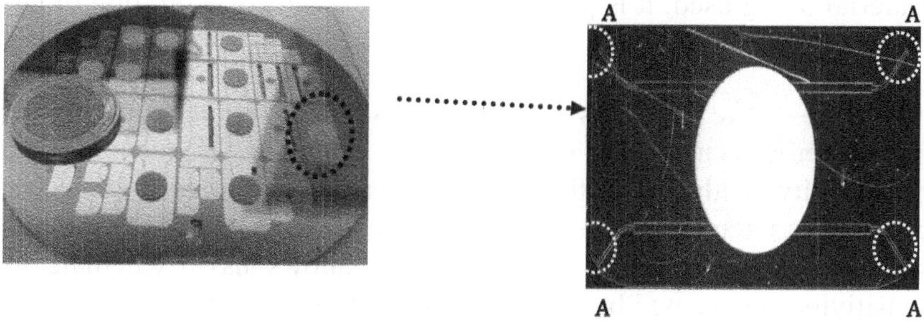

Figure 8.18. *Photos of the first gas sensors with electromagnetic transduction*

Whilst some of these sensors have been produced, there is an unavoidable difficulty in the assembly of these structures. Perfecting the assembly procedure is essential to guarantee:

– that the DR is well centered between the lines;

– the resonator's flatness and that intervals are uniform (and symmetrical) regarding the two lines;

– the good state of the structure, accounting for the fragility of membranes.

These new sensors, which are still being researched, use an innovative transduction based on a resonator's dielectric relaxation excited in gallery mode and wireless reading using classic RADAR.

Applying this to gas detection

In the design of the new sensor examined previously, the electromagnetic field is used to measure the resonant frequency of a gas concentration. The detection principle relies on measuring the resonant frequency of gallery modes (WGMs) in the hyper-frequency dielectric resonator (DR). The dielectric material used is a metal oxide (potentially TiO_2 or SnO_2) whose electrical properties depend on the gas concentration. The adsorption of gas molecules by this sensitive layer leads principally to a more or less significant modification in its dielectric permittivity. Based on the characteristics of the sensitive material being used, it is possible to calculate the concentration of gas from this measure.

NOTE.– The sensor's sensitivity is the combination of the transducer's sensitivity (variation in resonant frequency with permittivity) with that of the sensitive element (variation in dielectric constant (ε_r) depending on the concentration of gas). An example of a sensor's response, given in Figure 8.19, allows us to calculate its sensitivity represented by the derivative of this curve at any point. We can see that this response is almost linear on the frequency range considered and that this sensor's sensitivity is particularly high (0.2 GHz/unit of ε_r).

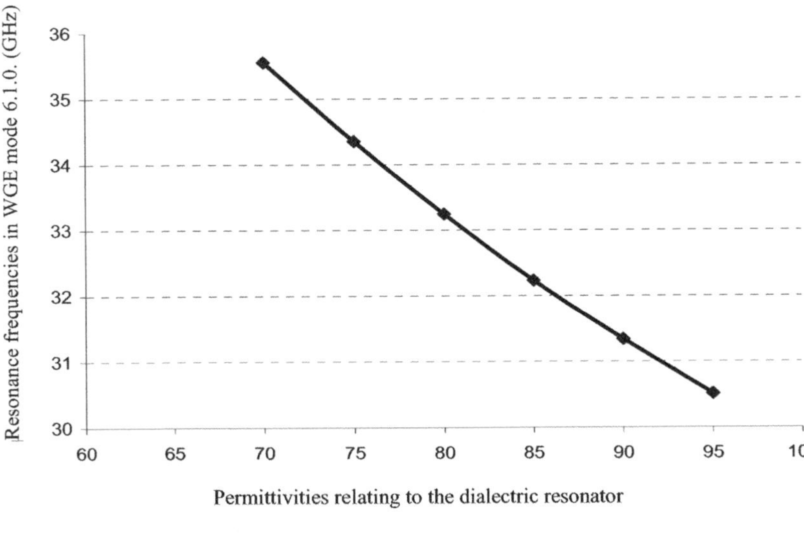

Figure 8.19. *Resonance frequency according to the DR's permittivity (transducer's sensitivity), example of WGH6.1.0 mode)*

8.4. Conclusion

In the near future and coming years, ambitious but primarily industrially oriented areas of research will focus on three main axes:

– the integration of new sensitive materials on mono and multisensor structures;

– the development of multisensor microsystems, enabling the selective detection of chemical (gas or biological) environments;

– the development of new generations of communicative sensors, which could be integrated into a sensor network for ambient intelligence, security or even the environment.

8.5. Bibliography

[CHA 07] H. CHALABI, Conception et Réalisation d'une plateforme multicapteurs de gaz conductimétriques...Vers le nez électronique Intégré, PhD Thesis, University of Aix Marseille III, 2007.

[HAL 10] H. HALLIL, Conception et Réalisation d'un nouveau capteur de gaz à transduction RF communicant sans fil, PhD Thesis, University of Toulouse III, 2010.

[YOB 09] P. YOBOUE, Etudes technologiques pour l'amélioration des performances d'un capteur de gaz à oxyde métallique : développement d'une plateforme chauffante haute température et intégration de couche sensible nanostructure, PhD Thesis, University of Toulouse III, 2009.

Chapter 9

Multisensors: Measurements and Behavior Models

9.1. Introduction

The aim of this chapter is the calibration, i.e. the design of a behavior model, of a system composed of multisensors and more specifically the measurement of concentration levels from raw signals emitted by chemical sensors.

General comments on behavior modeling with either single or multiple variables will be made before detailing the performance of chemical sensors as well as studying their interdependence.

Finally, we will go over an applied example where commercial sensors are used to detect air pollution.

Chapter written by Philippe BREUIL.

9.2. Modeling the behavior of multisensors

9.2.1. *General concepts*

A sensor is a component that allows the transformation of a physicochemical variable *y*, which is "difficult-to-measure", into an "easy-to-measure" variable *x*, usually an electrical signal.

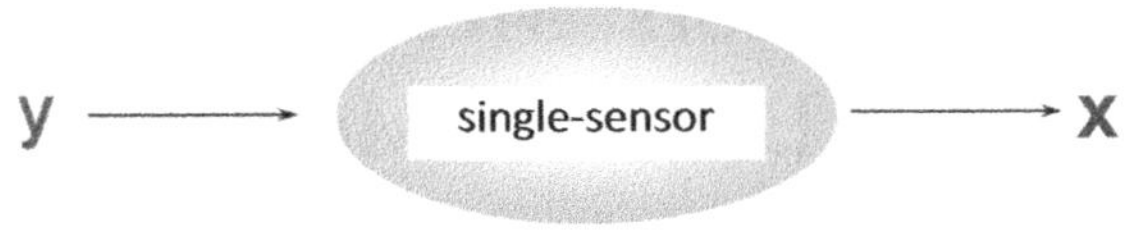

Figure 9.1. *Single-sensor*

For a "multisensor", variables X and Y, respectively named the X variable and Y variable, are vectors. The transfer function of a sensor is a vector function F such that Y = F(X).

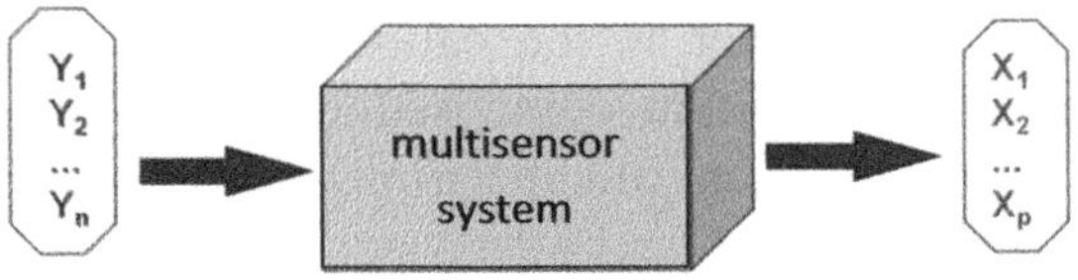

Figure 9.2. *Multisensor system*

The calibration of the multisensory system consists of calculating an "inverse" $Y = F^{-1}(X)$ model from experimental data that will generally be a "black box"-type behavior model (Figure 9.3):

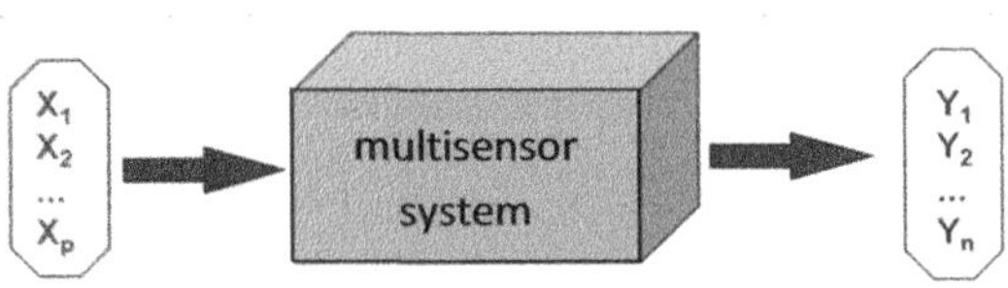

Figure 9.3. *Inverse model*

The "science" behind this behavior modeling is sometimes called "chemometrics" as it was initially developed for multi-wavelength spectrometry.

For example, UV-visible spectrometry consists of determining the concentration of absorbing species (Y variables) using the absorbances (X variables), which are optic quantities measured with the help of a spectrophotometer. These devices composed of a diffraction grating and a photodiode array, can process up to several thousand wavelengths.

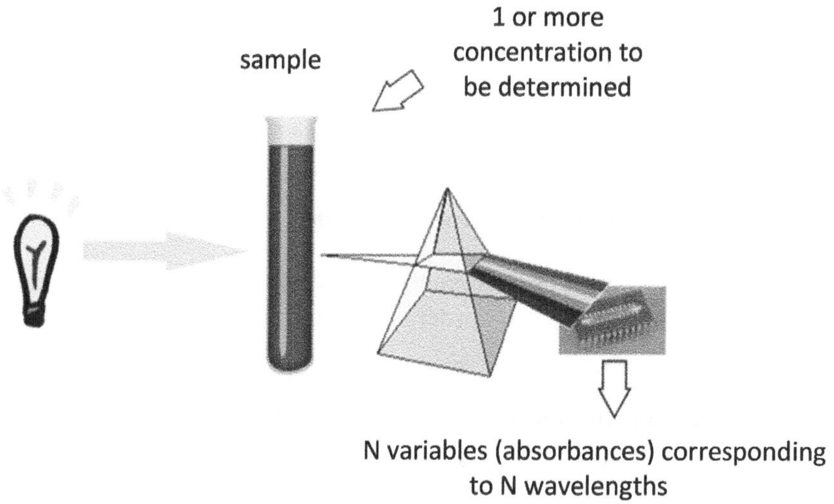

Figure 9.4. *Principle of visible UV spectrometry*

In the same way as it is possible to detect one or more gaseous species using a network of several gas sensors different semiconductors can also be used, even if the components have a reputation of having poor selectivity.

Before discussing the modeling methods for the behavior of sensor networks, a quick reminder of some of the basic fundamental principles is in order.

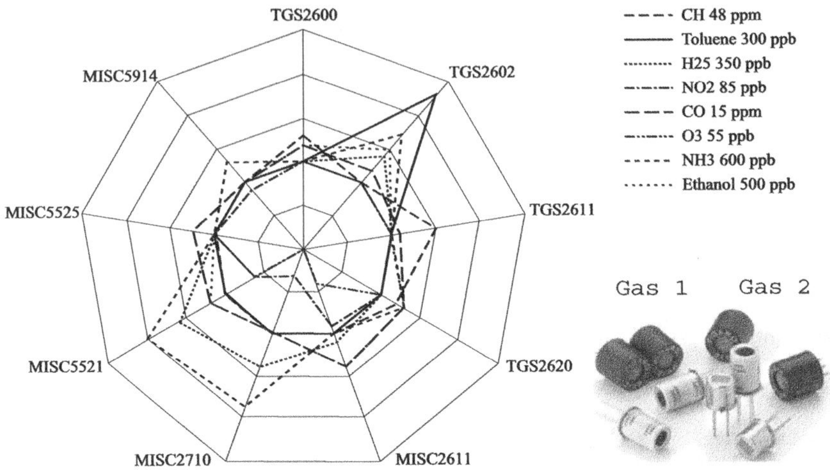

Figure 9.5. *Gas multisensors*

9.2.2. *Performance of measurement systems*

All measurement systems, including sensor networks, will have performance indicators in order to determine the measurement uncertainty.

The measurement uncertainty is an indication of the probability of error, for example:

$$C = 12.3 \text{ ppm} \pm 0.7 \text{ at } 95\%$$

shows that there is a 5% chance that the measurement has an error above 0.7 ppm. The error is usually calculated from the standard deviation of a measurement of the same value. For a probability of 95%, the uncertainty will therefore be about twice the standard deviation of a normal distribution for error (Gaussian property).

Another indicator will also be used, the limit of detection (LOD), which is defined as being the smallest non-zero concentration detectable (present or not) with a given probability for error. We can show that for a probability of 5%, the detection limit is equal to three times the standard deviation of the measurement close to zero.

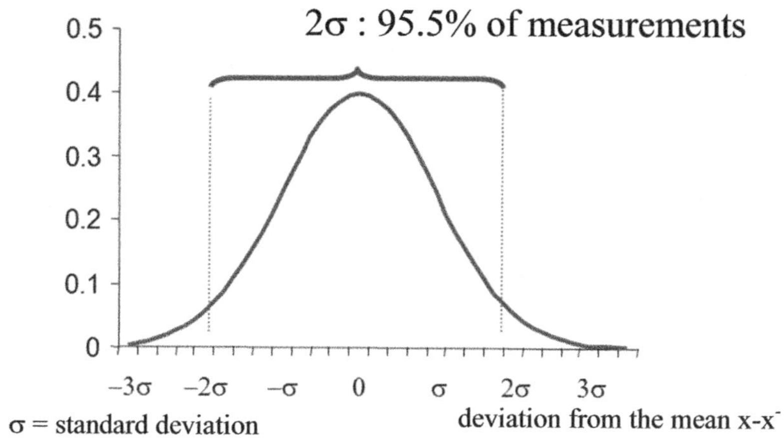

Figure 9.6. *Normal error distribution*

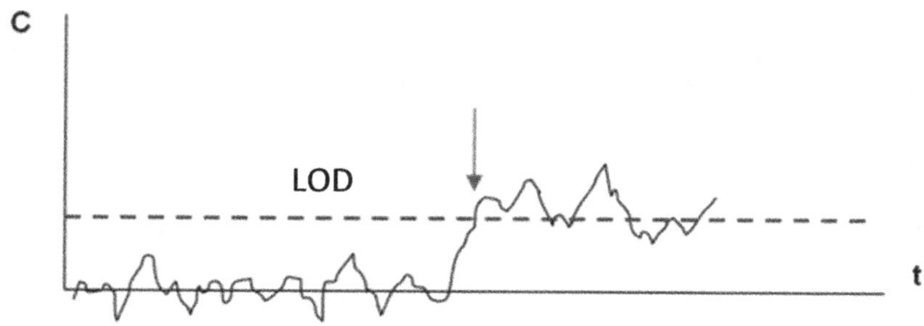

Figure 9.7. *Limit of detection*

9.2.3. *Modeling = calibration*

Calibration consists of building a behavior model describing, at best, a set of data where X and Y variables are known, from a mathematical method based on statistics. The "normal" use of the sensor, called "prediction", consists of evaluating Y variables from X variables and model G alone.

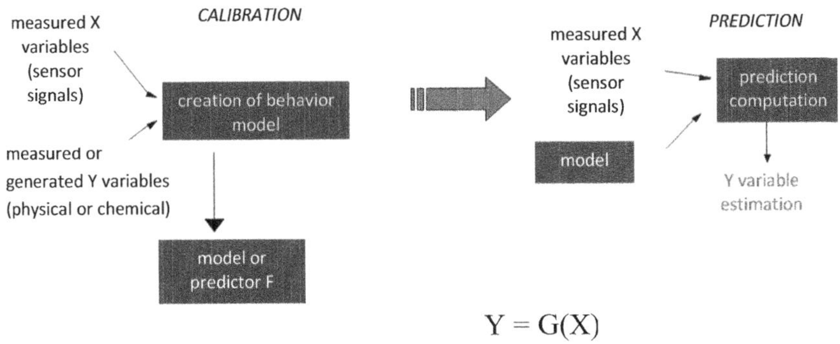

Figure 9.8. *Calibration principle*

It is possible to distinguish calibration, the calculation of a complete model that needs a large number of experiments, and gauging, which is "resetting" a model from one experiment ("zero") or more ("zero + gain") and which allows the correction of constant deviations. If the model is not linear, gauging will only allow the correction of small deviations close to the operating point.

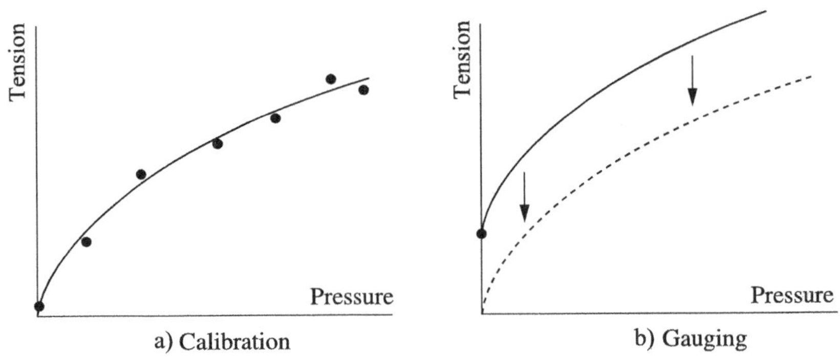

Figure 9.9. *Calibration and gauging*

9.2.4. *Principles of behavior modeling: least squares*

In the calibration process, a number of samples or experiments with Y variables and X variables are at our disposal.

$$Y = (y_1 \dots y_n) \qquad X = (x_1 \dots x_m)$$

(ex: concentrations) (ex: electrical signals)

Figure 9.10. *X and Y variables*

The aim is to model Y = G(X). This is generally impossible and in effect Y = F(X) + E is obtained where E is the residual term (unused information from X variables).

$$E = Y - F(X) = \left(y_j^i - f_j\left(x_1^i, \dots, x_j^i\right)\right) \qquad [9.1]$$

The objective is to minimize the residual term by using a distance taken as the Euclidian distance. This is the sum of the squares of the deviations, hence the "least square":

$$\sum_{i,j}\left(y_j^i - f_j\left(x_1^i, \dots, x_k^i\right)\right)^2 \qquad [9.2]$$

Later it will be seen that minimizing for the calibration samples alone is not enough, and that wherever possible it is necessary to do so for all possible samples in the range of multisensors.

It is usually necessary to take a hypothetical function G based on knowledge of the multisensor's operating principles. The most common is the linear hypothesis:

Monovariable y = a.x + b, linear regression.

Multivariables Y = K.X, (K = matrix) linear multivariable.

9.2.5. *Single linear regression*

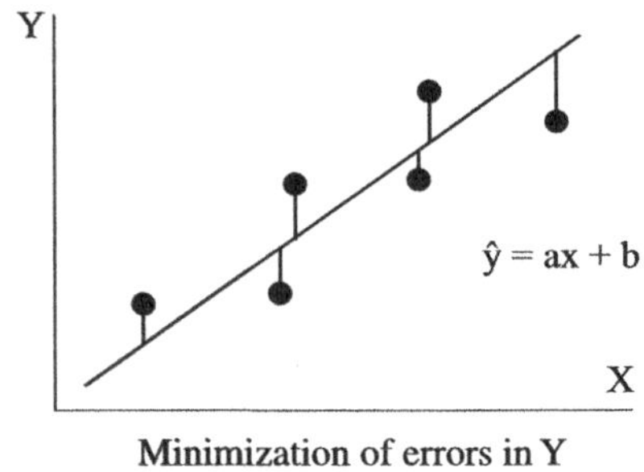

Figure 9.11. *Single variable linear regression*

In a linear regression, the aim is to minimize the deviations DΔy between experimental points and the straight line. This implies the hypothesis that X variables do not have a random error. Estimations of the gradient *a* and the intercept *b* can be calculated easily:

$$a = \frac{\sum_i (x_i - \bar{x})(y_i - \bar{y})}{\sum_j (x_i - \bar{x})^2} = \frac{Cov(x,y)}{Var(x)} \qquad [9.3]$$

$$b = \bar{y} - a\bar{x} \qquad [9.4]$$

For gas sensors, it is appropriate to take the X variable as the "relative conductance response": $(G\text{-}G_0)/G_0$, where G is the electrical conductivity and G_0 is the value for the reference gas.

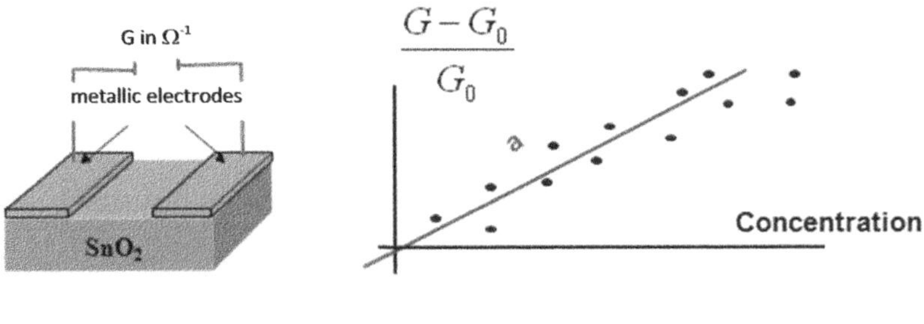

Figure 9.12. *Principle of semiconductor sensors*

If the sensor is nonlinear, an attempt to linearize it can be made. By doing this, semiconductor sensors often have a response to reducing gases proportional to the square root of the concentration.

9.2.6. *Multiple linear regression*

We now look at the case where we have several X variables, eventually leading to several Y variables.

$$\begin{cases} \text{n types} \neq \text{sensors} & \text{x = matrix } n\times k \text{ from } \dfrac{G-G_0}{G_0} \\ \text{c present compounds } (c \leq n) & \text{c = matrix } c\times k \text{ from concentrations c} \\ \text{k calibration experiments } (c \leq n \leq k) & \end{cases}$$

Figure 9.13. *Multiple variables X and Y*

"Multiple linear regression" (MLR) enables us to determine the calibration matrix nc A:

$$C = A.X \qquad \begin{pmatrix} c_{1,1} \cdots c_{1,k} \\ \vdots \;\; \ddots \;\; \vdots \\ c_{c,1} \cdots c_{c,k} \end{pmatrix} = \begin{pmatrix} a_{1,1} \cdots a_{1,n} \\ \vdots \;\; \ddots \;\; \vdots \\ a_{c,1} \cdots a_{c,n} \end{pmatrix} . \begin{pmatrix} x_{1,1} \cdots x_{1,k} \\ \vdots \;\; \ddots \;\; \vdots \\ x_{n,1} \cdots x_{n,k} \end{pmatrix} \qquad [9.5]$$

The solution using the ordinary least square method is:

$$A = C.X'(X.X')^{-1} \qquad [9.6]$$

Using Excel, the function LINEST (one single Y variable) can be used. With MatLab:

$$A = C.X'/(X.X') = X \backslash C \qquad [9.7]$$

This method, which is useful for simple problems, does have its shortcomings:

– it does not work for nonlinear problems;

– it cannot cope well with unquantified "interferants";

– it does not work well when X variables are not independent (colinearity).

Finally, there is a problem of "over-modeling", meaning that the model, *if too complex*, is in danger of modeling the particularities of calibration samples, such as measurement noise.

9.2.7. *Influence of the model's complexity*

When a calibration model is used in order to predict Y variables, the measurement uncertainty contains 2 components, here represented by their variance (square of the standard deviation *s*):

Variance of the error on prediction: $s^2 = s_{mes}^2 + s_{\text{mod}}^2$

Figure 9.14. *Two error components*

S_{mes} is the standard deviation of the "*error linked to measurements*", which is the propagation of the measurement uncertainty of X variables through the model. S_{mod} is the "*modeling error*", meaning the error due to the fact that the model is not perfect.

The modeling error logically tends to decrease with the complexity of the model. However the error linked to the measurement is amplified with complexity, as will be shown later.

An optimal complexity therefore exists, which enables minimization of the error on the prediction, hence the interest in methods that are able to adjust this complexity – that do not allow linear regression (MLR). These are factor analysis and neural network methods.

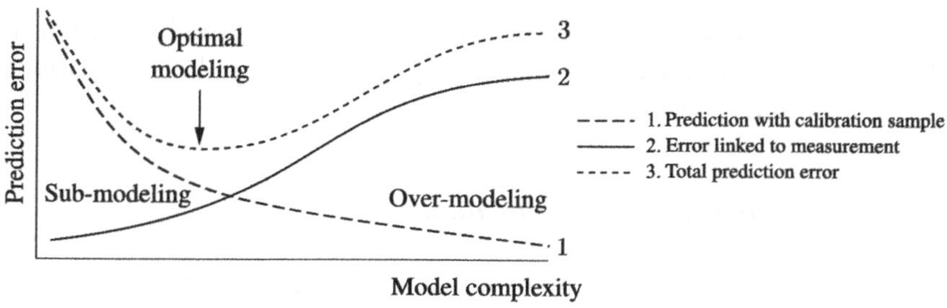

Figure 9.15. *Sub- and over-modeling*

9.2.8. *Methods of factor analysis (PCR, PLS)*

These methods consist of taking new X variables called *factors*, old linear combinations, in smaller numbers and that are independent of each other.

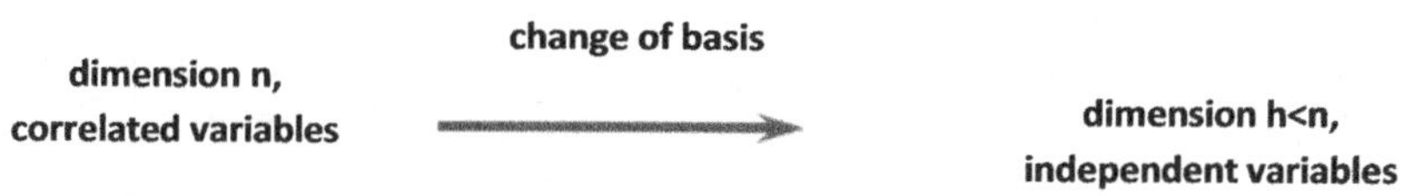

Figure 9.16. *Factorial analysis*

The number of factors therefore determines the complexity of the model. There are various methods available to define this, all based on a performance test with samples that have not been used in the design process of the model.

There are two main methods implemented in Matlab, Statistica, Unscrambler and Simca, etc.:

PCR: "principal component regression" consists of a linear regression on the principal components, the "factors" calculated from the X variables alone, maximizing their variance, and therefore information of the first factors.

PLS: "partial least squares" is a similar method to the former but in this case the factors are calculated by maximizing the correlations between X and Y variables.

9.2.9. *Neural networks, quantification*

Neural networks are algorithms allowing the calculation of "targets" using X variables. They are linear combinations where the coefficients are called "synaptic coefficients" with one or more intermediate layers. The difference with previous methods (based on linear algebra) is the presence of a nonlinear transfer function at the heart of the network after the calculation of linear combinations. This function gives superior modeling capacities to the algorithm for nonlinear phenomena, whereas factor analysis methods are usually quite limited.

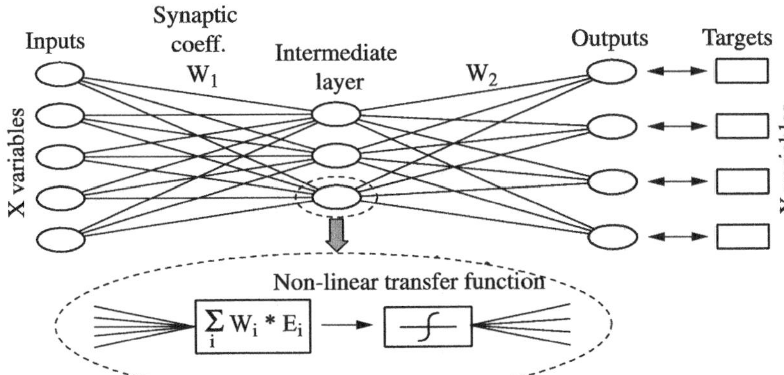

Figure 9.17. *Principle of the neuron network*

The synaptic coefficients are evaluated using iterative optimization algorithms in order to minimize, while still using the least square principle, the distance between the targets and Y variables for a series of calibration (or "learning") samples. The complexity of the model can be regulated when the optimization process ends early on. In this case it is necessary to test the performance during the learning phase with the help of samples not being used for optimization in order to stop the phase as soon as the performance stops improving.

9.3. Performance of gas sensors, influence on prediction computation

We will now discuss the influence of the metrological characteristics of gas sensors in a sensor network, on the performance of the prediction.

The main characteristics are sensitivity, stability and selectivity, sometimes referred to as the "3Ss".

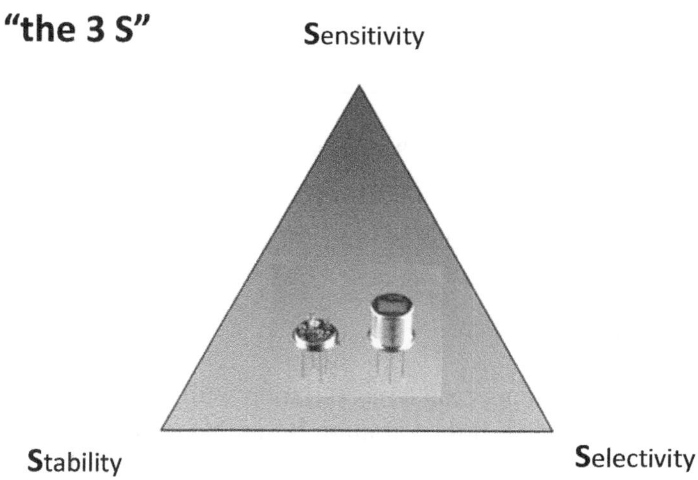

Figure 9.18. *3Ss of a gas sensor*

9.3.1. *3S: sensitivity*

By definition, this is the relationship between the variation of the electrical parameter X provided by the sensor and the parameter to be measured, generally the concentration C.

Example 1: X variable = electrical conductance,

$$\text{sensitivity: } S(C) = \frac{G(C) - G(0)}{C} \Omega^{-1}.\text{ppm}^{-1}$$

Example 2: Used X variable = relative conductance response,

$$\Re(C) = \frac{G(C) - G(0)}{G(0)} \qquad [9.8]$$

sensitivity is dimensionless and is given by the following figure.

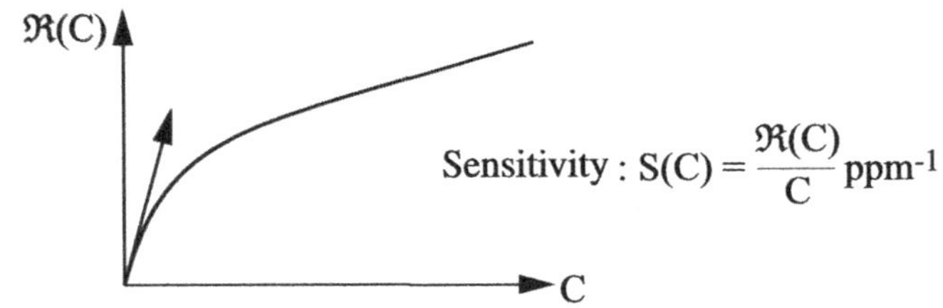

Figure 9.19. *Definition of sensitivity*

For semi-conductor sensors and certain gases S(C) > 1 ppm^{-1}.

9.3.2. *3S: stability*

Instability is an unpredictable variation with time. This means that it cannot be modeled. If the variation is monotonic, drift occurs.

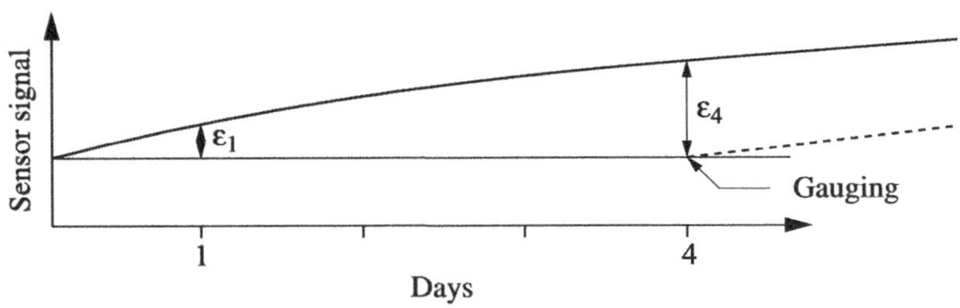

Figure 9.20. *Drift*

Instability can be considered as a supplementary source of uncertainty, especially as the measuring time between two gaugings is large.

9.3.3. *3S: selectivity*

This is defined as the ability of the multivariable device to measure the concentration of a gas independently of the concentration variations of other gases likely to be present for a given application. In this context, selectivity can range from "zero" to "absolute" with all possible intermediate states (low, medium, high, etc.)

Some authors have attempted to quantify selectivity (generally, $0 \leq \zeta \leq 1$). A sensor is said to be "specific" if it responds to only one gas.

9.3.3.1. *Selectivity in a multisensor system: an attempted definition*

Absolute selectivity, or specificity, is referred to if the sensor only responds to one gas for a given application.

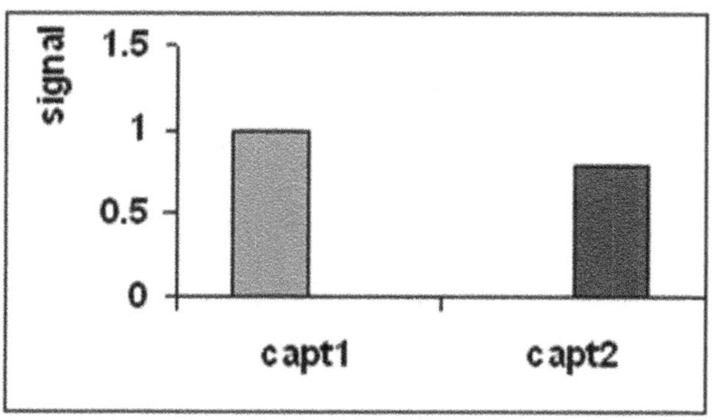

$\zeta = 1$

Figure 9.21. *Absolute selectivity*

Relative selectivity exists if sensors respond differently to all gases likely to be present: the problem can be resolved by a multivariable approach.

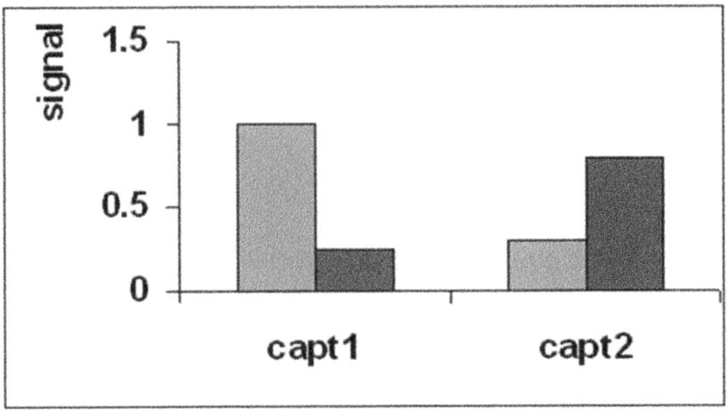

$0<\zeta<1$

Figure 9.22. *Partial selectivity*

Selectivity will decrease as the response of the sensors becomes similar. It will be zero if there is no mathematical solution.

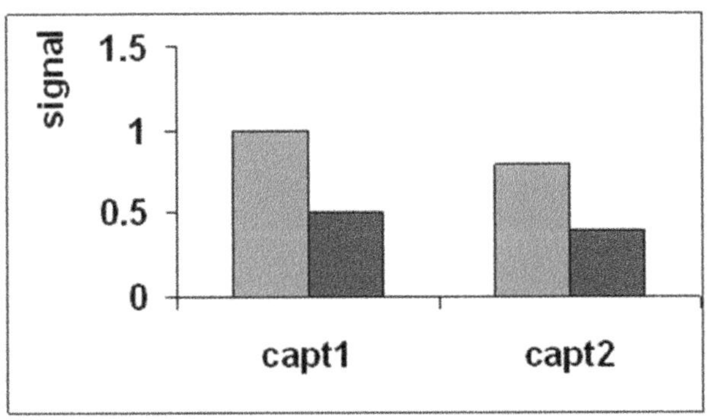

$\zeta=0$

Figure 9.23. *No selectivity*

9.3.3.2. *Trivial example: 2 linear sensors for two compounds*

There are two sensors of relative conductivity X_1 and X_2, such that:

$$X_i = \frac{G_i(C)-G_i(0)}{G_i(0)} \quad [9.9]$$

In the presence of two gas concentrations C_1 and C_2, the following can be written, if linearity is assumed:

$$\begin{cases} X_1 = a_{11}C_1 + a_{12}C_2 \\ X_2 = a_{22}C_1 + a_{21}C_2 \end{cases} \qquad [9.10]$$

a_{ij} is then the sensitivity of sensor i to gas j.

In the case of *absolute selectivity*: $a_{21} = a_{12} = 0$

$$\begin{cases} X_1 = a_{11}C_1 \\ X_2 = a_{21}C_2 \end{cases} \qquad C_1 = \frac{1}{a_{11}}X_1 \qquad C_2 = \frac{1}{a_{22}}X_2$$

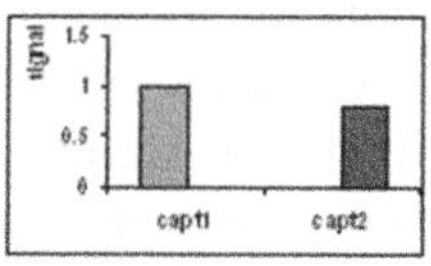

each sensor responds to one gas only

Figure 9.24. *Absolute sensitivity*

In the case of *relative selectivity*: a_{12} and $a_{21} \neq 0$ ("cross effects"). A system with two equations and two unknowns is obtained and can therefore be resolved:

$$\begin{cases} X_1 = a_{11}C_1 + a_{12}C_2 \\ X_2 = a_{22}C_1 + a_{21}C_2 \end{cases} \qquad [9.11]$$

$$\begin{cases} C_1 = \dfrac{a_{22}X_1 - a_{12}X_2}{D} \\ C_2 = \dfrac{a_{11}X_2 - a_{21}X_1}{D} \end{cases} \qquad [9.12]$$

$D = a_{11}a_{22} - a_{21}a_{12}$, is the determinant of the system. The determination of a_{ij} requires at least two calibration experiments but usually more are carried out with the use of the least square method. A mathematical solution exists as long as $D \neq 0$, and D tends towards zero as the selectivity decreases.

In the case of *"almost zero" selectivity*: the determinant D is zero when the selectivity is zero. There is therefore no mathematical solution. Physically, the model cannot be used if D is too small.

9.3.4. *Propagation of error in the model*

We have seen that raw signals from multisensors are "transformed" into concentrations through the use of the behavior model. In fact these signals are subject to errors, in the statistics sense of the term, and in which we can include a drift. These errors can be evaluated using standard deviation and there is a question as to what the repercussion of the standard deviation of the X variables is on the calculated Y variables, and therefore on the final uncertainty and detection limit. This is what we call the study of error propagation through the model. This issue is the object of extremely complex mathematical theories, including the Vapnik-Chervonenkis theory. We will limit ourselves to the "trivial" two linear sensor example mentioned previously.

If we come back to a case of absolute selectivity, the standard deviation S_C of the concentration can be easily calculated as a function of the signal S_X and we can see that the sensitivity is, predictably, of prime importance.

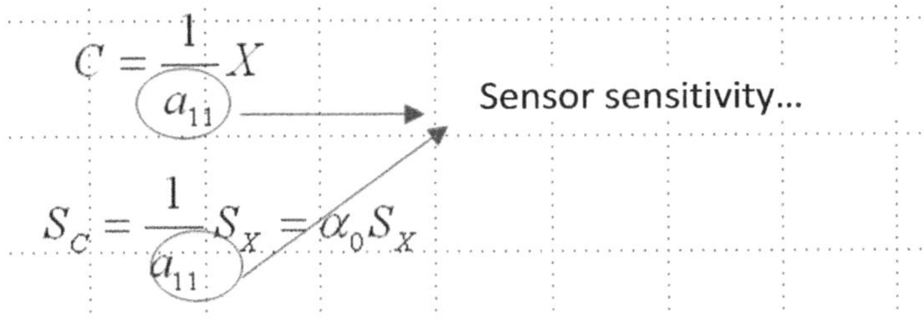

Figure 9.25. *Error sensitivity*

We can easily show that in this case the relative standard deviation of the signal and of the concentration are identical:

$$\frac{S_C}{C} = \frac{S_x}{X} \qquad [9.13]$$

In the case of relative selectivity, more complex calculations show that this is no longer the case: the relative standard deviation of the concentration is greater and the phenomenon of "amplification of the relative standard deviation" occurs:

$$\frac{S_C}{C} = \alpha.\frac{S_X}{X} \qquad \alpha > 1 \qquad [9.14]$$

The "amplification gain" α tends towards infinity when D the determinant of the system as mentioned previously, tends towards zero, therefore when selectivity decreases. The relative error is therefore amplified as selectivity decreases. The same applies for the limit of detection.

We can generalize this result to the complexity of the model and make the following statement.

> "The performance of a multisensory system will decrease with lower selectivity and/or the complexity of its behavior model".

9.3.5. *Conclusion on the 3Ss*

We can see that in terms of performance (uncertainty and limit of detection of multisensors) the 3Ss (stability, sensitivity and selectivity), are inseparable.

Stability will affect the standard deviation of the signal, which, spread through the calibration behavior model, will be amplified even with lower values of sensitivity and selectivity.

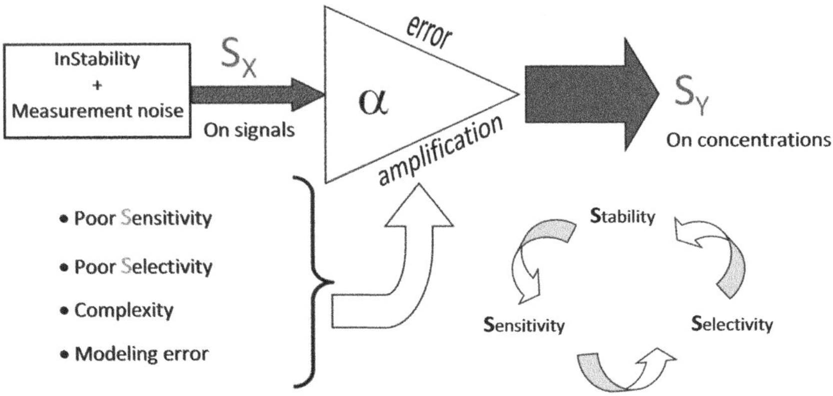

Figure 9.26. *Balance of the 3Ss*

9.4. Example with four sensors

For this application demonstration, four commercial sensors (one MICS and three Figaro, Figure 9.27) have been integrated into a cell (Figure 9.28a), insulated and in their nominal conditions. A relative humidity sensor (Humirel HS1101) and a temperature sensor enable the calculation of absolute humidity that, used as an X variable, allows us to improve the performances. Four identical cells have been used.

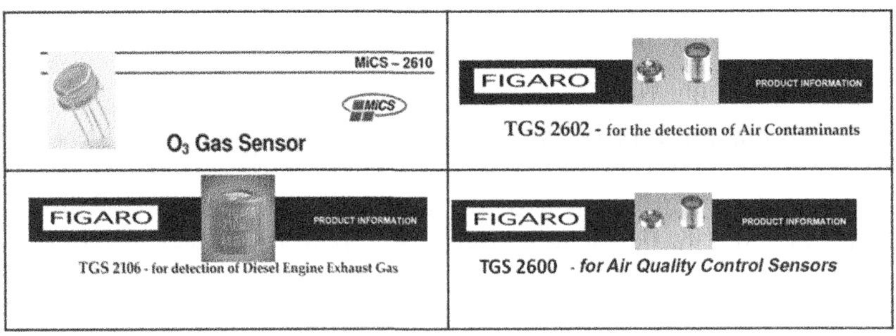

Figure 9.27. *Types of sensor used*

The X variables are therefore the four signals from the sensors (relative conductances) and absolute humidity. The Y variables, those to be predicted, are the pollutant concentrations. During calibration, they are measured with the help of two analyzers (ozone and nitrogen oxides) and a PID (photo ionization detector), which provides very sensitive but poor selective measure of all hydrocarbons (HC), n>3 (Figure 9.28b).

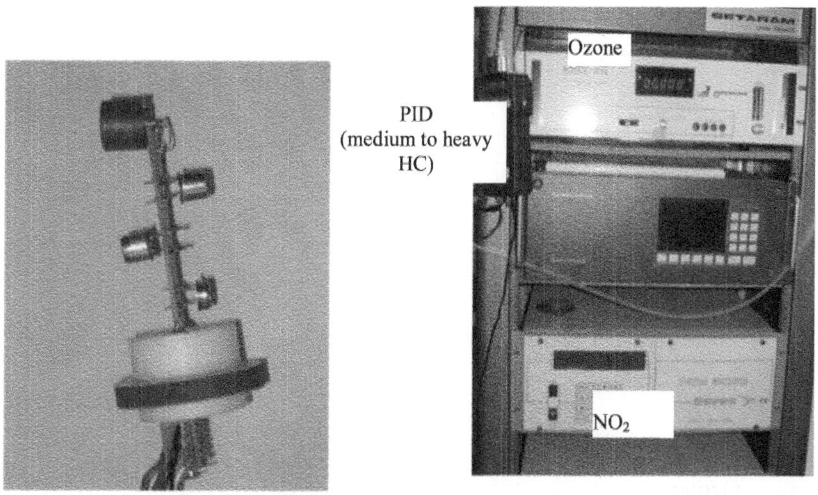

Figure 9.28. *a) Sensor network and b) analyzers*

A series of measurements (over five days) was used to calculate a model that was tested here in a second series of measurements lasting four days, one week later.

These tests were carried out with a "three-compound" model (ozone, NO_2 and HC) (Figure 9.29).

A second model, with two-compounds (ozone and HC), which is therefore less complex, was tested for the same range of X variables (Figure 9.30).

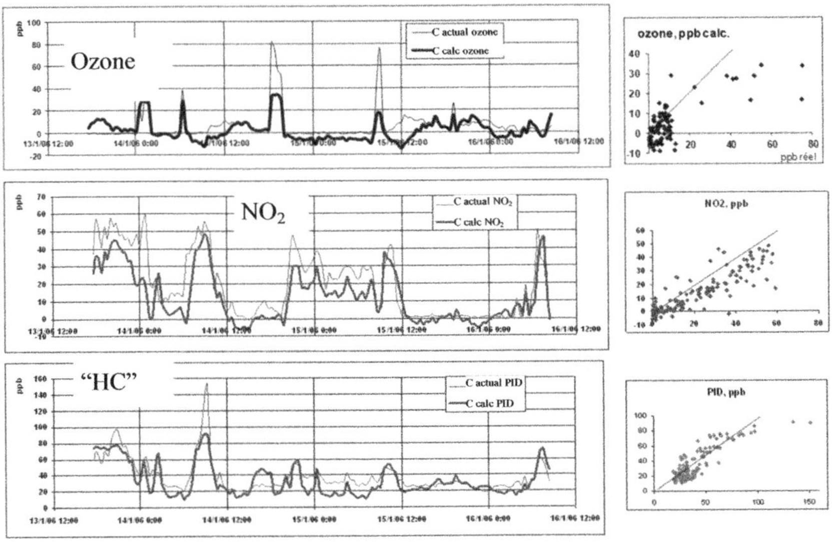

Figure 9.29. *Results for a model with three components*

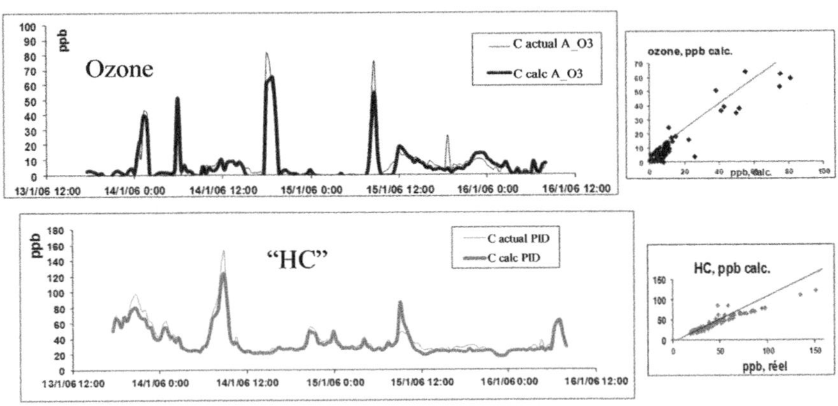

Figure 9.30. *Results for a model with two components*

We can clearly see that the three-compound model, which is naturally more complex, does not perform as well as the two-compound model (Figure 9.31).

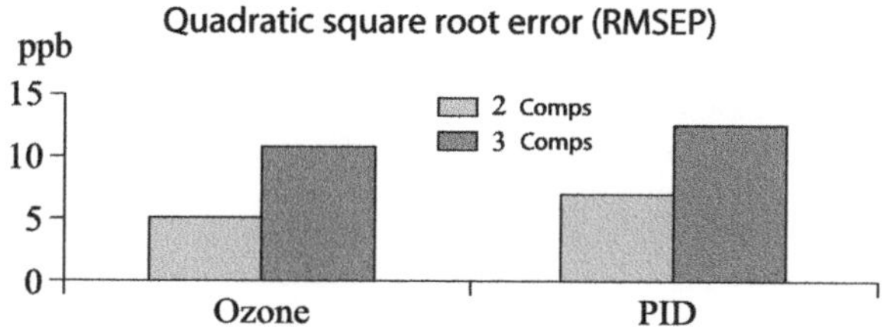

Figure 9.31. *Performance of models*

9.5. Bibliography

[ESB 02] K. ESBENSEN, Multivariate Data Analysis, in Practice. An Introduction to Multivariate Data Analysis and Experimental Design, 5th edition (reprint), Camo – Lavoisier, 2002.

[MIL 98] J.C. MILLER, J.N. MILLER, *Statistics for Analytical Chemistry*, 3rd edition, John Wiley & Sons, 1998.

[MON 86] B.F.J. MONLEY, *Multivariate Statistical Methods, A Primer*, Chapman & Hall, 1986.

[NEU 93] M. NEUILLY, "Modélisation et estimation des erreurs de mesure", *CETAMA*, Lavoisier, Paris 1993.

[PRI 00] J.C. PRINCIPE, N.R. EULIANO, W.C. LEFEBVRE, *Neural and Adaptive Systems*, John Wiley & Sons, 2000.

Chapter 10

Development of Microtechnologies for the Realization of Chemical, Biochemical and/or Biological Microsensors

10.1. Introduction

In the course of the last two or three decades, we have witnessed an evolution towards more reliable operations and decisions in all scientific areas. This tendency is linked, on one hand, to the sudden expansion of computer systems and, on the other hand, to the increasing complexity and miniaturization of electronic systems. The change in society's needs has meant that classic electronics now takes into consideration global systems and not only conventional electronic circuit boards. Microtechnologies came about for this reason. They are based on technology specific to microelectronics as well as technologies developed in various scientific areas such as physics, mechanics, thermodynamics, electromagnetism, optics, chemistry, biology, etc. Considerable advances have already been made in the design and manufacturing of intelligent systems including information detection, signal measuring, data processing, visualization and operation.

Chapter written by Pierre TEMPLE-BOYER.

Whether it is in the domain of physics, chemistry or biology, the development of analysis techniques has followed this trend. In order to answer the needs of miniaturization, mass production, autonomy, communication, performance, reliability and cost, the development of sensors has been carried out using three complementary approaches:

– technological integration for the manufacturing of transducers and microsensors with the use of tools derived from microelectronics;

– "system" integration for the creation of detection and/or analysis microsystems from technologies for on-board systems;

– software approach to setup efficient tools for diagnosis and data processing using informatics.

However, if these developments have helped the industrialization and/or "general public" commercialization of various physical microsensors (measurements on current, tension, temperature, pressure, flow, position, velocity, acceleration, radiation levels, etc.) in a wide scope of applications (military equipment, aeronautics, transport, home automation, leisure, health, etc.), it appears that this has not been the case for chemical, biochemical and/or biological analysis techniques. Besides rare exceptions such as the glucose sensor developed by the company Abbot, the development of chemical, biochemical and biological sensors has generally been limited to laboratory-level research and has not reached industrial levels. However if we consider only the fields related to health, food processing and the environment, the potential "general public" applications are more abundant.

It is therefore suitable to look more specifically at chemical, biochemical and biological sensors in order to identify the principal scientific, techno-economic and/or strategic challenges, and to eventually consider the industrial development of solutions in the context of "general public" applications. Subsequently, we will refer to all sensors interacting with chemistry, biochemistry and biology as "chemical sensors".

10.2. Chemical sensors

This section will present the world of chemical sensors by defining them briefly, by describing the obstacles associated with their development and by proposing solutions according to these different obstacles.

10.2.1. *Defining chemical sensors*

The field related to chemistry, biochemistry and biology is extremely large and therefore has a number of application fields: the chemical industry, fine chemicals, cosmetics, pharmaceutics, health, food-processing, the environment, home automation, civil protection, etc. However, whatever the field, it is possible to define the general outline of a chemical sensor (Figure 10.1). Potentially, it regroups the different elements or components characteristic of its operation, namely:

– the medium to be analyzed (gas, liquid, tissue, etc.) and its peculiarities;

– a detection/transduction principle in order to transform the chemical signal being analyzed into a measurable electrical parameter;

– a measuring and signal processing interface in order to format the desired electrical signal;

– a data processing software tool connected to a calibration system in order to guarantee the reliability of the measurements;

– an energy supply to ensure that the entirety of the detection/transduction process is autonomous.

Lastly, the general outline does not hide the diverse use of chemical sensors as it takes into account obtaining the qualitative (precise measurements) or quantitative (detection of thresholds) results according to a compromise between the performance and the cost linked to the application.

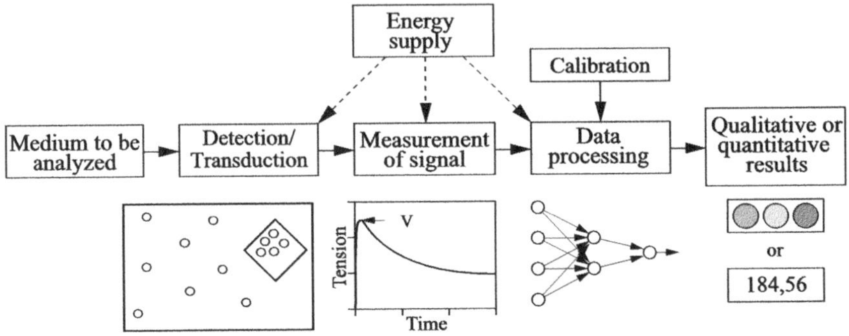

Figure 10.1. *General outline of a chemical sensor*

10.2.2. *The obstacles in developing chemical microsensors*

It becomes clear that the approaches related to technological integration, "System" integration and/or software approach all produce solutions for the development and improvement of the different elements and therefore of the chemical sensors as a whole. However, as mentioned before, the results do not meet expected standards. This failure is in fact linked to a number of obstacles related to the field of chemical sensors.

The first obstacle is technological. Reinforced by the success of physical and mechanical microsensors, the fundamentals of microelectronics have been rigorously applied to the development of chemical microsensors, by aiming to obtain optimal detection properties characterized by the "S" law: sensitivity, selectivity, stability, safety, self-calibration, sampling and speed, while ensuring a low-cost collective manufacturing process in order to cater for market forces ($). This strategy neglects one of the fundamental parameters in chemistry, also related to physics: entropy (S). This parameter, which represents the disorder and which is difficult to quantify, plays a part in all levels of chemical, biochemical and biological detection: diversity and complexity of the analyzed media, reaction principle, health and lifespan of the biological culture, tiredness of the experimenter, etc. It is therefore responsible for a certain number of chaotic phenomena that influence the detection properties: loss of sensitivity, non-selective interferences, temporal drifts, non-reproducibility of detection, etc. In principle these chaotic phenomena cannot be apprehended by the laws specific to microelectronics. For example, the main causes of the dysfunction of physical and/or mechanical microsensors are often linked to the accidental occurrence of turbulent and/or chaotic operating regimes, usually with serious consequences due to the confidence we generally have in the equipment.

The second obstacle has to do with economy. It appears that the market for chemical sensors is one of the most fragmented and in fact combines a multitude of niche technologies. This diversity is a consequence of the impact that entropy has on the field of chemical sensors. The development of a pH microsensor for medical analysis therefore does not have much in common with the development of a pH microsensor for water analysis: the former generally aims for a differential qualitative measurement over a period of several hours and highlights the use of low-cost collective manufacturing processes (~1€), the latter, on the contrary, implies a reliable quantitative measurement over the period of several months and highlights the optimization of encapsulation processes and autocalibration techniques, which have a much higher cost (~100€). In any case, these niche technologies are ultimately characterized by small production batches (from 10^3 to 10^6 sensors per year) of low profitability. They are also privileged by small and medium-sized enterprises that do not generally have the financial support to invest in the considerable and costly means associated with microtechnologies. On the other hand, big industrial groups that work in microelectronics do have the ability to involve themselves in chemical microsensor development, but choose not to: the profitability of their undertakings requires the consideration of a considerable production series ($>10^6$ sensors per year). Therefore there is yet a technico-economic compromise to be defined.

Lastly, the third obstacle is related to strategy. The development of chemical microsensors clearly leads to progress in matters of analysis but also causes parallel issues. First of all, it can question certain branches of industry that are extremely profitable today, for example in the areas of medical analysis and food processing. Next, it generally involves preventive measures that are not profitable in the long-term and that are unfortunately difficult to estimate the cost of. Finally it has the ability to expose certain dysfunctions of which the responsibilities are difficult to handle. To give an example, the sensor for temperature has been industrialized for many years now and dozens of different technologies coexist depending on the scope of application. Even though the technology is advanced enough, a smart label involving a temperature microsensor has still not yet been developed in order to systematically check the cold chain integrity in the case of perishable goods as the solution may create more problems than advantages for the food-processing industry.

10.2.3. *Development solutions*

As far as strategic obstacles are concerned, the solutions do not concern scientific, but political issues. It is a question of setting standards and regulations in order to encourage the development and use of chemical (micro)sensors, and therefore to progress towards improved analysis techniques, the prevention and control of industrial processes linked to chemistry, biochemistry and/or biology, all for the benefit of civil and industrial societies. This is an important debate that is outside the scope of this chapter and will therefore not be discussed here.

As for technico-economic obstacles that are more or less linked to entropy (see section 10.2.2), the solution concerns the approaches mentioned previously, by taking into account multidisciplinary aspects related to chemistry, biochemistry and/or biology. This strategy mainly concerns the approaches associated with technological integration. Microelectronic technology has to be adapted to the development of chemical sensors and not the other way round. However, this adaptation must not be carried out in an anarchic way. For example, the notion of "low-cost" collective manufacturing processes must be preserved. At the same time, it must be accepted that the chemical microsensors that have been developed do not fulfill the "S" law in its entirety but prioritize only certain criteria by taking into account the different aspects related to the analyzed system's entropy and to the application of interest.

Schematically, this comes back to developing multidisciplinary research in areas that concern fields of both "microelectronics" and "chemistry/biochemistry/biology" with the aim of setting up technological platforms for transducers, and lastly the development of integrated chemical microsensors (Figure 10.2). This adaptation of microelectronics will hereafter be termed "silicon and polymer microtechnologies".

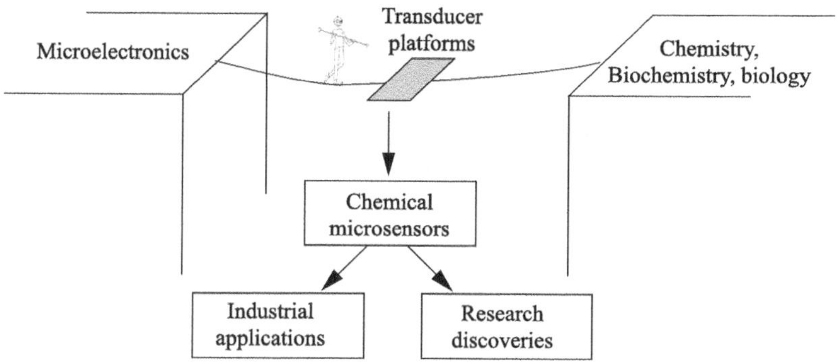

Figure 10.2. *Development of transducer networks at the crossing point between microelectronics and chemistry/biochemistry/biology*

Initially transducers and chemical microsensors created in such a way will not be designed to be absolutely perfect. However, in the scope of research discoveries related to chemistry, biochemistry and/or biology, they will be the best components to optimize themselves with regards to technological integration. In addition, they allow us to work on the industrial applications through more concrete achievements. In association with the system integration approach and the software approach, it will therefore be possible to remedy their imperfection in order to develop chemical detection microsensors that will always be optimized. By searching for systematic optimization through complementary iterations, this development strategy will have a tendency to mimic – as closely as possible – the nature from olfactory receptors having limited detection means, it succeeded in developing highly efficient senses, such as taste and smell, through the creation of devices or multisensory systems associated to *ad hoc* data processing.

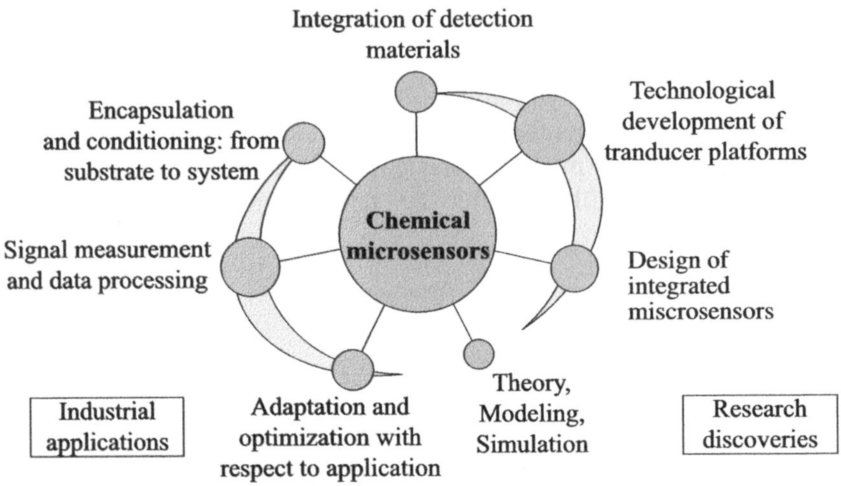

Figure 10.3. *Development strategy of chemical microsensors: from research discovery to industrial applications*

In order to further discuss the subject of solutions, it is necessary to determine the key points that come into play in the strategic development of chemical microsensors (Figure 10.3). From the discoveries related to chemistry, biochemistry and biology, materials science and processes and even industrial applications, they will combine:

– technical studies, modeling and/or simulation of detection/transducer principles and the operation of chemical microsensors;

– the design of integrated chemical microsensors;

– the design of generic technological procedures for transducers;

– the integration of detection materials;

– the encapsulation and conditioning starting from substrate to system level;

– signal measurement and processing of associated data;

– adaptation and optimization depending on the application.

This strategy has a number of advantages. First of all, the provision of generic transducer networks will allow the pooling of technology from the different industrial partners concerned. By sharing their needs and money, a series of "medium-sized" production series ($\sim 10^6$ sensors per year) can be created in silicon and polymer foundries, while preserving the "low-cost" aspect. Each industrial partner will then be able to recover their generic transducers according to their needs and to adapt them according to their application.

Depending on the case, the optimization work could concern the integration of detection materials, encapsulation methods, packaging techniques, measuring, data processing, etc. However they will generally be specific to each developed application and will enable each industrial partner to distinguish themselves from each other. This strategy will not only be a solution to the economic obstacle caused by the access to silicon and polymer microtechnologies, but will also allow each one to develop their own industrial property in chemical microsensors.

Hereafter we will focus on approaches specific to the technological integration in the scope of silicon and polymer microtechnologies and not in developments related to system integration and sensor software.

10.3. Development of silicon and polymer microtechnologies applied to chemical sensors

It has already been stated previously that the notion of silicon and polymer microtechnologies is related to the adaptation of microelectronic technologies for the design of chemical microsensors. It entails the respect of constraints due to low-cost collective manufacturing and regroups the technological developments of generic transducer networks in particular. The major principles of transduction, respectively connected to electrochemistry, electromechanics and optronics, can thus be identified and are related to microelectronics, micromechanics and photonics. In fact, it appears that the principles for optical detection/transduction have up to this point not been included in chemical sensor applications because of the use of technologies III-V and their specificities. This will not be detailed further at this point and we will therefore continue the discussion with transducer networks originating from components of microelectronics, such as resistors, microelectrodes and field effect transistors, as well as those resulting from micromechanical microdevices already developed to create physical microsensors, such as electromechanical microsystems and acoustic wave transducers.

10.3.1. *Technological platform based on chemically sensitive resistors*

The resistor is the simplest component in microelectronics. It can be inserted very easily by creating conductive patterns where semiconductors are placed in thin layers on an insulating surface. In order to develop a chemical sensor using a resistor, it is necessary to use a (semi-)conductive material for which the conductivity varies with the chemical, biochemical and/or biological parameter. By measuring the current/voltage it will therefore be possible to study the change in resistivity and conductivity in order to analyze the parameter of interest. It becomes apparent that this type of resistive transducer can only be used to analyze conductive media, such as aqueous solutions, due to the short-circuiting of the measurement resistance, and has therefore mainly been used for gaseous phases.

Two types of application have come from this. The first uses the principles of catalytic reactions of certain metals, for instance platinum or palladium, in the presence of gases such as dihydrogen H_2, alkanes C_nH_{2n+2} and/or their derivatives. In the presence of oxygen these exothermic combustion reactions are responsible for the increase in temperature and therefore a change in the resistance of the metal conductor. The second application uses the redox principles that take place on the surface of some semi-conductive metal oxides (SnO_2, ZnO, TiO_2, WO_3, etc.). In the presence of oxidizing or reducing gases, respectively acting as the electron acceptor or donor, these mechanisms will cause variations in the conductivity of conductive materials and therefore a change in the measurement resistor.

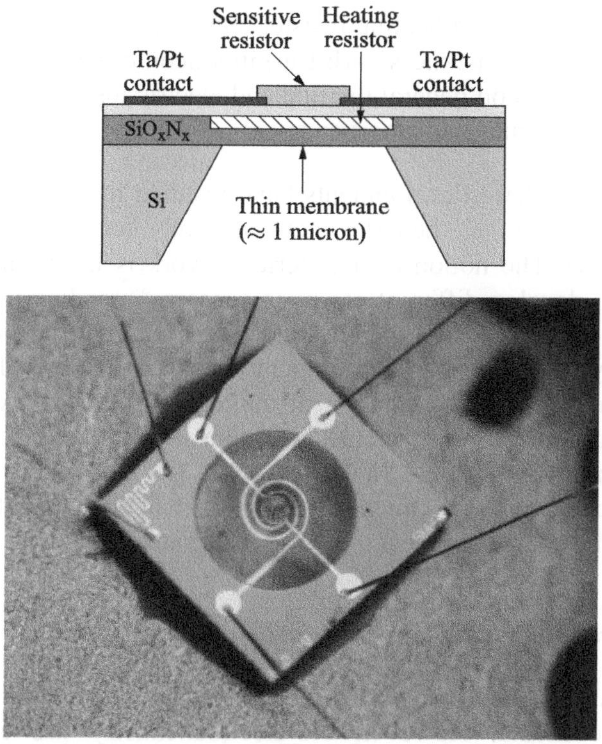

Figure 10.4. *Heating microplate for gas microsensors: diagram of layout and photo of the completed element*

These two detection principles are founded on different theories. Nonetheless similarities exist between them, such as the need to operate at temperatures above 400°C. This operating temperature is a critical parameter in order to master the detection properties and it is therefore necessary to have exact control of it. From the point of view of technological design, this leads to:

– the stable association of metal deposits and bonding layers at high temperature, for example a tantalum/platinum Ta/Pt bilayer;

– the inclusion of a heating resistor in order to thermally activate chemical reactions in a gaseous phase;

– the inclusion of a heating resistor and a measurement resistor on a thin membrane (thickness: ~1b µm) in silicon oxynitride SiO_xN_y in order to ensure good general thermal and electric insulation while still permitting good mechanic resistance.

In the end, these developments have resulted in the technological development of a generic network of heating microplates (Figure 10.4). The notion of a generic network is interesting as it is possible to develop different gas microsensors depending on the final sensitive materials used from the same initial investment in technology. The use of a measurement resistor composed of platinum or palladium allows the creation of microsensors for catalytic gases, also called pellistors. The inclusion of thin layers of semi-conductive oxide films using physical vapor deposition (PVD) in vapor phase, or nanoparticulates using microlithography methods, results in the creation of SGS (semiconductive gas sensors) microsensors.

10.3.2. *Technological platform based on microelectrodes*

As has been mentioned before, resistors cannot be used in a liquid conductive media. However, bio-electrochemistry has established various analysis techniques based on potentiometry, ampermetrics, conductometry, etc., with the help of miniature electrodes. For example, these include the measurement of conduction properties of a liquid, the analysis of redox reactions on the conductive surface placed in a liquid medium, or even the study of impedance of a solid/liquid surface using spectroscopy, etc.

In theory these detection/transduction principles can be technologically included with relative ease as microelectronics has for a long time been designing contact microelectrodes placed in thin layers on an insulating surface, while using a number of metallic materials such as aluminum, tin, chrome, gold, platinum, copper, etc. In reality the mistake is in taking this as fact and forgetting the necessity of adapting microelectronic technology to chemical microsensors, therefore to electrochemical microelectrodes in this particular case. The inclusion of thin metal layers on a silicon substrate generally requires metal bonding layers coupled with thermal annealing silicidation. If the processes ensure the adhesion of microelectrodes as a result of thermally activated atomic inter-diffusion, they are also responsible for the drastic deterioration of their electrochemical properties. In addition the active surfaces of the electrochemical electrodes must be precisely controlled and aim to include all of the components present in the electrochemical cell according to the specific geometry of the application in question.

All of the limitations must be taken into account to develop microelectrodes dedicated to electrochemical analysis in liquid phases. The technological process must therefore involve their integration by prioritizing the optimization of their electrochemical properties, including:

– the use of either metallic bonding layers, such as chrome (Cr), tantalum (Ta), titanium (Ti) or a barrier to prevent atomic inter-diffusion, such as platinum (Pt);

– the development of "lift-off"-type photolithographic methods and low temperature annealing methods in order to include, in a controlled way, the different levels of bonding while ensuring adhesion and electrochemical properties;

– the development of processes of "low temperature" surface passivation, i.e. under 100°C, either using organic layers like SU8-type epoxy resins, with photolithographic techniques, or using inorganic materials like silicon oxynitride SiO_xN_y, obtained by chemical-vapor-deposition with the help of plasma (called plasma enhanced chemical vapor deposition – PECVD).

It is now possible to setup a generic network of "electrochemical microcells" (Au – Pt – Ag/AgCl), called ElecCell for short, which contains a gold (Ti/Pt/Au) working microelectrode, an auxiliary platinum (Ti/Pt) microelectrode and a chlorinated silver (Ti/Pt/Ag/AgCl) reference microelectrode on the same chip. The chip as a whole, passivated to the substrate level by a layer of SU8 resin, allows the determination of the active surfaces of each electrode (Figure 10.5).

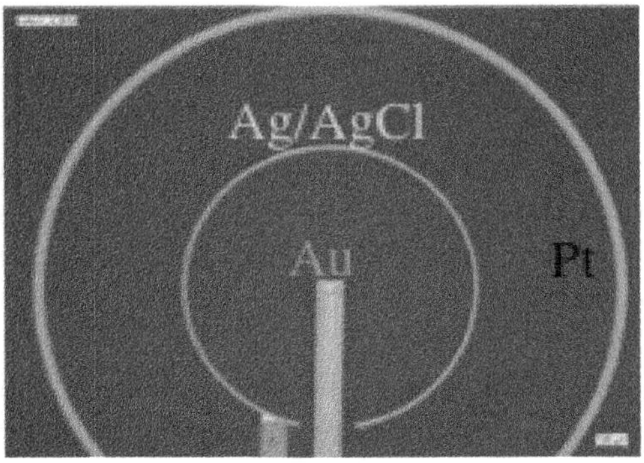

Figure 10.5. *Manufacture of electrochemical microcells (Au – Pt – Ag/AgCl) with silicon and polymer technologies*

10.3.3. *Technological platform based on field effect transistors*

The most common components in microelectronics are metal-oxide-semiconductor field effect transistors (MOSFET) and their use has been pushed to extremes in the area of complementary metal-oxide-semiconductor (CMOS) technology. It was therefore only natural to find parallel functions for them in detection applications.

Another component, the metal-insulator-semiconductor field effect transistor (MISFET) is comparable to a voltage-controlled switch. An applied voltage V_{GS} between gate and source contacts enables the control of the current I_{DS} going through the channel created between the drain and source points for a given drain-source voltage V_{DS} (Figure 10.6). The control law is generally given by the following equation:

$$I_{DS} = K(V_{GS} - V_T - 0{,}5V_{DS})V_{DS}$$

The field effect transistor is hence characterized by the threshold voltage parameter V_T, which dictates whether the drain-source switch is open or closed in relation to the applied gate-source voltage V_{GS}. The use of MISFETs in detection is usually based on the principle of charge trapping at the gate insulation level, as well as on the variation of the threshold voltage V_T with the parameter of interest. Even if

MISFETs are characterized by a variation in the current, their detection principles are purely potentiometric. MISFETs actually have the advantage of functionally incorporating an element of potentiometric detection at the structure level with an amperometric measurement element at the drain-source structure level.

Initially, the field effect transistor was developed for detection at the palladium (Pd)-silicon oxide (SiO_2) interface that appears to be responsible for the manifestation of electric dipoles and detectable variations in the transistor's threshold voltage V_T. In this context the setting up of a technological MOSFET network dedicated to the detection of hydrogenated gases was relatively simple as only a metal palladium gate (Figure 10.6) was required. In addition, the resulting Pd/SiO_2-GasFET allowed the industrial development of chemical microsensors as the technologies taken from microelectronics could be used directly. However the principle of detection is limited to the Pd/SiO_2 interface and its use has not been extended to applications such as the detection of hydrogenated gases.

Nonetheless it has been possible to adapt the MISFET component to detection in liquid media. The concept is to replace the metal-insulator-semiconductor gate control switch with an electrolyte-insulator-semiconductor (EIS) structure and to transfer the polarization of the gate to the electrolyte conductor (Figure 10.7). The physicochemical equilibrium of the insulator/electrolyte interface is maintained by trapping charges on the surface of the chemically sensitive insulator. The characteristic potential difference ψ_0 for the electrolyte's chemical properties therefore appears and directly influences the general operation of the electrolyte-insulator-semiconductor field effect transistor (EISFET), also known as a ChemFET (chemical field effect transistor):

$$I_{DS} = K(V_{GS} - V_T + \psi_0 - 0{,}5V_{DS})V_{DS}$$

The ChemFET transducer is therefore designed for the detection of charged species, such as ions, in aqueous phases. Although chemical microelectrodes (see section 10.3.2) are well adapted to measurements in liquid phases, the ChemFET transducer is characterized by a number of defects linked to its intrinsic operation. First of all, the use of potentiometric detection principles, functionally associated with an amperometric measurement, is sensitive to the risks of the inherent short-circuiting of liquid conductive phases. Secondly, the electric polarization of electrolytes requires the use of a reference gate to avoid interfering with the measurement of the interface voltage ψ_0. Lastly, these defects are responsible for the instabilities, temporal drifts and/or lifespan limits that must be minimized as much as possible.

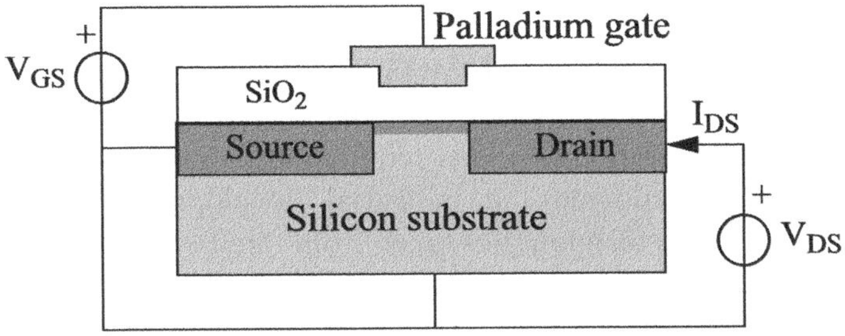

Figure 10.6. *Schematic diagram of the palladium/SiO$_2$-GasFET microsensor*

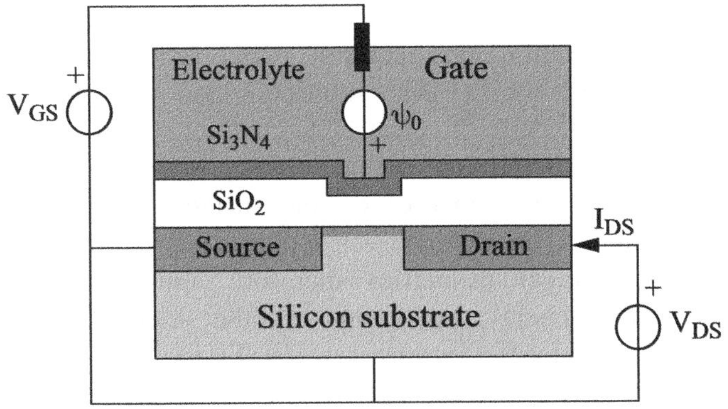

Figure 10.7. *Schematic diagram of an SiO$_2$/Si$_3$N$_4$-pH-ChemFET microsensor*

The ChemFET transducer was first used to detect ions in water, i.e. hydronium H_3O^+ and hydroxide OH^-, in order to measure the pH. An interest was then taken in the pH-ChemFET, while taking into account the numerous disadvantages linked to insulation with respect to the liquid medium and the use of a reference gate electrode. As a consequence, technological developments have been made in the following areas:

– the incorporation, using chemical or physical vapor distribution, PVD or CVD, of chemically pH-sensitive materials, for example silicon nitride Si_3N_4, aluminum oxide Al_2O_3 or tantalum oxide Ta_2O_5, to create insulating bilayers and to optimize the dielectric functions and pH measurement associated with silicon oxide SiO_2;

– the development of passivation and conditioning methods in liquid phases at the substrate scale, using both organic and inorganic technologies (see section 10.3.2), to ensure the electric insulation of the electrolyte;

– the inclusion of a reference (micro)electrode, either at the substrate level using Ag/AgCl microelectrodes (see section 10.3.2) or at the system level.

Figure 10.8. *Inclusion of microsensors SiO_2/Si_3N_4 pH-ChemFET in the SU8 resin fluid microstructure*

As a result the industrialization of pH-ChemFET electrodes, termed "solid", was successful for uses relating in particular to food-processing analysis. Similarly, generic networks have been developed for SiO_2/Si_3N_4-pH-ChemFET microsensors that incorporate a reference microelectrode made of chlorinated silver (Ti/Pt/Ag/AgCl) within the same chip, that have been passivated at substrate level and that are conditioned for microfluid analysis of SU8 technology (Figure 10.8).

The pH-ChemFET network perfected for pH measurement has been adapted to other types of liquid phase analysis by integrating ionophores, enzymes and/or biological molecules and through the use of microlithographics methods (photolithography, serigraphy, ink-jet printing, contact printing, etc.) It has enabled the development of microsensors using ion sensitive field effect transistor (ISFET), enzymatic field effect transistor (EnFET), biological field effect transistor (BioFET), etc., components.

10.3.4. *Technological platforms based on micro electromechanical systems*

Micro electro mechanical systems (MEMS) of the microbeam or micromembrane type have shown great potential for the industrial development of physical microsensors. It is therefore only natural that they should be used to create chemical, biochemical and biological microsensors.

There are two main transduction/detection principles that can be highlighted as far as beam-type microdevices are concerned: either for statics or dynamic modes. The static mode is related to the detection of the microbeam's displacement Δz along the vertical axis. On a microscale, the vertical displacement is dependent on the surface forces linked to the appearance of gradients of mechanical stress at the heart of the microbeam that is a multilayer structure, and not the force density linked to the mass. The dynamic mode, on the other hand, is based on the study of the microbeam's change in mechanical resonance frequency. Here the frequency variation is always directly related to the weights placed on the microbeam and is known to allow highly sensitive detection, even if the viscous friction and damping phenomena caused by the analyzed medium play an important role in the deterioration of the resonator's quality factor.

No matter what transduction mode is chosen, i.e. static or dynamic, it is necessary to include the detection of the microbeam's displacement. The aim is to measure the displacement strain $\Delta\sigma$ that is directly related to deflection Δz. The piezoresistivity of single-crystal silicon is used to convert the change in stress $\Delta\sigma$ into a change in electrical conductivity that can be measured directly. Once the technology had been successfully developed in the context of microsensor development, it was simply a matter of being able to include strain gauges, based either on piezoresistors or piezoelectric field effect transistors (PiezoFET), and to ensure the setup of networks dedicated to microbeam-type MEMS (Figure 10.9).

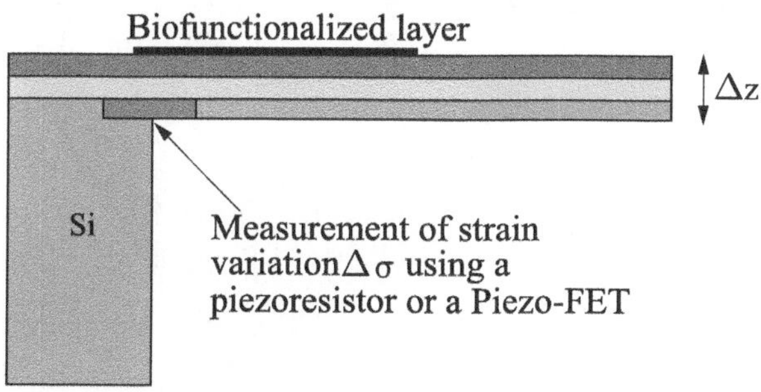

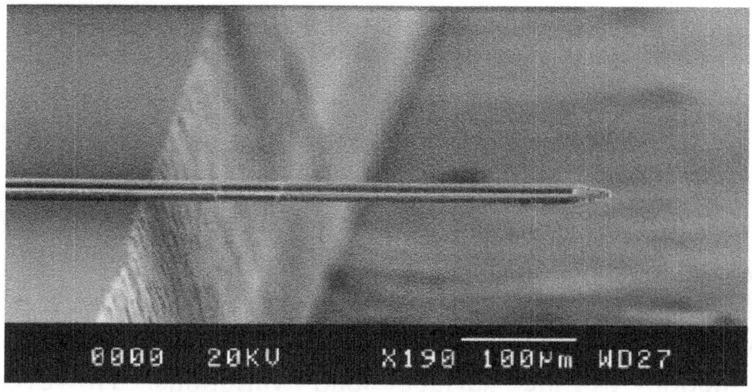

Figure 10.9. *Integrated microbeam platform: diagram and photo of the completed device*

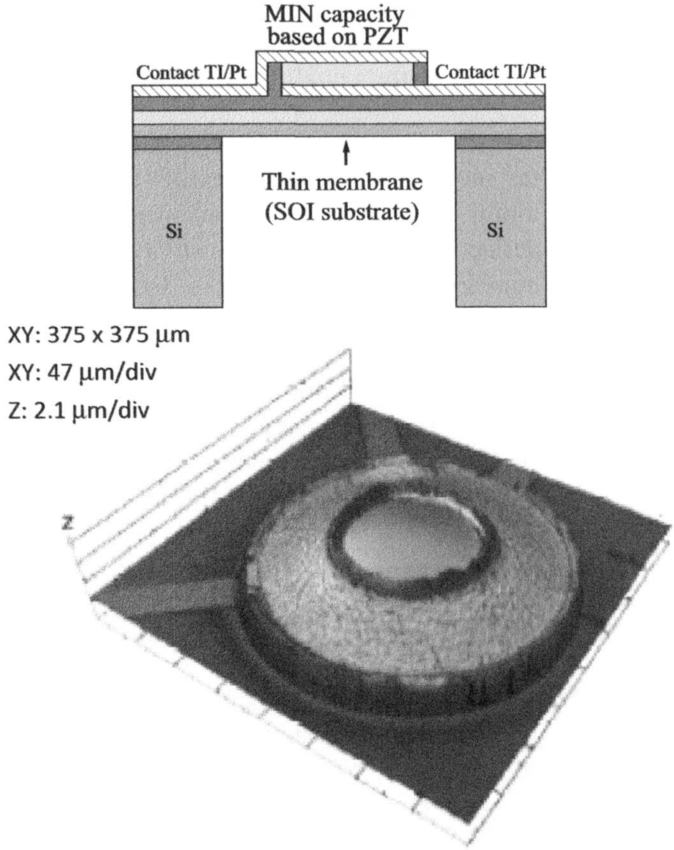

Figure 10.10. *Integrated piezoelectric membrane platform: diagram and photo of the completed device*

As far as membrane microdevices are concerned, the static method is difficult to use due to measurement interferences related to buckling. On the other hand the dynamic method is perfectly feasible. Similarly to microbeams, and with the same disadvantages, it consists of analyzing the changes in resonant frequency of the microstructure as a function of the surface load variations. However using micromembranes solves one of the major disadvantages of microbeams: the geometric configuration allows the co-integration of detection and dynamic control. To achieve this it is necessary to integrate a capacity based on a piezoelectric material like lead zirconate titanate $PbZrTiO_3$ (PZT) to ensure both functions simultaneously (Figure 10.10). The passive detection/transduction principles that are based on piezoelectricity are replaced by an active principle based on piezoelectricity. With these technological choices, the setup of networks dedicated to piezomechanic microsystems (Piezo-MEMS) required:

– the development of deep engraving methods for silicon-on-insulator (SOI) substrates;

– the thin-layering of PZT-type piezoelectric materials using "lift-off" photolithographic methods;

– the integration of thin membranes for metal-insulator-metal (MIM) capacities based on PZT-type piezoelectric materials.

Whether in static mode for microbeams or dynamic mode for micromembranes, the transducers based on MEMS mechanical microsystems are better adapted to biodetection, i.e. to the follow up of biological reactions or the detection of living organisms in the context of genomics, proteomics, immunology, virology, bacteriology, etc.

10.3.5. *Technological platform based on surface acoustic wave microsensors*

The last category of sensors presented here does not originate from microelectronics or micromechanics, but is nonetheless compatible with silicon technology. It is based on the use of acoustic waves developed in the context of quartz crystal microbalance (QCM) and bulk acoustic wave (BAW) devices, which have already been developed for chemical, biochemical and/or biological detection. The idea is to use silicon and polymer microtechnologies to facilitate their complete integration.

The operation of quartz microbalances is based on the disturbance propagation of an acoustic wave in the observed medium. Various categories of acoustic waves exist that can enable the development of sensors but they will not be discussed here. However, restrictions linked to their integration on semi-infinite substrates in planar technology, and the requirement of working in gaseous or liquid media means that the use of a shear horizontal surface acoustic wave (SH-SAW) is preferred. The SH-SAW microdevice is comparable to a delay line: in the presence of the observed medium, the acoustic wave will slow down, resulting in a time delay Δt of the transmission. This delay will be responsible for the change in the circuit's resonant frequency Δf within the oscillator that is easily measured. Studying the detection principles of sensors is complex and will not be discussed here. However it appears that the variation in frequency correlates with the loads acting on the propagation surface. As for piezo-MEMS transducers (see section 10.3.4), everything comes back to load detection using a dynamic frequency measurement, a principle known to be highly sensitive.

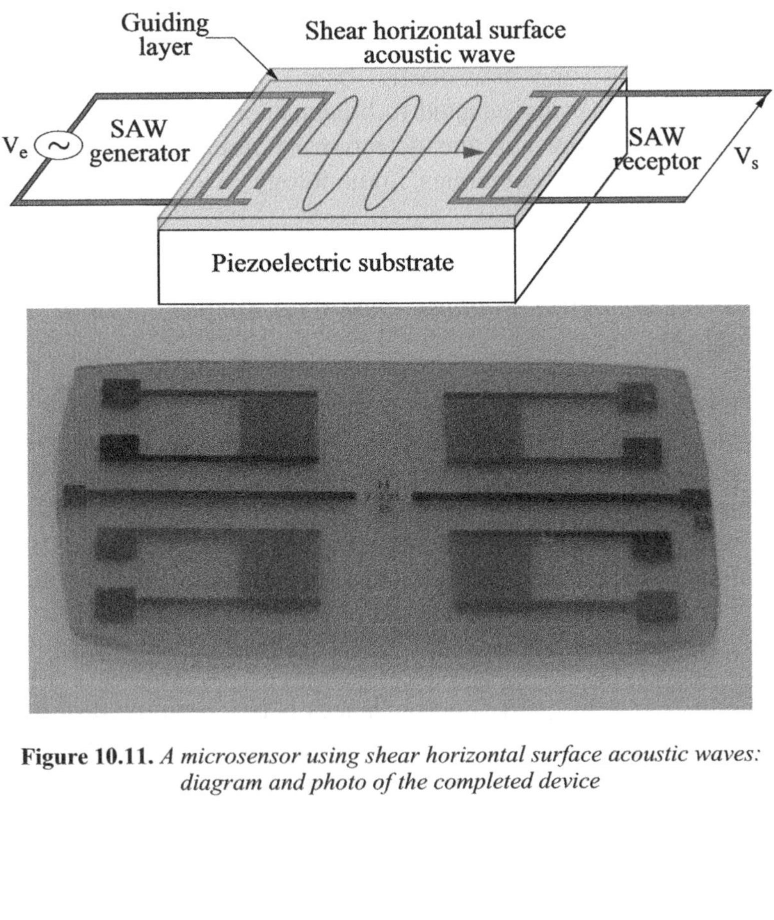

Figure 10.11. *A microsensor using shear horizontal surface acoustic waves: diagram and photo of the completed device*

SAW sensors are peculiar in that they need a piezoelectric substrate, such as quartz or single-crystal silicon c-SiO_2, for the propagation of surface acoustic waves to take place. Whether it is for the generation or reception of surface acoustic waves, transducers must also be created using interdigitated electrodes, the distance between digits being the determining wave vector for the surface vector, whereas the frequency is determined by the frequency of the applied incoming electric signal (Figure 10.10). It becomes apparent that the integration of surface acoustic wave microsensors can rely on the technological developments of microelectrodes integration (see section 10.3.2). They consist of:

– the use of piezoelectric substrates that have conformations compatible with silicon and polymer technologies;

– the development of "lift-off" type photolithographic methods in order to create interdigitated microelectrodes, traditionally aluminum (Al) based;

– the geometric alignment of interdigitated microelectrodes along the crystallographic axes of the piezoelectric substrate in order to optimize the surface generation and/or reception efficiency of surface acoustic wave transduction;

– the integration of guiding layers made from amorphous silicon oxide a-SiO_2, obtained by plasma enhanced chemical vapor deposit (PECVD).

By integrating different sensitive layers and by using microlithographic methods (photolithography, serigraphy, ink-jet printing, contact printing, etc.) the technological platform can be applied to various uses of chemical, biochemical and/or biological detection in gaseous and liquid media.

10.4. Conclusion

In this chapter the fields of chemical, biochemical and biological sensors have been presented, as well as the different obstacles related to their development and their industrialization. We have thus put forward the need to consider the constraints linked with entropy and more particularly those specific to chemistry, biochemistry and biology. As far as technological achievements are concerned, this has led us to suggest the development of silicon and polymer microtechnologies with the aim of integrating the different known detection/transduction principles, while promoting the following in particular:

– the development of generic transducer platforms;

– the adaptation and optimization of these transducers, depending on their application, through the integration of detection material, encapsulation methods, conditioning techniques, measurements, data processing, etc.

We have then presented different generic chemical transducers that are now possible to develop, namely:

– microhotplates used for the resistive detection of gases;

– electrochemical microcells;

– chemically sensitive field effect transistors;

– micro-electromechanical systems;

– surface acoustic wave transducers.

The development of chemical, biochemical and biological microsensors originating from these different technological platforms will be discussed in other chapters. However by highlighting the detection/transduction principles based on electrochemistry and/or electromechanics, this lists shows the lack of platforms using the integration principles of optical detection/transduction even though they are commonly used in a number of analytical systems. There are several reasons behind this lack of development. First of all, the integration of photonic and/or optical microcomponents is based on the use of III-V optoelectronics, which does not put forward the idea of technological networks and has excessive costs in comparison to the specifications of chemical, biochemical and biological microsensors. Secondly, optical scanning usually requires a complex environment, associating the emission, propagation in the observed medium and detection of light waves, which is difficult to minimize. However, this is currently changing.

The evolution of optoelectronics is in full bloom: photonic components are being developed beyond the needs of optical communication, giving rise to a great diversity of uses in spectral ranges going from ultraviolet to infrared and even microwaves. Their integration, which goes hand-in-hand with microelectronic technologies, has increased the potential of further integration and application. Lastly, flexible technologies now allow the creation of optical microcomponents such as organic light emitting diodes (OLED) and organic photo-detectors (OPD). All these factors create new opportunities for the low cost integration of optical detection/transduction principles, with the creation of opto-electo-fluid microsensors for microvolume analysis or laboratory development of chips. This is one of the future challenges for the development of silicon and polymer microtechnology.

Chapter 11

Development of Micro-preconcentrators for the Detection of Gaseous Species at Trace Level

11.1. Introduction

Public authorities are implementing increasingly restrictive standards and regulations regarding pollution emissions due to the concern of health issues related to air pollution. In order to determine whether they are applicable, it is necessary to be able to quantify emissions exactly, taking into account a threshold tolerance for certain gases of the order of one part per billion (ppb). There is therefore an increasing demand in the development of pollution control systems that are both more selective and more sensitive. Besides pollution control, the detection of harmful and explosive compounds is undergoing considerable development. This is in part explained by the current climate of permanent insecurity arising from terrorist threats. In addition to this, the challenge presented by miniaturization constitutes a significant advantage, not only for the attractiveness of these devices for on-site analysis, but also for their low energy consumption and the ease with which they can be integrated into other detection devices, thus broadening their field of application. Their

Chapter written by Jean-Paul VIRICELLE.

miniaturization may also favor the multiplication of measurement points owing to their offered spatio-temporal analyses.

Over the last two decades, a number of miniaturized systems have been developed to detect low gas concentrations, for example, micro-chromatographs (μ-GC) [ZEL 02], ion mobility spectrometry (IMS) [MAR 07] and electronic noises (E-noise) [CHO 06], which use a network of gas microsensors. The preconditioning of the gas is usually of high importance for these miniaturized devices. On the one hand it involves modifying the gas mixture, for example, to improve the selectivity of the detection device (the case of gas microsensors) and, on the other hand, the effect of preconcentrating in order to increase the detection system's sensitivity.

It is sometimes possible that the concentration of the gas needing detection is so low that it is below the detection limit of the sensors usually used. In this case it is necessary to provide a preconcentrating unit upstream of the sensor.

In this chapter, we will introduce the principle of preconcentration before going over some basics of adsorption phenomena and common adsorbing materials. We will then present the current developments of preconcentrators.

11.2. Preconcentration principle/preconcentration factors

The principle of preconcentration is based on a two-step process with one adsorption phase, where the trace gas is injected into a column containing an adsorbing material for a length of time called the "integration time"; followed by a fast desorption phase lasting only a few seconds. This second phase is caused by a sudden increase in temperature, which helps to generate a high concentration gas pulse and which is therefore easily detectable.

It is difficult to define an indicator for performance, such as the preconcentration factor, as the concentration amplification for a given preconcentrator is strongly related to the experimental conditions.

Some of the teams that work in the field of preconcentration, including those of P.R. Lewis *et al.* [LEW 06] and I. Gràcia *et al.* [GRÀ 08], define this preconcentration in terms of the relation between the detector's peak response areas with and without the preconcentration unit. This definition is ambiguous in the sense that it only expresses the quantity of gas absorbed, and not an amplification factor.

In addition, we can define another "theoretical" preconcentration factor, which is the relationship between the total volume of gas injected during the adsorption phase and the useful volume of the micro-preconcentrator. In fact, this would be a theoretical maximum that can be reached as the detection volume tends towards zero, as in the case of microchromatographs in a gaseous phase.

It is equally possible to define a "practical" preconcentration factor that is the relationship between the maximum concentration peak for desorption and the concentration of the injected gas.

For example, if we assume C_1, d_1 and t_1, C_2, d_2 and t_2, respectively concentration, flow rate and time during the adsorption and desorption phases (Figure 11.1), the laws of material conservation enables the following statement:

$$C_1 \cdot t_1 \cdot d_1 = C_2 \cdot t_2 \cdot d_2 ,$$

with a practical preconcentration factor of $k = \dfrac{C_2}{C_1} = \dfrac{t_1 \cdot d_1}{t_2 \cdot d_2}$.

The expression of the preconcentration factor shows that it is beneficial to have:

– a high adsorption flow rate (d_1);

– a low desorption flow rate (d_2);

– the longest possible integration time (t_1);

– the fastest possible desorption time (t_2).

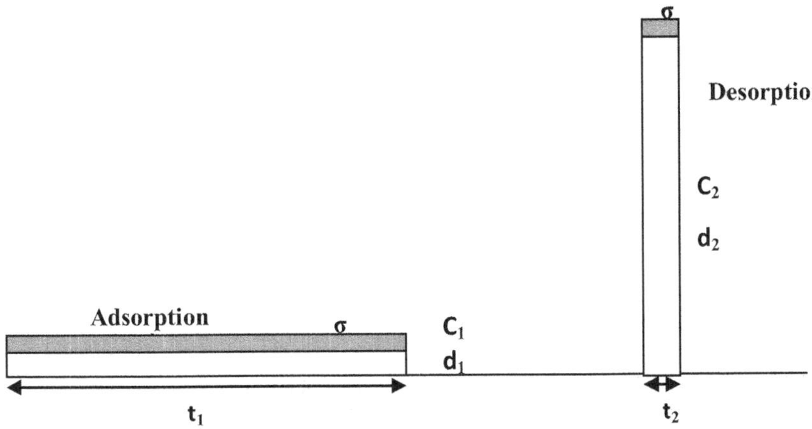

Figure 11.1. *Outline of a preconcentration cycle*

11.3. Adsorption phenomena

Adsorption phenomena are reversible processes on which the principles of preconcentration are based. They result from the attraction forces between individual gas molecules and atoms or ions on the surface of the solid. Two types of adsorption are possible, depending on the energy of applied forces:

– physisorption: van der Waals bonds of the order of 10 to 20 KJ/mol;

– chemisorption: chemical bonds, of the order of 100 KJ/mol or more.

In general, adsorption depends on both the adsorbing material's texture and its affinity with the target gas.

11.3.1. *Texture of materials: specific surface areas and porosity [ROU]*

The specific surface area (expressed in m^2/g) is an essential element for the characterization of finely divided solids and porous materials: it represents the surface accessible to gases. In homogeneous media consisting of spherical grains of similar size, the specific surface area S is related to the average diameter of the grains d by the following equation: $S = 6/\rho{*}d$, where ρ is the solid's specific gravity. We will later see that the most common method to measure the specific surface area is the "BET" method, which uses the adsorption isotherms of a gas (nitrogen or argon) for a temperature of liquid nitrogen of 77 K. These are determined using manometric and volumetric measurements that consist of measuring the equilibrium pressure for a given volume of adsorbed gas. It is also possible to find the adsorbed quantity using the gravimetric method (follow up of the adsorbed mass).

Fine-powder grains can contain cavities or pores and have a tendency to group together more or less solidly under the action of applied surface forces to form agglomerates (rigid structures) or aggregates (non-rigid grouping).

Porosity or a porous network will therefore be created and is characterized by porous volume and pore size distribution. Pores can be divided into three categories, based on the fact that there is a characteristic adsorption phenomenon for each size interval: (i) micropores with diameters less than 2 nm (considerable adsorbed volume for low relative pressures), (ii) mesopores with diameters ranging from 2 to 50 nm (phenomenon of capillary condensation), (iii) macropores with diameters larger than 50 nm (not used in areas applicable to physical gas adsorption).

Porosity can be categorized into two types: open porosity, where it is open to gas molecules and defines a specific surface area, and closed porosity, which is inaccessible to gases. The sum of these two types gives the total porosity. Most porosimetric techniques enable the measurement of porous volumes (helium pycnometer, buoyancy measurement) and pore size distribution for open porosity (mercury intrusion porosimetry). Total porosity is obtained by measuring the solid's true density.

Intrusion porosimetry consists of making a non-wetting liquid infiltrate a solid's pores, under the effect of pressure P of the liquid.

The applied pressure increases as the radius of the pores decreases, hence the following equation:

$$P.r = -2.\gamma.cos\Theta,$$

where γ is the liquid's surface tension and Θ is its wetting angle with the material. This technique allows the measurement of porosities ranging from 4 nm mesopores to macropores.

The mesoporous distribution can also be measured using the adsorption isotherms of a gas (nitrogen) with the BJH (Barrett Joyner and Halenda) [ROU] method, based on capillary condensation which is governed by Kelvin's law: a wetting liquid will condense in a capillary at a pressure P lower than its saturation vapor pressure P_0, which depends on the meniscus' radius of curvature r_K, the interfacial tension γ of the solid/liquid/gas/ interface, the wetting angle θ and the adsorbable liquid's molar volume V_l:

$$\ln\left(\frac{P}{Po}\right) = -\frac{2\gamma V_l \cos\Theta}{r_K RT}.$$

Specific operating methods for adsorption isotherms are available in the field of mircoporosity, especially for active carbons (t, α, Horvath and Kawazoe methods, etc.) but they will not be discussed here.

11.3.2. *Adsorption isotherms*

Measurements of the texture and affinity of an adsorbent to a particular gas are based on various physical gas adsorption phenomena. They consist of measuring the quantity of gas adsorbed at a given temperature as a function of the pressure of the gas above the solid. The obtained isotherm curves can be classified into five types [ROU] (Figure 11.2).

Type I isotherms are usually obtained for microporous adsorbents; they are characteristic of monolayer adsorption where micropores are filled. Type II and III isotherms occur when the adsorbent has a wide pore size distribution with a continuous transition from monolayer to multilayer adsorption and capillary condensation. Type IV isotherms are due to two successive adsorbate layers forming on the surface of the solid when the interaction between the adsorbed gas molecules and adsorbent's surface is stronger than the interaction between the gas molecules. For this type of isotherm, the adsorption sites of the second layer only start filling up once the first layer is nearly full. This is derived from a combination of isotherms of type I and V. Type V isotherms show the occurrence of important intermolecular interactions, keeping in mind that the interaction between gas molecules and the adsorbent is weak.

Furthermore, the existence of hysteresis during the desorption phase is evidence that vapor condenses in mesopores, forming a meniscus with a high curvature. Lastly, type VI isotherms are stepped curves. This is characteristic of multilayer adsorption on the surface of non-porous, highly homogeneous solids.

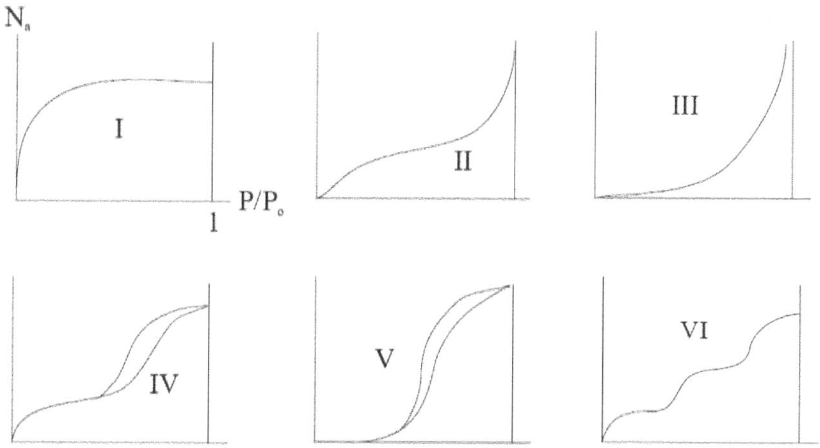

Figure 11.2. *Isothermal adsorption curves classification [ROU]*

11.3.3. *Physico-chemical adsorption-desorption models*

A great number of isotherm adsorption models have been put forward and are generally based on two distinct approaches: a surface approach that evaluates the adsorption in terms of site occupation, and a volume approach that considers adsorption as a volume filling phenomenon.

The Langmuir, BET and Dubinin models are most frequently used for the modeling of solid-gas adsorption-desorption. They all assume a stationary state and therefore the amount of gas adsorbed per unit area, according to the partial pressure and temperature, is known.

11.3.3.1. *Langmuir's theory*

The first model implemented for adsorption is the fundamental theory on the adsorption of gases by solids. It was put forward by Langmuir [ROU] in 1918 and is based on three main assumptions (Figure 11.3):

– monolayer adsorption localized on well-defined and equivalent sites;

– uniform surface adsorption;

– no interaction between adsorbed molecules.

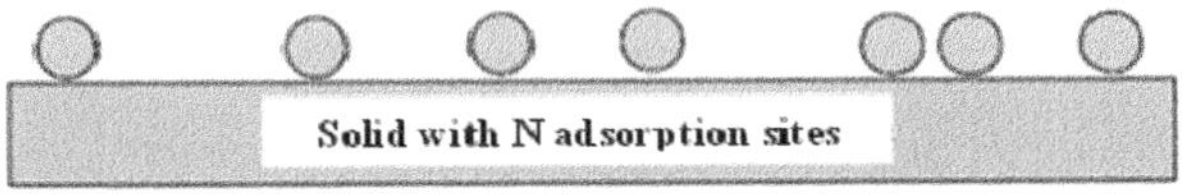

Figure 11.3. *Monolayer adsorption model representing adsorption sites and arrangement of adsorbate molecules*

The adsorption equation is represented by:

$$G + s \xrightarrow{\text{adsorption}} G - s$$

where s represents the adsorption sites, G the free state gas molecules and G-s the adsorbed gas molecules. For the surface coverage $\boldsymbol{\theta}$ of adsorption sites and pressure P of the gas, the equilibrium constant K is such that:

$$K = \frac{\Theta}{(1-\Theta)P} \text{ therefore: } \Theta = \frac{KP}{1+KP}$$

The Langmuir model describes type I adsorption isotherms with saturation being reached as pressure tends to infinity (Figure 11.4).

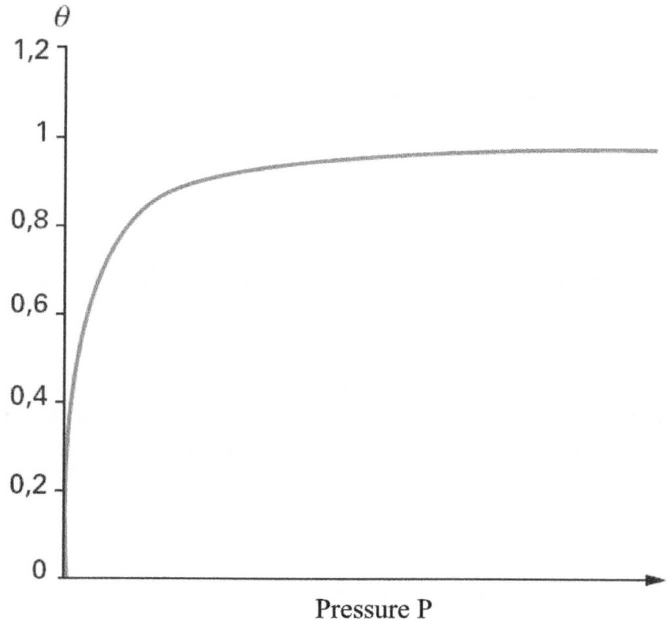

Figure 11.4. *Langmuir adsorption isotherm*

11.3.3.2. *BET theory*

This model was proposed by Brunauer, Emmett and Teller (BET) [ROU, SUN] and is founded on Langmuir's theory to describe the type II and III adsorption isotherms characteristic of multilayer adsorption. It considers an adsorbing surface S to consist of a set of surfaces S_0, S_1, S_2, S_3, etc. Adsorbate molecules are adsorbed on each of these surfaces by forming layers of molecules, respectively 0, 1, 2, 3, etc. (Figure 11.5).

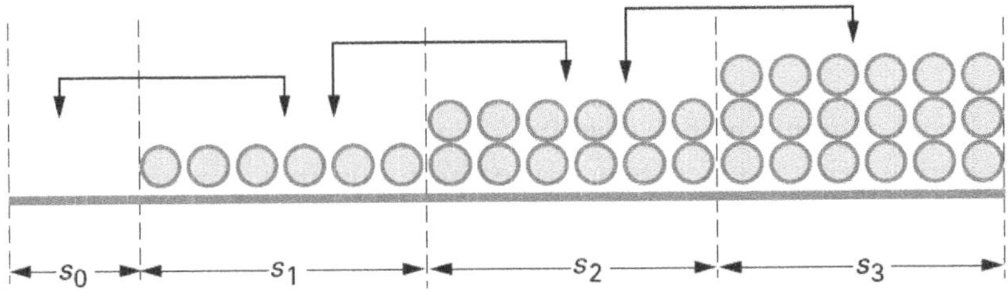

Figure 11.5. *Diagram of the adsorption of gas molecules on the surface of a solid, according to the BET model*

While taking the same hypotheses as Langmuir (one single independent adsorption site type and no interaction between adsorbed molecules), Brunauer, Emmett and Teller considered that for each equilibrium pressure *P*, his theory could be applied between the fraction of the non-covered surface S_0/S and the fraction of covered surface S_1/S, the fraction S_1/S and the fraction S_2/S, etc. They established an equation relating the amount adsorbed *n* to the relative equilibrium pressure *P*/*P*o as the number of adsorbed layers tends to infinity:

$$n / n_m = \frac{\cdot CP / P_o}{(1 - P / P_o)[1 - (P / P_o) + CP / P_o]} \quad \text{where}$$

$$C = \exp\left(\frac{E_1 - E_L}{RT}\right).$$

P_0, n_m, E_1 and E_L represent saturation vapor pressure, adsorbable amount needed to cover the solid's surface with a monomolecular layer, adsorption heat for the first layer and adsorbable's liquefaction heat.

The equation is more commonly used in linear form, called the "transformed BET equation". It contains two unknowns, C and n_m, that are found using the experimental measurement of adsorption isotherms ($n = f(P/P_o)$. The equation used is the following:

$$\frac{\cdot P/P_o}{n(1-P/P_o)} = \frac{1}{n_m C} + \left[\frac{C-1}{n_m C}\right](P/P_o)$$

11.3.3.3. *Dunbinin's theory*

Contrary to the two previous models, the Dubinin model is based on the volume filling approach by applying Polanyi's potential theory [DUB 89].

This model considers that the molecular interaction between the adsorbate and adsorbent surface is determined by a potential field and that the adsorbate's volume *Na* is solely a function of Polanyi's potential. According to the Dubinin-Astakhov equation:

$$Na = Nao \cdot \exp\left[-\left(\frac{A}{\beta E_0}\right)^n\right]$$

where *Nao* is the maximum adsorbable amount, *n* is the heterogeneity coefficient of the solid, β the affinity coefficient between adsorbate and adsorbent and E_0 is the solid's characteristic energy.

Polanyi's potential *A* is determined by the difference in Gibbs free energy ΔG, which is a function of the saturation vapor pressure P_0 and the partial pressure *P* of the phase:

$$A = RT \cdot \ln\left(\frac{P0}{P}\right) = -\Delta G$$

As for the solid's characteristic energy E_0, it characterizes the influence of the adsorbent's nature on the adsorption phenomenon and is inversely proportional to the size L of the pores [HUG 00]:

$$L = \frac{10.8}{E_0 - 11.4}$$

Lastly, the affinity coefficient β is equal to 1 for the benzene-active carbon system, taken as the adsorbate-adsorbant reference.

For specific cases of active carbons, the heterogeneity coefficient lies somewhere between 1.5 and 3. The coefficient is therefore taken as equal to 2 for a number of equations, including the Dubinin–Radushkevich equation.

11.3.4. *Study on the adsorbent's affinity to gases*

The adsorption of gas molecules by a gas is not only governed by the surface state of the material, but also, and most importantly, by its affinity to the target gas. The adsorbent's affinity can be estimated with the help of various techniques: thermogravimetry, thermodesorption or even breakthrough curves for an adsorbent bed.

11.3.4.1. *Thermogravimetry*

Thermogravimetric analysis (ATG) is an experimental technique for the macroscopic study of material. It consists of measuring the mass variation of a compound when exposed to an environment saturated with adsorbable gas. As the gas concentration, injection time and mass of the sample are known, it is possible to determine the adsorbance capacity of the solid with respect to the gas, or in other words, the solid's affinity towards the gas.

11.3.4.2. *Thermodesorption*

Thermodesorption, still referred to as thermal programmed desorption (TPD) consists of following the desorption of physisorbed molecules on a solid surface as the latter's temperature increases linearly. The process is followed using a mass spectrometer [NIC 06]. Two types of results are obtained using this method:

– the exact determination of desorption temperatures, and therefore of the activation energy required for desorption;

– the identification of desorbing species, particularly when several species desorb at the same temperature.

11.3.4.3. *Breakthrough curves*

When a polluted fluid goes through an adsorbing bed, the pollutant is stopped by adsorption. The portion of solid where the transfer takes place is called the transfer zone or adsorption front. The zone changes place within the adsorbent in order to reach the bed exit. At the end of the cycle, the output concentration is equal to the input concentration and the bed is completely saturated. Breakthrough curves are characterized by the breakthrough time, which corresponds to the time needed for the concentration of the pollutant in the fluid at the filter exit, to be non-zero.

The shape of the curves and the breakthrough time depend on the physico-chemical properties of both the adsorbent and gas adsorbed, the limiting step of the hooking mechanism of the gas on the solid, the mass of the adsorbent and the speed of the gas in the bed. For active carbon, the limiting step is more often linked to gas transfers than to reactions between gas and solid molecules.

11.3.4.3.1 General equation for the breakthrough time

Breakthrough time is the time at which a polluting gas will “start” to be detected at the preconcentrator’s exit. It can be measured using various models, including the Wheeler–Jonas model [JON 73, LIN 03, LOD 04] which is the most common. It is a function of the mass M(g) and adsorption capacity $W_e(g/g_{carbon})$ of the adsorbent bed at equilibrium, the volume flow Q (cm^3/min), the concentration $C_{in}(g/cm^3)$ of the pollutant inserted into the adsorbent, the chosen breakthrough concentration C_{out} (g/cm^3), the adsorbent’s density $\rho_b(g/cm^3)$ and, lastly, the adsorption speed coefficient $k_v(min^{-1})$:

$$tb = \frac{M \cdot We}{Q \cdot Cin} - \frac{We \cdot \rho b}{kv \cdot Cin} \ln\left(\frac{Cin - Cout}{Cout}\right)$$

Where the adsorption capacity *We* is given by:

$$We = W_0 \cdot d_L \cdot \exp\left(-\frac{BT^2}{\beta^2}\left(\frac{C_s}{C_{in}}\right)\right)$$

where W_0 (cm^3/g) is the adsorbent's microporous volume, d_L (g/cm^3) is the adsorbate's liquid density, B (K^{-2}) is the adsorbent's structural constant, T (K) is the experimental temperature, β (-) is the adsorbent's affinity coefficient for the adsorbent, and C_s (g/cm^3) and C_{in} (g/cm^3) are the adsorbate's concentrations at saturation vapor pressure and in air respectively.

As for the adsorption rate coefficient k_v (min^{-1}), it is expressed using the VOC's molar mass M_w (g/mol), the adsorbent's average particle diameter d_p (cm), the VOC's affinity coefficient β (-) and the speed V_L through the adsorbent bed:

$$k_v = 800 \cdot \beta^{0.33} \cdot v_L^{0.75} \cdot d_p^{-1.5} \cdot \sqrt{\left(\frac{W_e}{M_w}\right)}$$

The breakthrough model is only applicable for concentration ratios $C_{out}/C_{in} < 4\%$ and is therefore deemed more appropriate for devices of short length and large loading sections, for example gas masks. It is therefore necessary to design a simpler, more appropriate model for micro-preconcentrators, using thin adsorbent layers while ensuring a sufficiently long residence time.

As mentioned before, the breakthrough time depends on the adsorbent's adsorption capacity and on the polluting gas' concentration and injection rate. Strictly speaking, it also depends on the system's detection limit, placed upstream, as well as the device's geometry. If the gas has a high affinity for the solid, a "brutal" breakthrough will usually occur and the breakthrough time, as an initial guess, can be taken to be equal to the time necessary for the polluting gas to occupy all available adsorption sites in the adsorbent.

Therefore, if $N(t)$ represents the number of molecules injected into the adsorbent's bed, at a flow rate d for a time t, and all the gas molecules injected are adsorbed by the adsorbent, the following applies:

$$N(t) = d \cdot C_{in} \cdot t$$

Now, let us consider N_0, the maximum number of pollutant gas molecules that can be adsorbed by the bed for a given temperature, i.e. $\theta = 1$. When time t is equal to the breakthrough time t_p, there is an equality between the injected amount and the adsorbent's adsorption capacity, given by the following equation:

$$N_0 \cdot \theta = d \cdot C_{in} \cdot t_p \quad \Rightarrow \quad t_p = \frac{N_0 \cdot \theta}{d \cdot C_{in}}$$

For a perfect breakthrough, i.e. if all adsorption sites are occupied before breakthrough, we have:

$$\theta = 1, \text{ therefore } t_p = \frac{N_0}{d \cdot C_{in}}.$$

In addition, it should be noted that the breakthrough time is often shown by means of an output concentration C_{out} that represents the concentration threshold from which breakthrough is considered to have happened. This concentration C_{out}, linked to the detection limit for the detector being used, shows that the hypothesis of complete adsorption of all injected gas molecules is in actual fact an approximation: the same is true for perfect breakthrough.

11.3.4.3.2. Thermodynamic equilibrium hypothesis

Supposing that the system is in thermodynamic equilibrium, it is possible to use the law of adsorption isotherms $\theta = f(C)$, such that, if one uses the "Langmuir" hypothesis,

$$\theta = \frac{K\,Cin}{1 + K\,Cin}.$$

The breakthrough time is therefore given by:

$$t_p = \frac{KN_0}{d \cdot (1 + K\,Cin)}.$$

Two limit cases can be deduced from this equation:

– *Case 1*:

$$K.Cin \gg 1 \quad \Rightarrow \quad \theta \cong 1 \quad \Rightarrow \quad t_p = \frac{N_0}{d \cdot Cin} \propto \frac{1}{Cin}$$

This case is for either relatively high concentrations or high capacity adsorbents, such as active carbons or carbon nanotubes. The breakthrough time in this case is an inverse function of the concentrations.

– *Case 2*:

$$K.Cin \ll 1 \quad \Rightarrow \quad \theta \cong K.Cin \cong 0 \quad t_p = \frac{KN_0}{d} = Const.$$

This is the case for either low concentration or low capacity adsorbents for which the breakthrough time is constant and is only dependent on the flow rate and the length of the adsorbent bed.

The determination model for the breakthrough time thus implemented is much simpler to use than that of Wheeler–Jonas. The latter requires more information such as the adsorbent's capacity at equilibrium *We* (g/g_{carbon}) and the speed coefficient of adsorption *kv* (min^{-1}), as well as the adsorbent and adsorbate's specificities.

11.4. Adsorbing materials

There are many types of absorbents in existence as adsorption processes are used in a great variety of applications: VOC adsorption (in gas or liquid phases), separation of oxygen and nitrogen present in the air, hydrogen purification, desulfuration of natural gas, discoloration of sweetened juices, purification of effluents, separation of molecules in pharmaceutics, waste water treatment [GRÉ 04, LEC].

Four different types of adsorbent are mainly used for the adsorption of VOC: active carbons, carbon nanotubes, zeolites, and polymer-based adsorbents. However active carbons are by far the most commonly used adsorbent for the adsorption of organic compounds.

11.4.1. *Active carbons*

Activated carbons, which are activated via oxidative attack, are porous materials for which the specific surface area can reach up to 2,500 $m^2.g^{-1}$, with porous volumes ranging from 0.2 to 1.2 $cm^3.g^{-1}$ and bulk weights between 0.2 and 0.6 $g.cm^{-3}$. They are also characterized by a large pore size distribution ranging from micro to macropores. The pore size disparity is caused by thermic activation, the source of the micropores, and the presence of defects within the material which are synonymous to macro and mesoporosity.

11.4.2. *Carbon nanotubes*

Carbon nanotubes were discovered in 1991 by S. Iijima in Japan (NEC, Tsukuba). Iijima identified them for the first time in a sub-product of fullerene synthesis using electronic microscopy. Carbon nanotubes (CNTs) are composed of one or more concentric walls. The carbon atoms in these walls are organized in hexagonal networks [ESW 99]. CNTs are one of the four known organized states of carbon, the others being graphite, diamond and fullerene (football-shaped molecules). Two varieties of carbon nanotubes have been identified (Figure 11.6):

– single wall carbon nanotubes (SWCNTs) consisting of one single hollow tube;

– multi wall carbon nanotubes (MWCNTs) consisting of concentric tubes, separated by a few nanometers.

The internal diameter of these CNTs is of nanometer scale (0.4–3 nm) with micrometer lengths. The extremities of CNTs are often closed by carbon pentagons, characteristic of fullerenes.

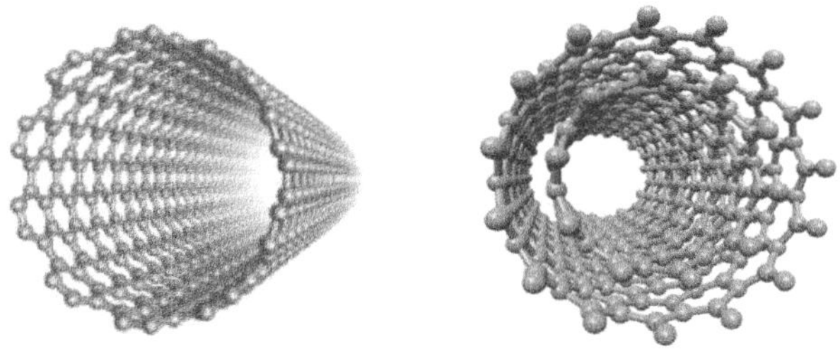

Figure 11.6. *Computer-generated images of SWCNT (right) and MWCNT (left) [ESW 99]*

The use of carbon nanotubes for the adsorption of organic molecules is attractive due to their large specific surface area (150-3,000 m^2/g) and the uniform diameter of their mesopores. They also offer three main sites for adsorption to take place: inside the nanotubes, in the interstitial chains between nanotubes, in the peripheral grooves where two nanotubes meet and on the other peripherals of the surfaces of nanotube bundles (Figure 11.7).

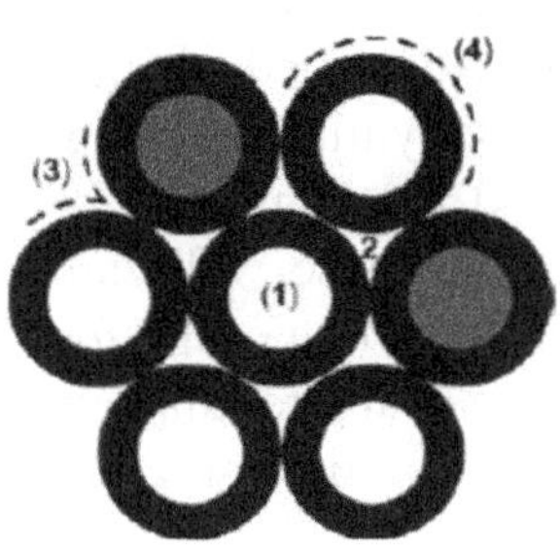

Figure 11.7. *Cross-sectional diagram of the different adsorption sites of a carbon nanotube bundle [AGN 06]*

It is also possible to functionalize CNTs: by opening their extremity through thermic treatment, their adsorption capacity can be increased [BIT 03]. Moreover thermic treatment leads to a loss of oxygenated groups that can block the nanotube's pores. The loss of these groups is a considerable advantage when it comes to making the CNTs hydrophobic.

11.4.3. *Other adsorbents*

11.4.3.1. *Zeolites*

Zeolites are hydrated metal aluminosilicates (calcium, magnesium, potassium, etc.). Their crystalline structure provides them with adsorption and catalytic properties, and has led to their use as molecular sieves, ion exchangers or as phosphate substitutes in detergents [PLE].

Zeolites are classed according to their structure into different groups. Some properties can vary within these groups, in particular the Si/Al ratio. Forty naturally occurring types have been identified, usually of volcanic origin (tuff, glass, clays).

The characteristics common to all zeolites are a low density (between 1.3 and 1.9 g/cm^3) and a large porous volume, a crystalline stability in dehydrated state, cationic exchange properties, uniform pore size, gas and vapor adsorption properties and, finally, catalytic properties.

11.4.3.2. *Polymers: Tenax TA*

Polymers are either organic or inorganic macromolecules consisting of a chain of identical elements called monomers that are linked together by covalent bonds. They can be found in both liquid and solid states at room temperature. Particular attention is drawn to them in the field of preconcentration where the ability to synthesize them, and even to functionalize *in situ*, opens up many perspectives [ROT 91, CAO 93].

Tenax TA is a 2.6-diphenylene oxide based porous polymer resin. It is widely used in a number of applications, including the preconcentration of airborne VOCs, or VOCs in moist media; because of its weak affinity to water vapor, its low impurity content, its porous and granular aspects which do not hinder the flow of fluids, and its thermic stability at temperatures under 350°C.

However it has two disadvantages that cannot be ignored: its low specific surface area (< 50 m^2/g) in comparison to other adsorbents like carbon nanotubes or zeolites, and its large grain size, which slows down its use for micro-preconcentration.

11.4.3.3. *Porous silicon*

As we will see later, the manufacturing processes for preconcentrators rely on silicon technology. The adsorbents described previously are introduced in micromachined silicon channels. The desire to improve the adhesion of adsorbent particles on the silicon support has resulted in the creation of the latter's porous surface via electrochemical anodization. Porous silicon is a crystalline "sponge", obtained by electrochemical attack of a mono-crystalline silicon substrate, usually in a concentrated hydrofluoric acid solution (HF). Silicon forms the anode and the inert metal (platinum) immersed in the electrolyte which acts as the cathode [FER 06, BET 02].

11.5. Development of preconcentrators

11.5.1. *State-of-the-art microcomponents/materials*

Most of the micro-preconcentrators mentioned in the references have either spiral or planar shapes.

The one developed by B. Alfeeli *et al.* [ALF 09], which is planar in shape, consists of a group of vertical pillars used to level the distribution of the fluid (Figure 11.8).

The adsorbent used in this micro-preconcentrator is Tenax, sprayed after its dissolution in dichloromethane. This micro-preconcentrator has been successfully used to detect traces of VOCs in high concentration alcohol.

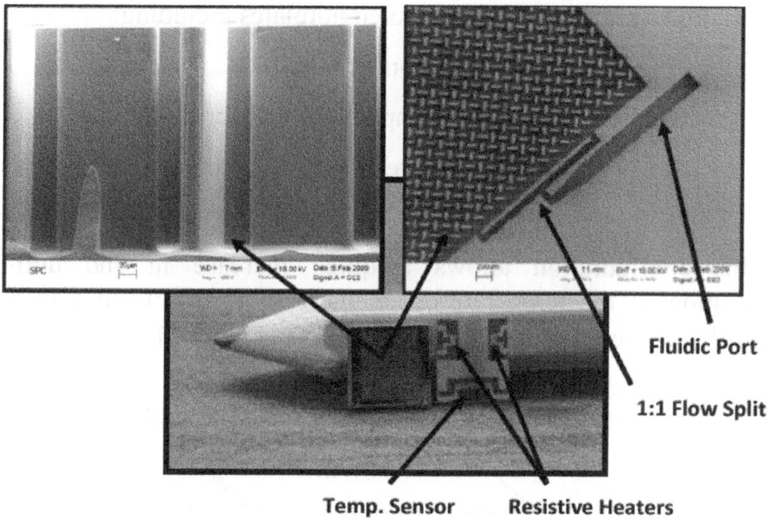

Figure 11.8. *Photos of planar-type micro-preconcentrators [ALF 09]*

The micro-preconcentrator proposed by I. Gràcia *et al.* [GRÀ 08] is an example of a spiral micro-preconcentrator. Its main channel is 300 μm deep and is filled with carbopack X before being sealed with a Pyrex cap (Figure 11.9).

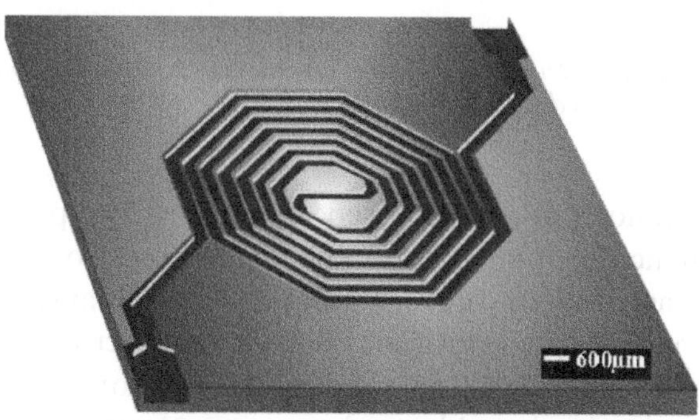

Figure 11.9. *Photo of a spiral micro-preconcentrator [GRÀ 08]*

The search for increasing sensitivity has generated other types of preconcentrators, such as microhotplates [VER 05, RUI 07]. R. McGill *et al.* [VOI 06] have created a microhotplate system based on a multilayer structure of assembled hotplates including:

– a heating coil coated in silicon oxide;

– a metal enabling uniform temperature distribution;

– an adsorbing layer to capture the gas molecules that need to be detected.

This configuration allows a quick increase in and uniform distribution of the temperature, i.e. increase to 200°C in a time of 35 ms, but a decrease in the adsorption time (Figure 11.10).

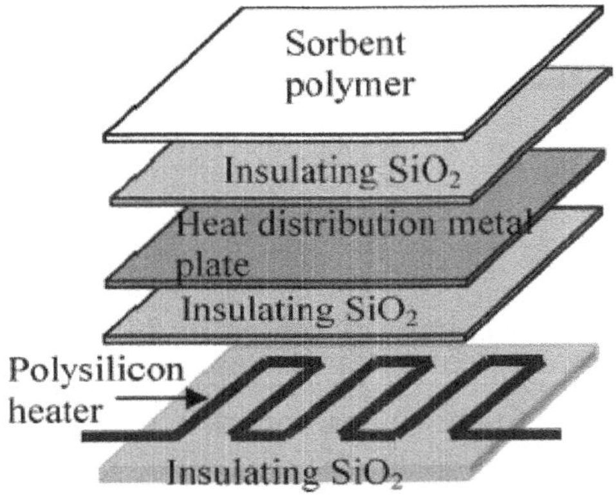

Figure 11.10. *Diagram of the different layers in a microhotplate used as micro-preconcentrators [VOI 06]*

The microhotplate is coated in a thin polymer layer (polycarbosilane functionalized with hexafluoroispropanol) and has been tested as a micro-preconcentrator. The obtained results are satisfactory as the signal obtained without the micro-preconcentrator is smaller by a factor of 100 for a collection time of the order of 60 s. The microsystem is also able to detect concentrations equivalent to 0.2 $\mu g/m^3$ (Figure 11.11).

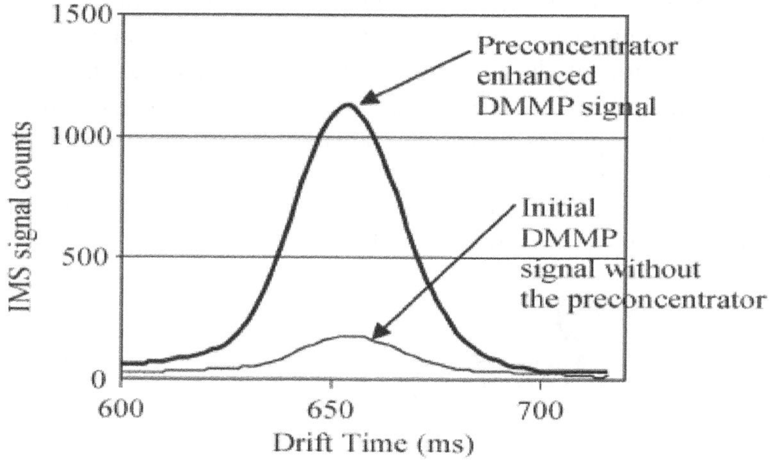

Figure 11.11. *Response of IMS to DMMP with and without a preconcentrator [VOI 06]*

It is also possible to combine different adsorbing materials in order to improve the preconcentrator's selectivity. For example, the micro-preconcentrator developed by E. T. Zellers *et al.* [ZEL 02] contains a column of adsorbents with three chambers filled with carbopack B, X and carboxen (carbon derivative), each with different specific surface areas and a total mass of 5 mg. The combination of several adsorbents enables a wide range of VOCs or explosives to be catered for as each adsorbent is specified for one or more gases (Figure 11.12).

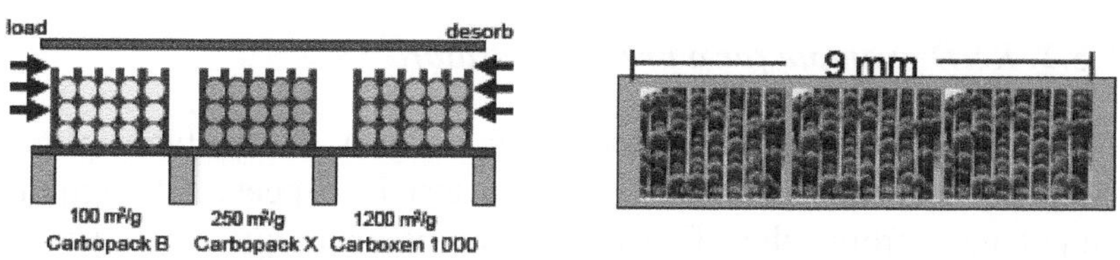

Figure 11.12. *Diagram (left) and image (right) of a micro-preconcentrator with three-phase preconcentration using different adsorbents [ZEL 02]*

The presence of water vapor in ambient media is another problem: it can cause a premature saturation of the adsorbent, leading to a lower preconcentration of the target gas. The use of other adsorbents like carbon nanotubes is promising in this case because of their hydrophobic properties, large specific surface area and low desorption temperature. F. Zheng *et al.* [ZHE 06] have developed an efficient detection system based on a preconcentrator where the adsorbent consists of a stack of carbon nanotube sheets (Figure 11.13).

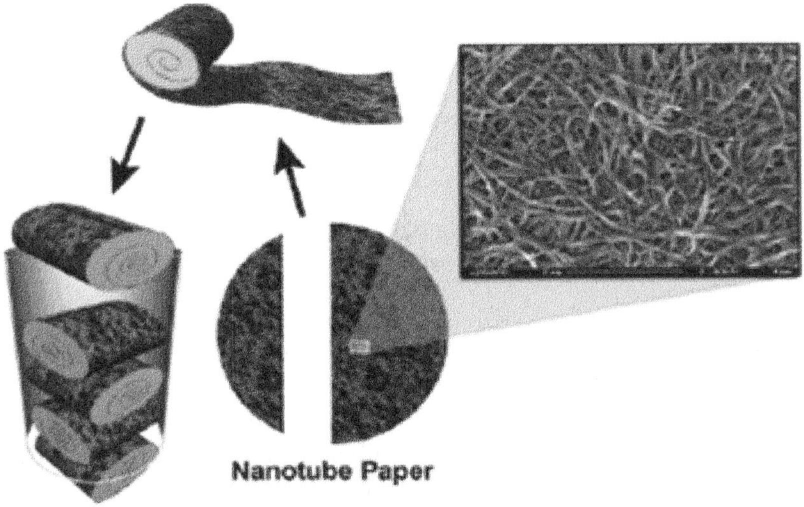

Figure 11.13. *Diagram of the carbon nanotube compaction process for preconcentration [ZHE 06]*

11.5.2. *R&D example for a preconcentrator*

Developing preconcentrators is the subject of multidisciplinary research: beyond the adsorption/material aspects it requires competence from the fields of microtechnology and process engineering, in order to obtain complete devices. As an example, the doctoral work of M. Camara [CAM 09] (co-supervised by EMSE Saint-Etienne – EPFL-IMT Neuchatel), uses microcomponents that have been created using silicon microtechnologies based on a series of lithographic and dry etching stages. These have then been closed using a glass cap and sealed with an anodic weld. Thermofluid models have been designed in order to define the geometries that allow the adsorbent to be homogeneously scanned by the gas, and which do not cause a loss in redhibitory charges of the device's mechanical resistance (wafer fracture or glass cap).

Figure 11.14 illustrates the microcomponents (30 × 15 mm^2) created using 525 μm thick wafers. The internal cavity (10 × 15 mm^2) made to a depth of 325 μm is capable of containing parallel microchannels (500 μm wide) and other patterns (staggered). The entrance and exit section allow the insertion of a capillary for fluid connections and are structured in such a way that they homogenize fluid distribution within the component.

Figure 11.14. *Geometry of silicon components: a) cavity; b) microchannels; c) staggered [CAM 09]*

As mentioned before, some devices have undergone electrochemical anodization (silicon etching in hydrofluoric acid-based electrolyte solution) in order to get porous surfaces. It is possible to obtain macroporous (Figure 11.15, 1–2 μm pore diameter and 90–100 μm deep) or microporous structures depending on the anodization parameters. This could respectively encourage the adsorbent's insertion and capture, and allow the silicon itself to behave like an adsorbent.

Devices thus obtained are equipped with a heating element (serigraphy of platinum resistance) that allows desorption to occur through rapid heating as well as electric and fluidic connectors (Figure 11.16).

Figure 11.15. *Porous aspects of silicon after electrochemical anodization [CAM 09]*

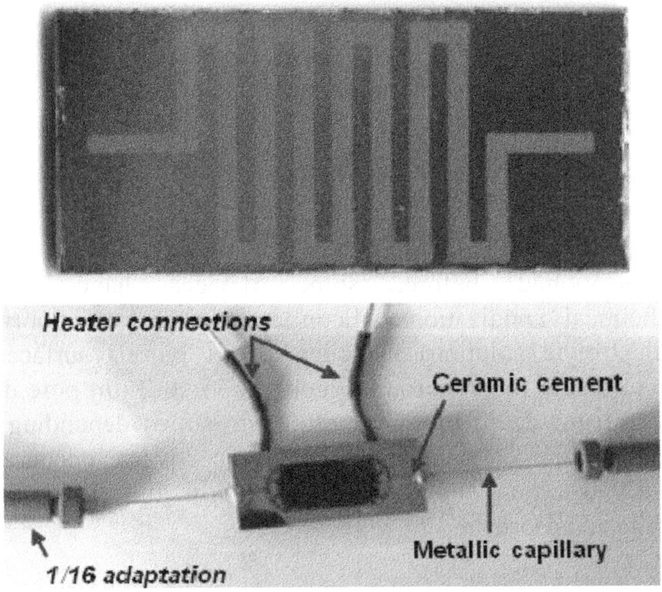

Figure 11.16. *Heating element in the microdevice (30 × 15 mm^2) and global view including connectors [CAM 09]*

The chosen adsorbent must be inserted into the microdevice. If the adsorbent's granular size allows it, it is possible to insert it as a fluid through the capillary once it has been suspended by means of a pumping or syringe system. It is equally possible to deposit the adsorbent in powder form into the open components before closing with the cap and sealing. There are pros and cons to all of these solutions (blocking the capillaries in the fluidic insertion method, difficulties in sealing the cap and adsorbent's thermal stability for post-deposit treatment in solid methods, etc.). Different technological solutions are still being studied and depend on the adsorbent's nature.

A benzene preconcentration example is presented in Figure 11.17: two devices filled with carbon nanopowder (roughly 0.9 mg) and consisting of dense or porous silicon. For five minutes, benzene (250 ppb) is injected at room temperature. The preconcentrator's temperature is then increased to 200°C and benzene is detected at the exit by a photon ionization detector (PID). The non-porous silicon component shows a peak desorption at 196°C. The porous silicon component has a maximum at 170°C with a concentration of around 15 ppm, i.e. a "practical" preconcentration factor of 60. Porous silicon therefore has a lower desorption temperature, which can be of interest for less volatile compounds.

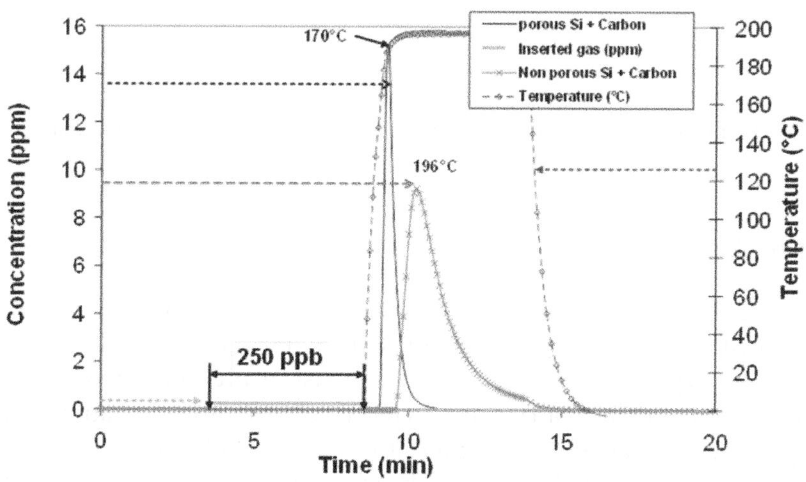

Figure 11.17. *Benzene preconcentration tests using carbon nanopowders in dense and porous silicon microcomponents [CAM 09]*

11.5.3. *Example of a commercial machine*

The marketing of machines using miniaturized preconcentrators has only just begun and to date, it seems that only one device exists: the MicroHound®, developed by SANDIA [PDDE]. It is a portable detector used for the detection of explosives, that combines a miniature preconcentration system with sensitive chemical microsensors capable of detecting quantities of explosive in the order of a nanogram (Figure 11.18) on the same support. This lightweight system (around 2 kg) can detect explosives in either vapor or particle state.

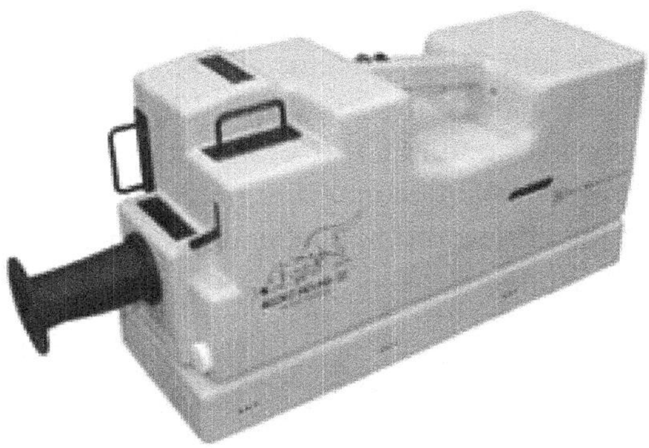

Figure 11.18. *Photo of an explosives detector based on preconcentration developed by SANDIA National Laboratories (USA) [PDDE]*

11.6. Conclusion

The development of micro-preconcentrators is now in full bloom and benefits from the technologies developed in microelectronics. Even if very few commercial devices exist at present, many studies are currently being carried out. The aspect of miniaturization goes beyond the scope of preconcentrators and includes whole systems such as separation columns and detectors placed on the same wafer, particularly for the development of micro-gas-chromatographs (μGC).

11.7. Bibliography

[AGN 06] S. AGNIHOTRI, J.P.B MOTA, M. ROSTAM-ABADI, M.J. ROOD, "Theoretical and experimental investigation of morphology and temperature effects on adsorption of organic vapors in single-walled carbon nanotubes", *J. Phys. Chem. B*, vol. 110, pp. 7640-7647, 2006

[ALF 09] B. ALFEELI, M. AGAH, "Selective preconcentration by temperature manipulation of cascaded micro preconcentrators", *Transducers 2009 Conference*, pp. 1662-1665, Denver, CO, USA, June 2009.

[BET 02] P. BETTOTTI, L. DAL NEGRO, Z. GABURRO, L. PAVESI, "P-type macroporous silicon for two-dimensional photonic crystals", *Journal of Applied Physics*, vol. 92, no. 12, pp. 6966-6972, 2002.

[BIT 03] E.W. BITTNER, M.R. SMITH, B.C. BOCKRATH, "Characterization of the surfaces of single-walled carbon nanotubes using alcohols and hydrocarbons: a pulse adsorption technique", *Carbon*, vol. 41, pp. 1231-1239, 2003.

[CAM 09] M. CAMARA, Développement de micro-préconcentrateurs pour la détection de substances chimiques à l'état de trace en phase gaz, PhD Thesis, Ecole des Mines de Saint-Etienne (France), 11 dec. 2009.

[CAO 93] X.-L. CAO, C N. HEWITT, "Evaluation of Tenax-GR adsorbent for the passive sampling of volatile organic compounds at low concentrations", *Atmospheric Environment. Part A. General Topics*, vol. 27, pp. 1865-1872, 1993.

[CHO 06] S.M. CHO, Y. JUN KIM, G.S. HEO, S-M. SHIN, "Two step preconcentration for analysis of exhaled gas of human breath with electronic nose", *Sensors and actuators B*, vol. 117, pp. 50-57, 2006.

[DUB 89] M.M. DUBININ, "Fundamentals of the theory of adsorption in micropores of carbon adsorbents: characteristics of their adsorption properties and microporous structures", *Carbon*, vol. 27, pp. 457-467, 1989.

[ESW 99] M. ESWARAMOORTHY, R. SEN, C.N.R. RAO, "A study of micropores in single-walled carbon nanotubes by the adsorption of gases and vapors", *Chemical Physics Letters*,vol. 304, pp. 207-210, 1999.

[FER 06] P. FERRAND, "Le Silicium nanoporeux: microstructuration diélectrique et application aux structures photoniques avancées", *Annales de Physique*, vol. 27, no. 2, 2006.

[GRÀ 08] I. GRÀCIA, P. IVANOV, F. BLANCO, N. SABATÉ, X. VILANOVA, X . CORREIG, L. FONSECA, E. FIGUERAS, J. SANTANDER, C. CANÉ, "Sub-ppm gas sensor detection via spiral µ-preconcentrator", *Sensors and actuators B*, vol. 132, pp. 149-154, 2008.

[GRÉ 04] G. GRÉVILLOT, Traitement d'air chargé en composés organiques volatils par adsorption sur solides microporeux, ENSIC-Nancy, www.culturesciences.chimie.ens.fr, 2004.

[HUG 00] D. HUGI-CLEARY, F. STOECKLI, "On the use of a standard DRK isotherm in Dubinin's t/F method", *Carbon*, vol. 38, pp. 1309-1313, 2000.

[JON 73] L.A. JONAS, J.A. REHRMANN, "The rate of gas adsorption by activated carbon", *Carbon*, vol. 12, pp. 95-101, 1973.

[LEC] P. LE CLOIREC, "Adsorption en traitement de l'air", *Technique de l'Ingénieur*, G 1770-1 to 13.

[LEW 06] P.R. LEWIS, R.P. MANGINELL, D.R. ADKINS, R.J. KOTTENSTETTE, D.R. WHEELER, "Recent advancements in the gas-phase MicroChemLab", *IEEE Sensors Journal*, vol. 6, no. 3, June 2006.

[LIN 03] M.J.G. LINDERS, E.P.J. MALLENS, J.J.G.M. VAN BOKHOVEN, F. KAPTEIJN AND J.A. MOULIJN, "Breakthrough of shallow activated carbon beds under constant pulsating flow", *AIHA Journal*, vol. 64, pp. 173-180, 2003.

[LOD 04] P. LODEWYCKX, G.O. WOOD, S.K. RYU, "The Wheeler–Jonas equation: a versatile tool for the prediction of carbon bed breakthrough times", *Carbon*, vol. 42, pp. 1351-1355, 2004.

[MAR 07] M. MARTIN, M. CRAIN, K. WALSH, R. E. MCGILL, E. HOUSER, J. STEPNOWSKI, S. STEPNOWSKI, H.-D. WU, S. ROSS, "Microfabricated vapor preconcentrator for portable ion mobility spectroscopy", *Sensors and actuators B*, vol. 126, pp. 447-454, 2007.

[NIC 06] S.I. NICHOLL, J.W. TALLEY, "Development of thermal programmed desorption mass spectrometry methods for environmental applications", *Chemosphere*, vol. 63, pp. 132-141, 2006.

[PDDE] PORTABLE DEVICE FOR DETECTING EXPLOSIVES: http://sine.ni.com/cs/app/doc/p/id/cs-700.

[PLE] D. PLEE, *Zéolites*, Techniques de l'ingénieur, J 6675.

[ROT 91] H. ROTHWEILER, P. A WÄGER, C. SCHLATTER, "Comparison of Tenax Ta and Carbotrap for sampling and analysis of volatile organic compounds in air", *Atmospheric Environment. Part B. Urban Atmosphere*, vol. 25, pp. 231-235, 1991.

[ROU] F. ROUQUEROL, L. LUCIANI, P. LLEWELLYN, R. DENOYEL, J. ROOUQUEROL, "Texture des matériaux pulvérulents ou poreux", Technique de l'Ingénieur, p. 1050-1 to 24.

[RUI 07] A.M. RUIZ, I. GRACIA, N. SABATE, P. IVANOV, A. SANCHEZ, M. DUCH, M. GERBOLES, A. MORENO, C. CANE, "Membrane-suspended microgrid as a gas preconcentrator for chromatographic applications", *Sensors and Actuators A*, vol. 135, pp. 192-196, 2007.

[SUN] L.M. SUN, F. MEUNIER, "Adsorption: aspect théorique", Technique de l'Ingénieur, J2730-1 to 15.

[VER 05] A. VERGARA, E. LLOBET, J. BREZMES, X. VILANOVA, P. IVANOV, I. GRACIA, C. CANÉ, X. CORREIG, "Optimized temperature modulation of microhotplate gas sensors through pseudorandom binary sequences", *IEEE Sensors Journal*, vol. 5, no. 5, December 2005.

[VOI 06] I. VOICULESCU, R. A. MCGRILL, M. E. ZAGHOUL, "Micropreconcentrator for enhanced trace detection of explosives and chemical agents", *IEEE Sensors Journal*, vol. 6, no. 5, October 2006.

[ZEL 02] E.T. ZELLERS, K.D. WISE, K. NAJAFI, R.B. BROWN, Q.J. CAI, J. DRISCOLL, "Determination of complex vapor mixtures in ambient air with a wireless microanalytical system: vision, progress, and homeland security applications", *Technical Digest of the IEEE Conference on Technologies for Homeland Security*, pp. 92-95, Waltham MA, IEEE, Boston, November 13-14, 2002.

[ZHE 06] F. ZHENG, D.L. BALDWIN, L.S. FIFIELD, N.C. ANHEIER JR., C.L. AARDAHL, J.W. GRATE, "Single-walled carbon nanotube paper as a sorbent for organic vapour preconcentration", *Anal. Chem.*, vol. 78, pp. 2442-2446, 2006.

Chapter 12

Microfluidics: Manipulation of Nanovolume Samples

12.1. Introduction

Microfluidics is understanding, manipulating and exploiting the flow of fluids in microchannels. Channels are defined as microchannels if the smallest flow quantity typically lies in the range of 1 to 100 μm.

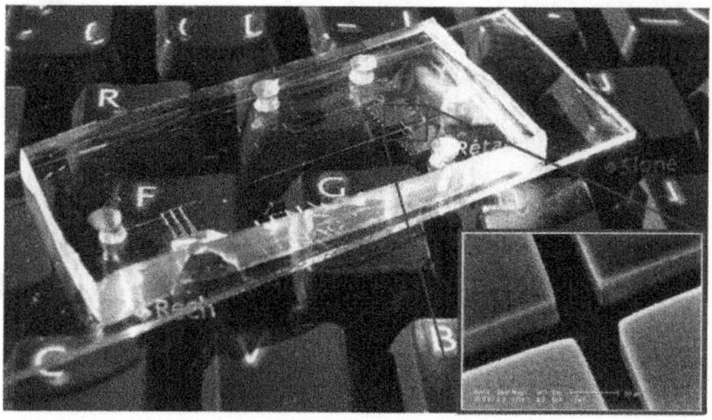

Figure 12.1. *Microfluidic chip and detail of microchannel*

Chapter written by Louis RENAUD.

The development of "Lab-on-Chip" or "microTAS" is one of the most promising aspects of microfluidics. These devices will allow the future combination of a multitude of conventional chemistry or biology laboratory functions on the same microfluidic chip: purification, extraction, mixing, preconcentration, separation, quantification, etc. In the long run, analysis will therefore become portable, faster and cheaper.

At microfluidic scales, the fundamental principles of fluid mechanics, i.e. Navier–Stokes equations, and the concept of macroscopic viscosity are still valid. For example, water maintains its "classic" Newtonian fluid properties in flows that are considerably more confined than those seen here, typically channels up to 50 nm deep [CHE 02]. However microfluidic flows have their own characteristics that are of interest both fundamentally and for their applicability: for example, microfluidic flows are always laminar unlike classic macroscopic flows (Figure 12.2).

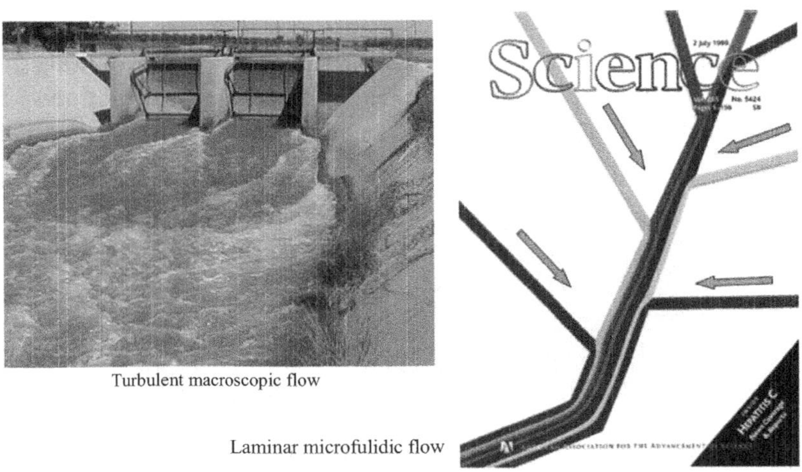

Figure 12.2. *Difference between microfluidic and macrofluidic flow*

This purely laminar aspect can have considerable advantages. For example, it is possible to make highly efficient "fuel cell" type microfluidic chips for energy production [ZEB 09]: the laminar flow replaces membranes standardly used at larger scales without ohmic loss. On the other hand, a mixture is sometimes required. Obtaining microfluidic mixtures can be problematic. A considerable amount of research is now being carried out in order to find out more about this problem.

Microfluidic flows are unique as they differ greatly from large scale flows. However, they obey the same macroscopic laws for fluid dynamics. In the context of a publication dedicated to sensors and microsensors, this chapter aims to demonstrate these flows and give readers a basic understanding of the physics behind them. It will also discuss the means by which small sample volumes are brought to the sensor quickly in simpler and more controlled ways than for macroscopic levels.

We will start by defining the fundamental characteristics of microfluidic flows: highly simplified Navier–Stokes equations. We will then see how to create simple hydrodynamic flow models in any complex microchannel network, before presenting the concept of electro-osmosis and its advantages for the displacement of matter by fluid displacement. Finally we will see how it is possible to use the advantages of microfluidics for innovative uses in chemistry and biology through the presentation of a two-phase microfluidic flow.

12.2. The physics of microfluidic flows: simplified Navier–Stokes equations

The Navier–Stokes equations describe flows in fluid mechanics and are a set of equations containing partial derivatives of the following form [OUZ 94]:

$$\mathrm{div}\,(\vec{v}) = 0 \qquad [12.1]$$

which is the continuity equation (for incompressible fluids).

$$\frac{\partial \vec{v}}{\partial t} + \left(\vec{v} \cdot \vec{\nabla}\right).\vec{v} = -\vec{\nabla}\left(\frac{P}{\rho}\right) + \upsilon\,.\,\Delta\vec{v} \qquad [12.2]$$

is the equation for the conservation of momentum.

In these equations, $\vec{v}$ is the flow velocity, with components (u, v, w), and is a function of position, time, absolute pressure P, density of the fluid ρ and kinematic viscosity υ. Equation [12.1] and [12.2] do not have a general solution and this problem is one of the seven greatest mathematical challenges of our time [CAR 00].

Fortunately equation [12.2] is considerably simplified in the case of microfluidic flows. By decreasing the flow dimensions, the key factors are surface effects and not volume effects. This is a concept which is widely used and generally approved of: the law of scales in physics often involves drastically modifying the behavior of physical objects simply by decreasing their dimensions. An innocuous grain of wheat can thus become an explosive when in powdered form. Similarly, when one macroscopically inserts gold as nanoparticulates, it becomes very reactive [PAN 09]. The number of examples is infinite and the same is applicable to fluid mechanics.

The channel dimensions in microfluidics is always much smaller than the flow's maximum laminar layer. Microfluidic flow is therefore invariably laminar and the calculation of Reynolds Number always leads to a value of around 1 and under. In all cases, this is very far from the turbulence transition limit which is 2,300. It follows that the rotational terms in equation [12.2] disappear and the current lines in the fluid are only in the flow's longitudinal direction. The velocity of the fluid therefore only has one component (arbitrarily in the x-axis direction, for example) and becomes a 2D function that only depends on the position in the cross-section (y-z). The loss of loads becomes linear and the relative pressure difference ΔP between the entrance and exit of a microchannel of length L can simply be considered as:

$$\frac{\partial u}{\partial t} = -\frac{1}{\rho}\frac{\Delta P}{L} + v\left(\frac{\partial^2 u}{\partial y^2} + \frac{\partial^2 u}{\partial z^2}\right) \qquad [12.3]$$

For constant flow, the time-related derivative term cancels out and a simplified version of the velocity field is obtained:

$$\frac{\partial^2 u}{\partial y^2} + \frac{\partial^2 u}{\partial z^2} = \frac{\Delta P}{\eta L} \qquad [12.4]$$

here η is the fluid's dynamic viscosity defined as the product of kinematic viscosity and density. For example, the dynamic viscosity of water at a temperature of 20°C is 10^{-3} Pa.s and typically varies by 2% per degree at room temperature. This type of equation is frequently used in Physics and results in solving a Laplacian constant. Even though it has few written forms, it can be solved quickly and simply using numerical computation software, by discretizing the problem in finite form:

$$\frac{u_{i+1,j} - 2u_{i,j} + u_{i-1,j}}{\Delta y^2} + \frac{u_{i,j+1} - 2u_{i,j} + u_{i,j-1}}{\Delta z^2} = \frac{\Delta P}{\eta.L} \qquad [12.5]$$

In the next two sections, we will see how equation [12.5] ends in flow modes, which differ considerably depending on the fluid's actuation mode: hydrodynamic and electro-osmosis modes.

12.3. Hydrodynamic flow: concept of microfluidic resistance

The hydrodynamic flow mode is encountered daily: fluids move when a pressure is applied. For microfluidic channels, the relationship between pressure and velocity is described by equation [12.4] stated in the previous chapter. The equation alone cannot describe flow completely. There is no condition for the wall flow velocity, which we will consider to be zero for a vast majority of materials encountered in microfluidics [MAA 09]: PDMS (PolyDimethylSiloxane), COC (CycloOlefinCopolymer), glass, silicon, etc.

The velocity profile develops even more as as we come away from the walls as the fluid is forced to move by the pressure; at the same time the zero wall velocity "information" applies to all fluids because of viscous Newtonian friction, which "slows down" the fluid as it approaches the walls. For example, the solution to equation [12.4] for a tube gives the following velocity profile:

$$V(r) = \frac{\Delta P.(R^2 - r^2)}{4.\eta.L} \qquad [12.6]$$

where R is the tube's radius and r is the distance between a fluid element and the centre of the tube. The velocity profile has a parabolic shape and this does not affect load transport, as will be seen later.

The integral of the velocity field is used to calculate the flow rate of the fluid and the flow's laminar appearance guarantees a perfectly linear relationship between the pressure and the flow rate. This linear relationship is analogous to Ohm's law, U = R.I, used in electrical engineering and it is possible to write a microfluidic version of Ohm's law valid for all microfluidic channels:

$$\Delta P = R_{\mu f}.Q \qquad [12.7]$$

where Q is the fluid's flow rate (analogous to electric current), ΔP is the applied pressure (analogous to voltage, potential difference), and $R_{\mu f}$ is the microchannel's microfluidic resistance

Therefore to answer the question: "What is the value of the hydrodynamic flow rate at any point in a complex microfluidic chip?" it is simply a matter of replacing each microchannel – or each section of microchannel – with its equivalent microfluidic resistance, replacing the flow sources with their electrical equivalent (voltage source for pressurized systems, current source for pump-syringes) and solving Kirchhoff laws for continuous electrical networks.

Only the problem of calculating a microchannel's microfluidic resistance remains. For capillary tubes, we simply integrate equation [12.6] to obtain a relationship between the pressure and flow rate:

$$Q = \frac{\pi.\Delta P.R^{4}}{8.\eta.L} \qquad [12.8]$$

After this, determining the microfluidic resistance of a capillary tube of radius R and length L is straightforward:

$$R_{\mu f,tube} = \frac{8.\eta.L}{\pi.R^{4}} \qquad [12.9]$$

For example, a capillary tube of diameter 20 µm, length 10 cm, filled with water at a temperature of 20°C has a microfluidic resistance of 14 µl/h/bar. Similarly, for the flow for a rectangular section of length L, width w and height h (for example, silicon etching by DRIE – deep reactive ion etching – followed by anodic bonding Si/glass), H. Bruus proposes an equation which takes into account a factor for the shape of the microchannel's cross-section [BRU 07]:

$$R_{\mu f,rectangle} = \frac{12.\eta.L}{w.h^{3}.(1-0.630.\frac{h}{w})} \qquad [12.10]$$

We have put forward a series of microfluidic resistances, determined through systematic numerical resolution of equation [12.5], for a wide range of sizes regularly encountered in microfluidics, and for common sections obtained in microtechnologies: rectangular sections (DRIE etching of silicon) and isotropic section (HF glass etching). This is already available in the literature with "instructions" that enable the rapid and simple determination of a microchannel's microfluidic resistance [REN 08].

It is necessary to give an example in order to better understand the method: a complex microfluidic network analogous to a multi-phased microfluidic mixer, used to generate controlled molecular gradients [STE 07]. Each channel in the microfluidic chip has the same section (50 μm wide by 70 μm deep). Only the length of the channels varies (Figure 12.3). The chip is fuelled by two syringe pumps assimilated to the electric current generators, and the microfluidic chip's exit, at air pressure, is defined as the system's conductive part.

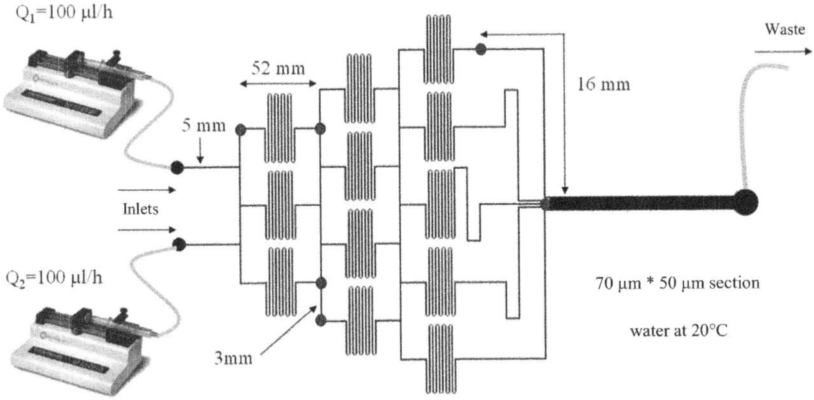

Figure 12.3. *Diagram of a chip used to generate a microfluidic gradient*

To solve this complex-looking problem, a simple system of linear equations needs to be solved. In practice, simple linear equation softwares are used which are dedicated to calculating electrical networks, such as PSPICE, the education version of which is free. The results shown in Figure 12.4 are thus obtained and give the flow rate at any point in the microfluidic chip by showing the equivalent system's electric currents. The syringe pumps placed at the system's entry, and working at a flow rate of 100 μl/h, are replaced by 100 A current generators and each ampere calculated in the electric circuit corresponds to a flow rate of 1 μl/h at the equivalent point in the microfluidic chip.

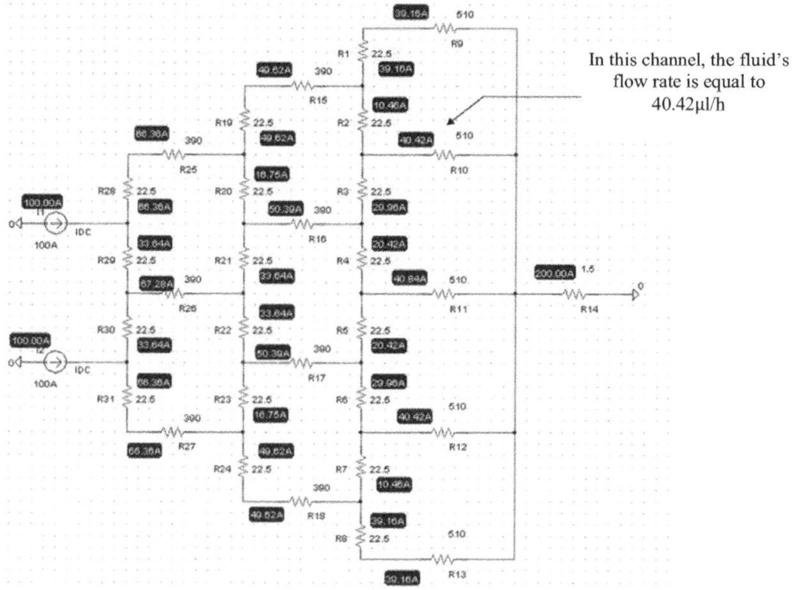

Figure 12.4. *Microfluidic flow rates calculated using equivalent electric circuits*

Using the same software it is possible to know the pressure at any point in the microfluidic chip. This is done by showing the voltage at each junction in the circuit and converting them into equivalent pressure levels. Here, we must take care with the unit of microfluidic resistance used. For example if the microfluidic resistance unit used is mbar/(µl/h), and the flow rate of 1 µl/h corresponds to an electric current of 1 A, each volt in the electric circuit will represent a pressure of 1 mbar in the microfluidic chip. For the microfluidic chip shown in Figure 12.3 it is therefore possible to determine a maximum theoretical pressure of 69 mbar.

12.4. Electro-osmotic flow

When a solid is placed in direct contact with an electrolyte, a surface charge sometimes appears at the liquid/solid interface. This is the case for glass and PDMS, which are far from being the most commonly used materials in microfluidics and both become negatively charged when put into contact with aqueous buffers [KIR 04a, KIR 04b]. The charge is a consequence of two main physico-chemical phenomena: it can be caused by the ionization of surface atoms upon contact with an electrolyte through ionic dissociation, or through the means of a physical interaction between the solid and the electrolyte, for example, in the case of the adsorption of charged ions on the solid's walls. Numerous models describe the charged liquid/solid surface, namely the physical behavior of ions present in the electrolyte, close to the surface. For simplicity and educational reasons, we will use a very simplified model that more than amply describes electro-osmotic flow (see Figure 12.5).

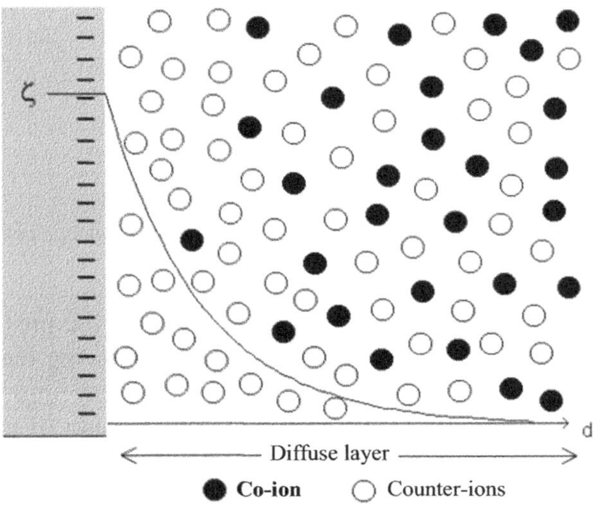

Figure 12.5. *Simplified Gouy–Chapman model*

Each surface charge has a corresponding electrical potential, called the zeta potential (ζ). It represents the potential difference between the wall and the electrolyte at an infinite distance from the surface. The surface charge is counterbalanced in the liquid by a layer composed mainly of counter-ions that shield the electrostatic field created by the surface charge: this layer is called the spread. It penetrates the electrolyte to a depth called the Debye length, which is usually in the tens or hundred nanometers scale (depending on the concentration of the electrolyte filling the microchannels), which is therefore much smaller than the microchannel's smallest dimension.

As an example, a microchannel with negatively charged walls is used (Figure 12.6):

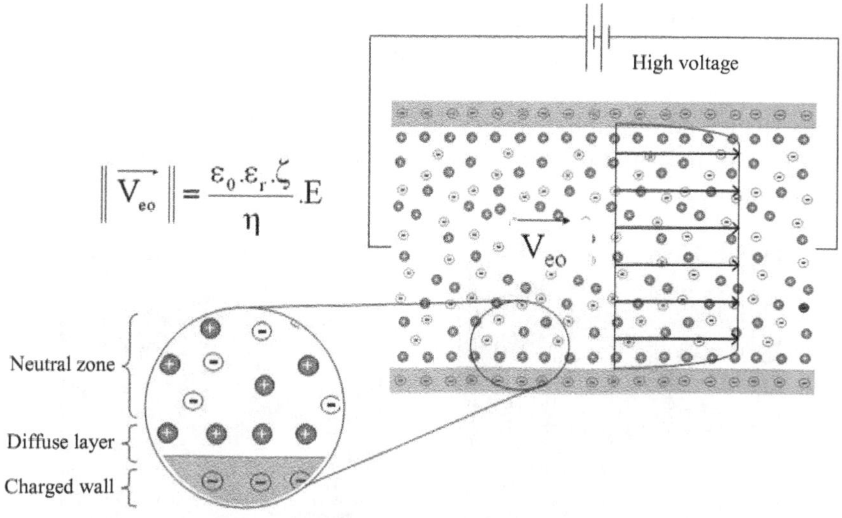

Figure 12.6. *Outline of electro-osmosis*

If a voltage is applied to the microchannel's entry and exit, an electric field will cause the charges to move in the spread, of which a majority is of the opposite sign to that of the charged surface. The fluid is displaced by these charges via viscous friction: this displacement of fluid is called electro-osmosis. The velocity of the fluid is constant over the microchannel's cross-section as in the case of equation [12.4] which is solved using zero applied pressure and a non-zero wall velocity: this is called "plug" flow. The electro-osmotic speed is given by the Schmoluchowski equation:

$$\overrightarrow{V_{eo}} = -\frac{\varepsilon_0.\varepsilon_r.\zeta}{\eta}.\vec{E} \qquad [12.11]$$

where E is the applied electric field, ε_r is the electrolyte's relative dielectric permittivity (equal to 80 for water), η is the electrolyte's dynamic viscosity and ζ is the channel wall's zeta potential.

Electro-osmosis has two advantages over the hydrodynamic mode for fluid displacement. The first relates to the fluid's velocity, which is constant throughout the channel's cross-section. This property is remarkable as it enables the transport of matter without convection dilution, as shown by the images in Figure 12.7 [ROS 01]. Material is only dispersed through diffusion in a way that is completely controlled if there is information on the diffusion coefficients of the pertinent molecules.

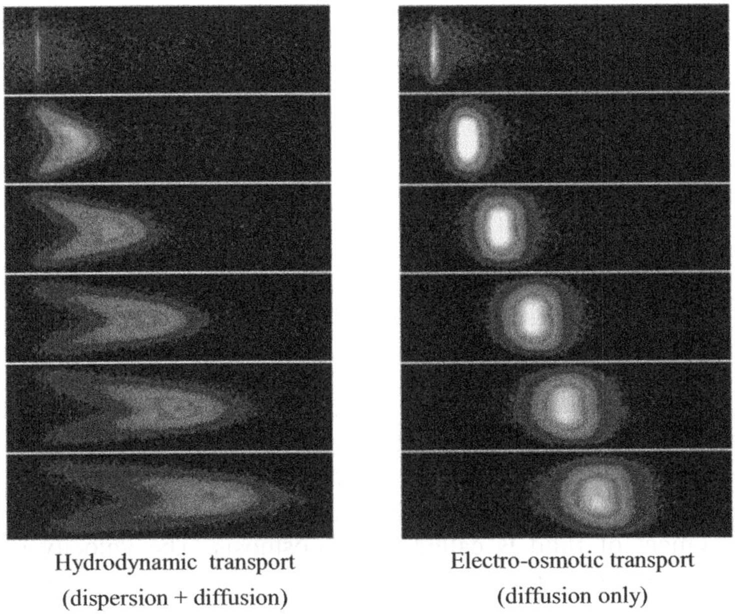

Figure 12.7. *Advantage of electro-osmosis for the material transport (excerpt from [ROS 01])*

The second advantage uses this transport mode and is the reason why its use is so widespread: electro-osmosis enables the availability of flows in complex 2D microchannel networks. As has been seen previously, the current lines in the fluid follow the same lines as those of the electric field.

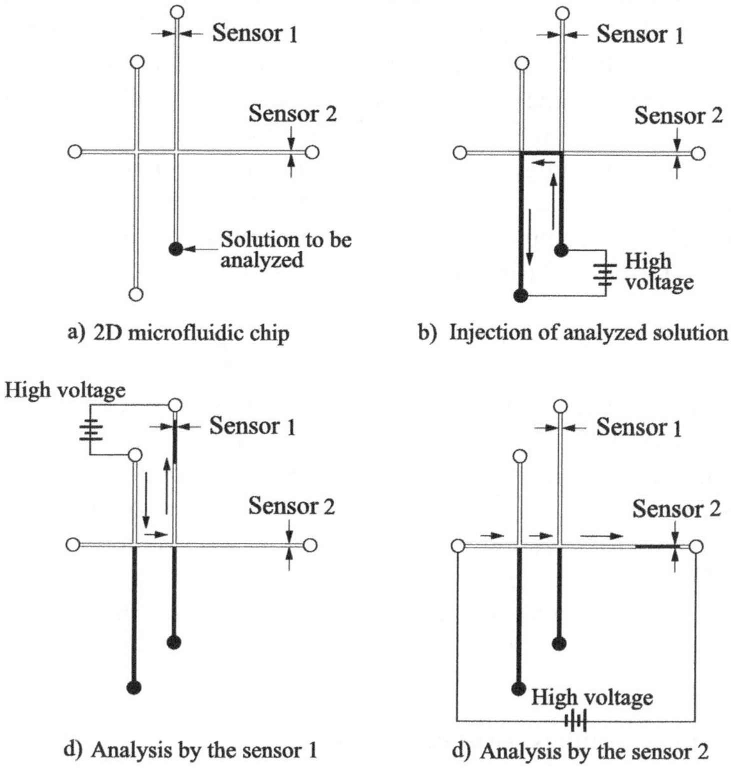

Figure 12.8. *Use of electro-osmotic flows for the control and manipulation of sample nanovolumes*

Therefore, if we place a complex microfluidic chip at n inlets/outlets, only some of them can be "activated" at any time simply by controlling their voltage with electrodes. Figure 12.8 is an example of what can be achieved simply using electro-osmosis. Figure 12.8a shows a complex 2D channel network with one of the entrances containing the desired molecule for analysis. If the entry is linked to one of the exits the shortest distance between entry and exit is gradually filled with analyzed solution in a time which depends on the electro-osmotic speed and that can be controled with the applied voltage value (Figure 12.8.b). Therefore if a section of the channel (in this case the central section) is filled, the volume can be controlled using the microfluidic chip's manufacturing technology. As stated previously, a cube with 10 µm sides contains a volume of 1 picoliter whereas a cube with 100 µm sides contains a volume of 1 nanoliter. It is easy to obtain volume ranges with relatively cheap clean room technologies that we unfortunately will not discuss here [STE 07]: PDMS soft molding using dry resins, etc. Once the volume has been determined, the applied voltages between entries and exits are changed to bring the sample towards the desired channel, for example from sensor 1 to sensor 2 (Figure 12.8c and 12.8d).

In conclusion, electro-osmotic flows can be simply summarized: it is the conversion of an electric energy (applied electric field, which releases heat in the channel by the Joule effect as well as generating electro-osmosis) into mechanical energy corresponding to the transport of the desired matter (and an energy loss due to viscous friction in the solution). The fluid's velocity is constant throughout the microchannel's cross-section and the flows are easy to find in complex 2D networks. These properties explain why electro-osmosis is one of the most commonly used fluid displacement methods in microfluidics.

12.5. Drop microfluidics: two-phase flows

Take a "Y" shaped set of macroscopic channels with two entries and one exit. Two immiscible fluids, for example water and oil, are passed through each of the Y's branches, from "top" to "bottom". What we are trying to observe is what happens at the junction of the two top branches, and in the exit channel. Most of the time, a perfectly inhomogeneous emulsion is obtained. Sometimes we see joint flows. The importance of surface effects in microfluidics means that these observations are invalid: the decrease in flow induces a new functionality (Figure 12.9). If the channel walls are even slightly hydrophobic, i.e. they have a preference for oil, then the system will immediately result in water droplets forming, which will be identical in volume and imperviously separated from one another for a given volume of oil [SON 03]. The scale of the channels' dimensions means that the water droplet volume is of the order of 1 nm.

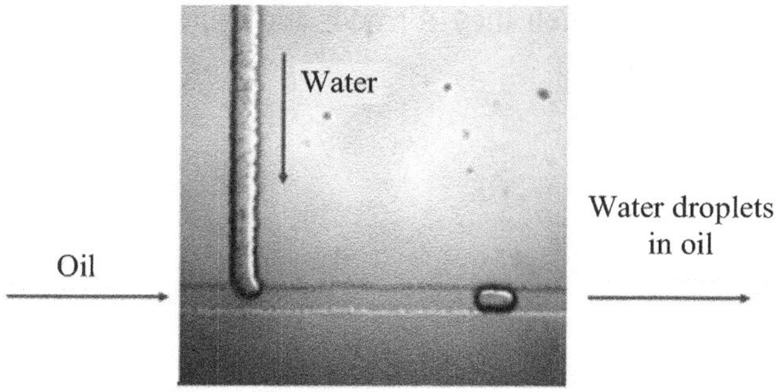

Figure 12.9. *Generation of nanodrop streams in a two-phase flow*

This particular type of microfluidics is noteworthy as each drop can be considered as a unique chemical nanoreactor, perfectly isolated both physically and chemically from droplets that are formed sequentially before or after it. Using the flow mechanics inside each droplet, in the shape of a double tank track, the question related to the mixture is answered by the creation of zigzag channels [SON 03] (Figure 12.10).

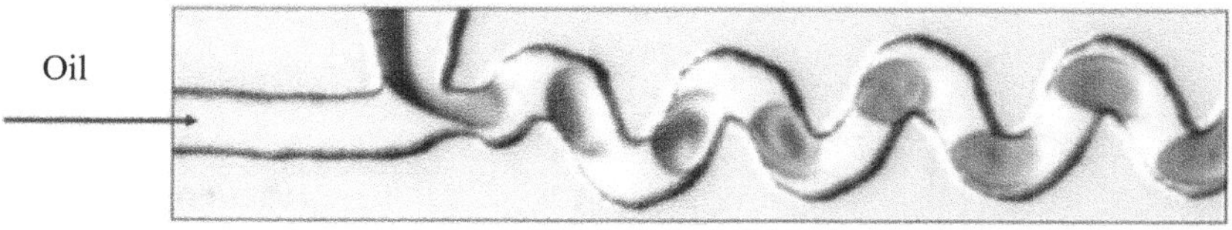

Figure 12.10. *Mixing in drop microfluidics [SON 03]*

The sequential creation of droplets means that the relationship between the lifetime and the location is perfectly established which, as an example, allows the efficient determination of the kinematics of chemical reactions in very small sample volumes [SHE 04]. In this particular case, every drop is identical. The only difference between them is the time at which they formed, and therefore their lifetime at the observation time.

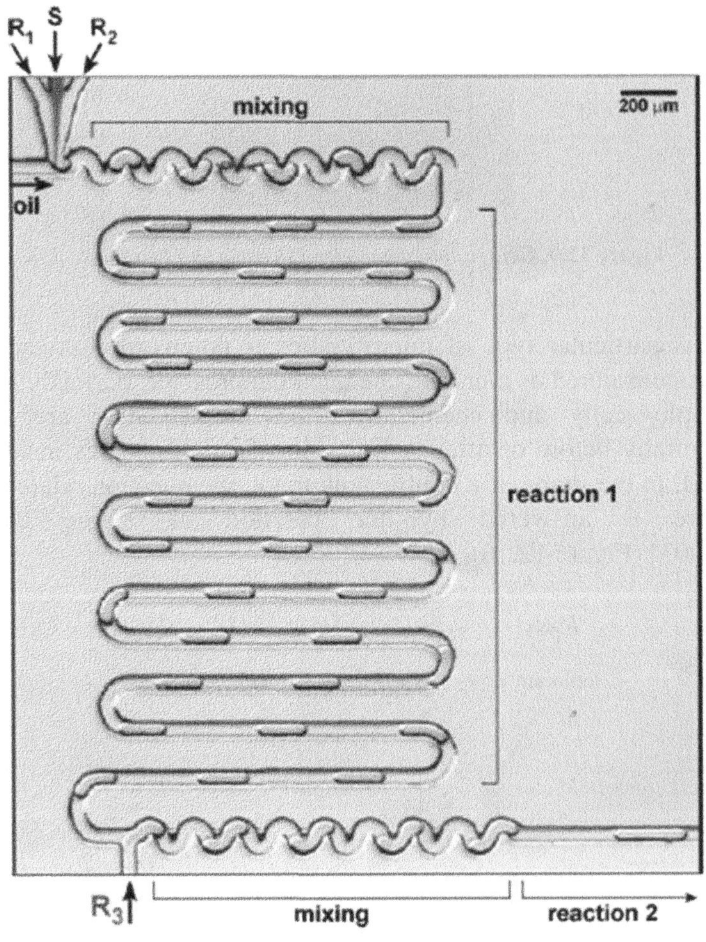

Figure 12.11. *Kinematic study of quick reactions [SHE 04]*

As the drop volume is always identical, it is possible to vary the composition of each drop with time if the microfluidic chip has multiple entries for aqueous solutions by changing the relative flow of these entries with time. This has been used for the remarkable determination of protein crystallization parameters by screening at high flow rates without knowing *a priori* the crystallization conditions in extremely small protein solution volumes: only a few microliters [ZHE 03]. This type of product is only starting to emerge industrially and be commercialized to biology laboratories.

Lastly, we can quote some very interesting studies which consist of manipulating drops individually by laser beams [BAR 07] to fractionate, fuse and orientate them in different secondary channels, or to stock them in a given location of the chip for future use, etc.

A very modest insight into drop microfluidics has been given here. From both scientific and industrial points of view, drop microfluidics has a very promising future as it brings together an increasing number of innovative functionalities for uses in biology and chemistry.

12.6. Conclusion

Microfluidics not only means decreasing the size of macroscopic fluidic systems to recreate homothetic flows at a smaller scale. Microfluidics offers new, innovative perspectives and these few introductory pages can clearly not cover all the research that has been done in this discipline. Readers will have been able to discover the physical fundamentals of microfluidic flow and there is an abundance of literature in this domain in which numerous applications are proposed for mainly chemistry and biology.

From a physics point of view, microfluidics is very close to electrical engineering, at least for the resolution of common problems and those related to classical Newtonian liquids. In fact, the distribution of flow rates in a complex microchannel system is similar to DC supplied electrical network problems. The distribution of velocities by electro-osmosis follows, via intuition, the distribution of the electric field in a network, i.e. the shortest path going from an entry to an exit in the microchip.

It is therefore not necessary to have done a degree in the field of fluid mechanics in order to understand the vast majority of the microfluidic systems put forward in literature. This is probably the reason why microfluidics is so popular in multidisciplinary teams. Microfluidics is still a relatively new subject area, hence the interest from larger and multidisciplinary teams; this will most probably lead to the emergence of attractive and innovative applications in the future.

12.7. Bibliography

[BAR 07] C.N. BAROUD, M.R. DE SAINT VINCENT, J-P. DELVILLE, "An optical toolbox for total control of droplet microfluidics", *Lab on Chip*, vol. 7, pp. 1029-1033, 2007.

[BRU 07] H. BRUUS, *Theoretical Microfluidics*, Oxford University Press, 2007.

[CAR 00] J. CARLSON, A. JAFFE, A.WILES, *The Millenium Prize Problems*, Clay Mathematics Institute, 2000.

[CHE 02] J.T. CHENG, N. GIORDANO, "Fluid flow through nanometer-scale channels", *Physical Review E*, vol. 65, pp. 1063-1065, 2002.

[KIR 04a] B.J. KIRBY, E.F. HASSELBRINK JR., "Zeta potential of microfluidic substrates: Part 1. Theory, experimental techniques, and effects on separations", *Electrophoresis*, vol. 25, pp. 187-202, 2004.

[KIR 04b] B.J. KIRBY, E.F. HASSELBRINK JR., "Zeta potential of microfluidic substrates: Part 2. Data for polymers", *Electrophoresis*, vol. 25, pp. 203-213, 2004.

[MAA 09] A. MAALI, Y. WANG, B. BHUSHAN, "Evidence of the no-slip boundary condition of water flow between hydrophilic surfaces using atomic force microscopy", *Langmuir*, vol. 25, no. 20, pp. 12002-12005, 2009.

[OUZ 94] R. OUZIAUX, J. PERRIER, *Mécanique des fluides appliquée*, Dunod, 1994.

[PAN 09] N.R. PANYALA, E.M. PEÑA-MÉNDEZ, J. HAVELL, "Gold and nano-gold in medicine: overview, toxicology and perspectives", *Journal of Applied Biomedicine*, vol. 7, pp. 75-91, 2009.

[REN 08] L. RENAUD, C. MALHAIRE, P. KLEIMANN, D. BARBIER, P. MORIN, "Theoretical and experimental studies of microflows in silicon microchannels", *Materials Science and Engineering C*, vol. 28, pp. 910-917, 2008.

[ROS 01] D. ROSS, T.J. JOHNSON, L.E. LOCASCIO, "Imaging of electroosmotic flow", *Analytical Chemistry*, vol. 73, pp. 2509-2515, 2001.

[SHE 04] I. SHESTOPALOV, J.D. TICE, R.F. ISMAGILOV, "Multi-step synthesis of nanoparticles performed on millisecond time scale in a microfluidic droplet-based system", *Lab On Chip*, vol. 4, pp. 316-321, 2004.

[SON 03] H. SONG, J.D. TICE, R.F. ISMAGILOV, "A microfluidic system for controlling reaction networks in time", *Angewandte Chemie International Edition*, vol. 42, no. 7, pp. 768-772, 2003.

[STE 07] K. STEPHAN, P. PITTET, L. RENAUD, P. KLEIMANN, P. MORIN, N. OUAINI, R. FERRIGNO, "Fast prototyping using a dry filmphotoresist: microfabrication of soft-lithography masters for microfluidic structures", *Journal of Micromechanics and Microengineering*, vol. 17, pp. N69-N74, 2007.

[THO 01] T. THORSEN, R.W. ROBERTS, F.H. ARNOLD, S.R. QUAKE, "Dynamic pattern formation in a vesicle-generating microfluidic device", *Physical Review Letters*, vol. 86, no. 18, pp. 4163-4166, 2001.

[ZEB 09] A. ZEBDA, L. RENAUD, M. CRETIN, F. PICHOT, C. INNOCENT, R. FERRIGNO, S. TINGRY, "A microfluidic glucose biofuel cell to generate micropower from enzymes at ambient temperature", *Electrochemistry Communications*, vol. 11, pp. 592-595, 2009.

[ZHE 03] B. ZHENG, L.S. ROACH, R.F. ISMAGILOV, "Screening of protein crystallization conditions on a microfluidic chip using nanoliter-size droplets", *Journal of the American Chemical Society*, vol. 125, pp. 11170-11171, 2003.

Chapter 13

Electrochemical Biosensors

13.1. Defining biosensors

Biosensors are defined as measuring devices that use biological bodies (enzymes, antibodies, antigens, cell tissues, RNA, DNA, etc.) to detect or quantify a specific analyte. According to IUPAC, the definition is limited to the use of biological bodies that have catalytic activity (enzymes): "a biosensor enables the continuous measurement of an analyte's concentration". This definition has now been extended to single-use biosensors (antibodies, antigens, DNA based, etc.). The analyte can be of chemical or biological origin.

13.2. Principle behind biosensors

Biosensors are created by combining a biological element (bioreceptor) with a transducer (detection mode) to ensure that the biological phenomenon (catalysis, recognition) is converted into an electrical signal (Figure 13.1). A conversion or amplification step may be necessary.

Chapter written by Chantal GONDRAN.

Transduction of the biological phenomenon can be electrochemical, optical, piezoelectric, thermic, etc. Only electrochemical transduction will be discussed in this chapter.

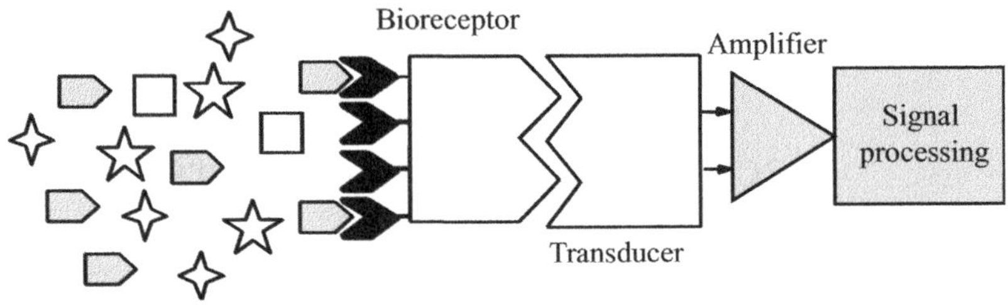

Figure 13.1. *Principle of biosensors*

13.3. Characteristics

Electrochemical biosensors can be compared to each other through different parameters: their sensitivity, response time, lifespan, etc., whether they are commercial or not. It is possible to use the biosensor correctly if the characteristics are known.

13.3.1. *Accuracy*

Measurements are accurate when the values obtained for several determinations are approximately the same as the real value, including error, and when the error is small.

The error on measurements is quantified by the value Δm: $m = \overline{m} \pm \Delta m$ where $\overline{m}$ is the average measurement. It can equally be expressed by the relative value $\frac{\Delta m}{\overline{m}}$ as a %.

13.3.2. *Repeatability*

In general, repeatability is the measuring device's ability to give similar responses to the repeated application of an entry signal. For biosensors this is the ability to give the same response for repeated measurements.

13.3.3. *Reproducibility*

Reproducibility is the ability of a series of sensors, made under the same conditions, to give the same results.

13.3.4. *Limit of detection*

The sensitivity range is indispensable for the use of biosensors. The limit of detection (lowest detectable value) is essential for analytical applications. This limit is set by the magnitude of the electrical background noise which interferes with the signal. This definition is only valid for linear response biosensors (for example, amperometric biosensors), background noise must not go over one third of the signal's magnitude (Figure 13.2).

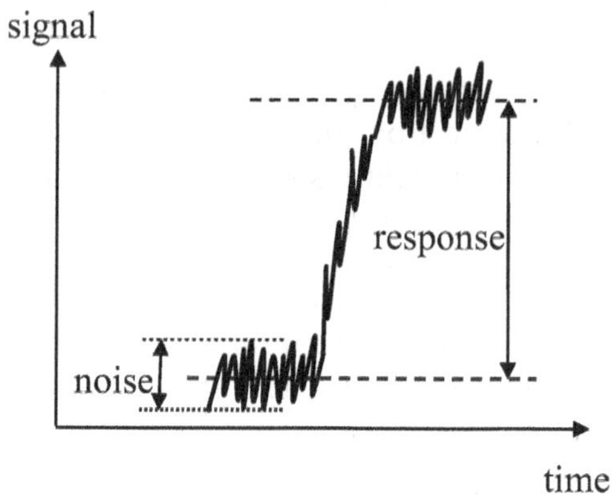

Figure 13.2. *Signal-to-noise ratio*

IUPAC has defined a limit of detection for potentiometric sensors as being the intersection of Nernstian and insensitivity asymptotes (Figure 13.3).

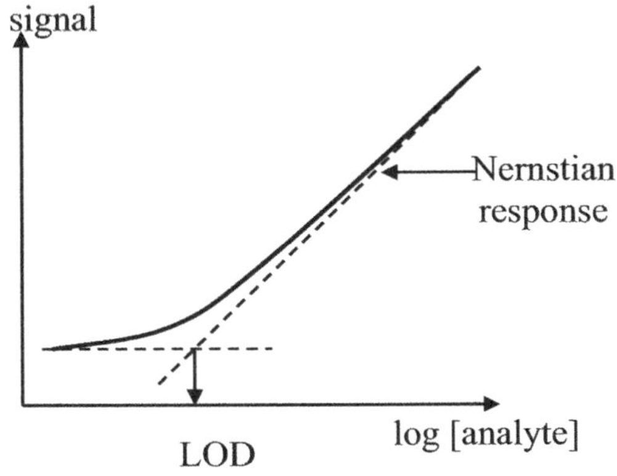

Figure 13.3. *Limit of detection for potentiometric biosensors*

13.3.5. *Sensitivity*

Sensitivity is linked to the variation in the signal's response with the variation in the analyte's concentration. It corresponds to the signal's derivative with respect to concentration.

If the function is linear then the sensitivity S is constant for all concentrations. A deviation from linearity (expressed in %) between the real response and the straight response line obtained can therefore be defined using the method of least squares, for example.

When the response obeys logarithmic laws, as is the case for potentiometric biosensors, a linearized function can be obtained using the logarithm of the concentration. The same deviation from linearity is therefore defined.

13.3.6. *Selectivity*

Selectivity is an essential characteristic of biosensors regardless of the transduction mode. A biosensor is not meant to be used in the presence of only one analyte. It is usually in contact with a mixture of species, which can provide an additional signal. These species are called interferants (Figure 13.4).

Selectivity is defined as a sensor's ability to recognize one species amongst others and to give an undisturbed quantitative signal.

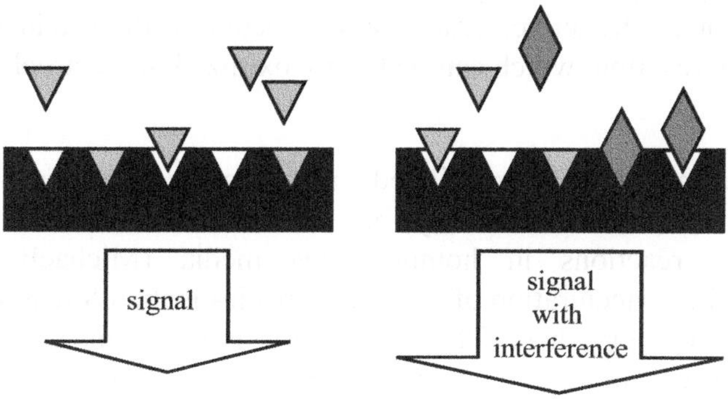

Figure 13.4. *Diagram of the interference phenomenon*

For enzymatic biosensors, another phenomenon occurs: inhibition. We talk of inhibitors when the signal is diminished by species which inhibit enzymatic reactions.

13.3.7. *Response time*

Response time is the time necessary for the sensor to deliver a proportion of the signal's full amplitude. The proportion needs to be chosen, for example 90 or 99%: written as t0.9 or t0.99. Response time varies incrementally in the analyte's concentration so that different response times are obtained for different measurement ranges.

13.4. Enzymatic biosensors

Electrochemical transduction is separated into three different types: amperometrics, potentiometry and conductometry (or impedance measurement). Examples of these three transduction methods are found in enzymatic biosensors even though the most widespread is amperometrics.

13.4.1. *Amperometric enzyme biosensors*

The target species (enzyme substrate) for enzymatic biosensors is usually not electroactive. Transduction occurs on the products of the enzymatic reaction, which can be highly oxidized or electrochemically reduced.

If the enzyme is immobilized and unconfined in a matrix, the reaction kinetics can initially be modeled in a similar way to enzymatic reactions in homogeneous media (Michaelis–Menten model). The concentration of the target species is therefore assumed to be the same everywhere. This assumption is relatively well-founded for biosensors used to analyze solution microfilms (blood droplet deposited on sensor). Nonetheless it can be fine-tuned by taking into account the diffusion of the target through the medium where the enzyme is immobilized (kinetic model limiting support layer).

13.4.1.1. *Simplified kinetic model*

There are several phases in enzymatic reactions. The first can be represented by equilibrium between enzyme, substrate and the enzyme-substrate complex. The enzyme is then regenerated by a reagent called the co-substrate, R_{CS}, which leads to the generation of two products: one originating from the transformation of the substrate, and the other from the transformation of the co-substrate. Generally, one of the two products is electroactive in amperometric biosensors. The reaction's global kinetic progress, in the case of enzymes in the Michaelis–Menten model, is limited to the ES complex concentration. Therefore the reaction scheme is:

$$E + S \leftrightarrow ES \quad K = k_{-1}/k_1 = [E][S]/[ES]$$

$$ES + R_{CS} \rightarrow E + P_1 + P_2 \text{ slow phase } v = k_2\,[ES]$$

where k_1, k_{-1} are the kinetic constants of the first reactions (direct and inverse), and k_2 is that of the second reaction, considered non-reversible.

Analysis of one of the reaction products (P_1 or P_2) is carried out according to amperometric transduction:

$$P \rightarrow P' + z\,e^- \quad \text{or} \quad P + z\,e^- \rightarrow P'$$

In the Michaelis–Menten model, the co-substrate is not generally mentioned as the enzymatic reaction is negligible in comparison to the reagent (constant concentration during reaction). This means that the $[R_{CS}]$ is therefore included in the constant k_2.

If $[E]_{tot}$ in stationary operation is the total enzyme concentration, the progress velocity can be shown to be:

$$v = \frac{v_{max}[S]}{K_M + [S]} \quad \text{where} \quad v_{max} = k_2[E]_{tot}$$

The Michaelis constant K_M is a function of the kinetic constants and is given by:

$$K_M = \frac{k_{-1} + k_2}{k_1}$$

At the electrode, P is used as it is being produced and a limit operation is reached where the electrode's concentration [P] becomes zero. The limit current I, is then proportional to the velocity v:

$$I = \alpha_{corr}\, zFA \frac{v_{max}[S]}{K_M + [S]}$$

where z is the number of electrons, F is Faraday's constant and A is the electrode's surface area. The correction factor α_{corr} is included for several reasons: firstly, P species diffuse in all three directions and not only towards the electrode; enzyme activity can be modified by its conformation to the electrode or film texture. K_M obtained through the analysis of biosensors' amperometric curves is called apparent K_M as it differs greatly from that of free enzymes in the solution (Figure 13.5).

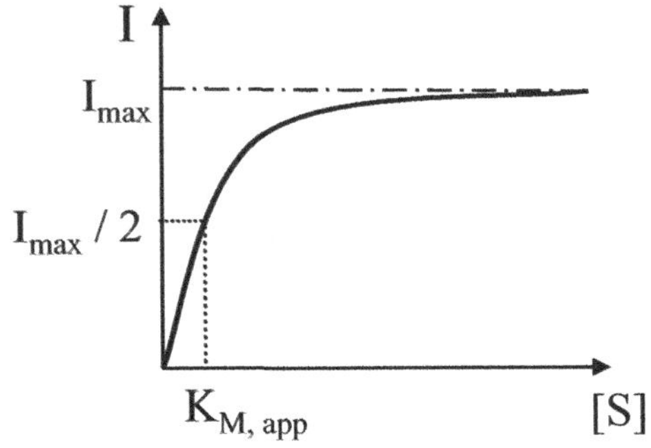

Figure 13.5. *Enzymatic biosensor's current response to enzyme's substrate concentration*

From this it is apparent that if [S] remains small compared to K_M, the response becomes linear. Sensitivity of the response is therefore given by a line going through the origin such that:

$$S = \alpha_{corr} zFA \frac{v_{max}}{K_M}$$

It is a function of the apparent Michaelis constant and is proportional to v_{max}, i.e. the total enzyme concentration.

13.4.1.2. *Kinetic model with limiting support layer*

When an immobilization matrix is used that is sufficiently permeable, the previous model is acceptable as long as concentrations in the solution and in the matrix are identical.

In some cases, the kinetic model should be completed to take into account diffusion inside the layer, to enzyme contact (Figure 13.5).

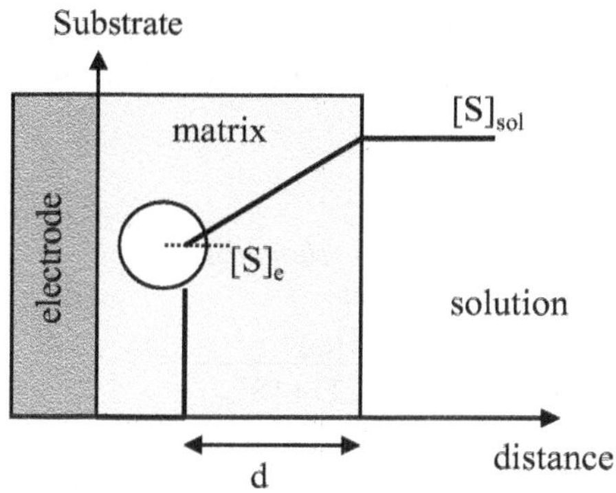

Figure 13.6. *Diagram of the concentration profile for an enzymatic biosensor with limiting diffusion in the matrix*

For stationary operation, the mass balance between the S input through diffusion in a thickness d (to the enzyme) and the consumption by enzymatic reactions is zero, leading to the following relationship:

$$\frac{d[S]_e}{dt} = \frac{D}{d}([S]_{sol} - [S]_e) - k_1[S]_e[E] + k_{-1}[ES] = 0$$

In this relationship, the partition coefficient between the membrane and the solution is assumed to be 1 for simplicity.

With:

$$[E] = [E]_{tot} - [ES]$$

$$k_1[S]_e[E] - (k_{-1} + k_2)[ES] = 0 \quad \text{(stationary ES concentration)}$$

The following relationship is obtained:

$$k_1k_2[ES]^2 - \left[k_1k_2[E]_{tot} + k_1\frac{D}{d}[S]_{sol} + (k_{-1} + k_2)\frac{D}{d}\right][ES] + k_1\frac{D}{d}[S]_{sol}[E]_{tot} = 0$$

The analytic law for [ES] is complex. However if there is a high enzyme concentration and the condition $[ES] << [E]_{tot}$ is assumed, then:

$$v = \frac{v_{max}[S]_{sol}}{K_M + v_{max}d/D + [S]_{sol}} \textit{ therefore}$$

$$I = \alpha_{corr} zFA \frac{v_{max}[S]_{sol}}{K_M + v_{max}d/D + [S]_{sol}}$$

For low values of $[S]_{sol}$, the linear part gives an equation for sensitivity (line through the origin) equal to:

$$S = \alpha_{corr} zFA \frac{v_{max}}{K_M + v_{max}d/D}$$

This shows that sensitivity decreases due to the presence of the term $v_{max}d/D$.

When $v_{max}d/D << K_M$, the sensitivity of the previous simplified model is obtained.

$$S = \alpha_{corr} zFA \frac{v_{max}}{K_M}$$

On the other hand, if $v_{max}d/D >> K_M$, sensitivity is given by:

$$S = \alpha_{corr} zFA \frac{D}{d}$$

For this last case, sensitivity no longer depends on enzyme concentration $[E]_{tot}$, but on the membrane parameters that limit diffusion: thickness of the matrix to the enzyme and the analyte's diffusion coefficient in the matrix. As has been shown, the sensitivity is reduced in this model. However linearity is more extensive.

13.4.1.3. *Inhibitors*

Enzymatic sensors are sensitive to species which poison them, called inhibitors. Inhibitors latch onto reaction sites thus making enzymes inactive.

Just like substrates, inhibitors form a complex with the enzyme but do not create reaction products. The sensitivity is therefore diminished and is reversible if the constant $K_I = [E]\,[I]/\,[EI]$ is not too low.

For competitive inhibition, the measured current equation becomes:

$$I = \alpha_{corr} zFA \frac{v_{max}[S]}{K_M(1+[I]/K_I)+[S]}$$

The law here is no longer linear with respect to [S] if $[I]/K_I >> 1$, in particular when K_I is very low. The sensitivity decreases and depends on the inhibitor's concentration. This leads to a default error on the substrate concentration.

The effect of inhibition can be used advantageously in amperometric biosensors by fixing a constant concentration of the enzyme's main substrate S, in the analyzed solution and by checking the effect of the inhibitor concentration I, on the current.

For competitive inhibition, the response decreases hyperbolically (shown below): the higher the inhibitor concentration [I], the lower the measured current becomes. It is convention to take the inverse of the current, $I^{-1} = f([I])$, to obtain a linear function for the inhibitor concentration.

$$\frac{1}{I} = \frac{1}{\alpha_{corr} zFAv_{max}} + \frac{K_M}{\alpha_{corr} zFAv_{max}[S]} + \frac{K_M[I]}{K_I \alpha_{corr} zFAv_{max}[S]}$$

This inverse current equation does not go through the origin and the sensitivity obtained increases as the surplus [S] decreases.

When inhibition is said to be uncompetitive, the following current equation is obtained:

$$I = \alpha_{corr} zFA \frac{v_{max}[S]}{K_M + [S](1 + [I]/K'_I)}$$

By taking the inverse function, I^{-1}, it becomes apparent that the sensitivity observed does not depend on the concentration [S].

In the case of non-competitive or mixed inhibition, as there are several inhibition reactions taking place in parallel, the equation is more complex:

$$I = \alpha_{corr} zFA \frac{v_{max}[S]}{K_M(1 + [I]/K_I) + [S](1 + [I]/K'_I)}$$

The inverse function is still linear and the gradient depends on [S].

13.4.2. *Potentiometric enzymatic biosensors*

The basic principle consists of keeping an enzyme within a matrix on the surface of the potentiometric sensor (ISE), which is sensitive to one of the enzyme reaction products. The principle of potentiometric biosensors relies on a stationary state between what is produced (and/or used) in the medium where the enzyme is kept, and the outside exchange, governed, for example, by diffusion. After a transitory period which depends on reaction speed constants and outside exchange coefficients (diffusion coefficients), the product concentration P therefore remains constant during measurement (Figure 13.7).

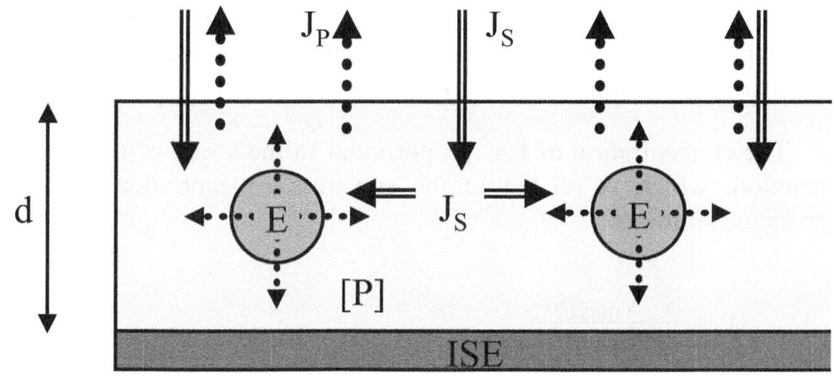

Figure 13.7. *Diagram of substrate diffusion flows and enzyme products in a matrix of thickness d*

The production (or consumption) of species P by the reaction is given by:

$$\frac{d[P]}{dt} = \frac{v}{Ad}$$

where A is the film's surface area and d its thickness. The input (or output) by diffusion can also be written if the concentration of one medium is negligible in comparison to that of the other:

$$\left|\frac{d[P]}{dt}\right| = A\ D_P\ \text{grad}[P] \approx D_P \frac{[P]}{d} A$$

where d is the diffusion thickness in the media (which overlaps the matrix thickness) and D_P is P's diffusion coefficient. Balance gives the following expression:

$$[P] = v/\ k_P$$

where parameter k_P is the coefficient for kinetic displacement and exchange of species P between the inside of the matrix and the outside (it includes the matrix's geometric parameters and coefficient for the diffusion of P within the matrix).

The concentration of P is proportional to the speed of the enzyme reaction, which is related to the substrate concentration S in the analyzed solution:

$$[P] = \frac{1}{k_P} \frac{v_{max}[S]}{K_M + [S]}$$

As long as there is an ISE which reacts to P, it is possible to create an assembly for a microsensor which gives a logarithmic response to the concentration of substrate S:

$$E = E^\circ + \frac{RT}{zF} \ln[P]$$

$$E = E^\circ + \frac{RT}{zF} \ln \frac{v_{max}}{k_P\ K_M} + \frac{RT}{zF} \ln[S] = E'^\circ + \frac{RT}{zF} \ln[S]$$

As the constant k_P only depends on geometric parameters, E'° needs to be determined by calibrating each sensor.

This type of potentiometric device is well-adapted to miniaturization techniques, particularly for the creation of field effect sensors (ENFET).

13.4.3. *Conductimetric enzymatic biosensors*

Conductometric, or impedance measuring, biosensors require enzymes that use and/or produce ionic species which have the effect of altering the medium's conductivity.

In stationary operation, the enzyme production of product P is compensated by the losses caused by diffusion out of the matrix, as in the case of potentiometric sensors. The conductivity of the medium where the enzyme is confined depends on the concentration of mobile species and is linked to the enzymes reaction speed, therefore to the substrate concentration.

For enzyme sensors, the enzyme is contained in the matrices placed on inert supports composed of two in-plane electrodes. In a coplanar system, the measured resistance is approximately the same as that of the film, depending on its thickness, placed in series with the resistance of the solution in the volume beyond these two electrodes (Figure 13.8).

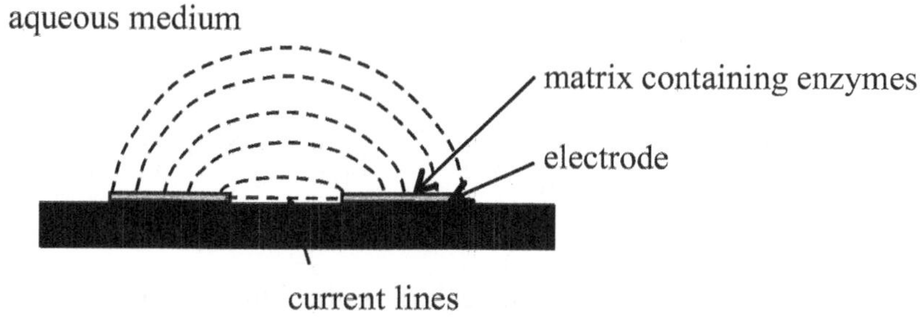

Figure 13.8. *Diagram of the current lines' passing zones for a conductimetric enzyme biosensor*

An example of this is the dosing of glutamate using glutamate-oxidase (GLOD). In aqueous solution, glutamic acid, which is weak, partially dissociates itself depending on the reaction:

$$C_5H_9O_4N \Leftrightarrow C_5H_8O_4N^- + H^+$$

When GLOD is present, oxo-2-glutarate is obtained:

$$C_5H_8O_4N^- + 1/2\ O_2 \rightarrow C_5H_4O_5^{2-} + NH_4^+$$

As the latter is a complete reaction, equilibrium shifts to the right and the acid is entirely used. The global balance of the reaction is therefore given by:

$$C_5H_9O_4N + 1/2\ O_2 \rightarrow C_5H_4O_5^{2-} + NH_4^+ + H^+$$

The initial conductivity is set by ions originating from the weak acid's dissociation ($pK_a = 4.3$), i.e. $C_5H_8O_4N^-$ and H^+, which are low in concentration if that of the acid is high, or similar if diluted. As the enzyme reaction is complete, H^+, NH_4^+ and $C_5H_4O_5^{2-}$ are produced in concentrations similar to that of the acid. The reaction therefore increases the initial conductivity.

The preservation of each ionic species depends on the reaction speed, and the relationship can be expressed as:

$$\left[C_5H_4O_5^{2-}\right]=\frac{v}{k_{C_5H_4O_5^{2-}}} \quad \left[H^+\right]=\frac{v}{k_{H^+}} \quad \left[NH_4^+\right]=\frac{v}{k_{NH_4^+}}$$

where parameter k_c (for example k_H) is the kinetic coefficient for displacement and exchange between the inside and outside of the matrix. We assume that the initial proton and glutamic acid concentration in these equations is negligible. The film's conductivity therefore equals:

$$\sigma=\sigma^\circ+\frac{F^2}{RT}\left(4D_{C_5H_4O_5^{2-}}\frac{v}{k_{C_5H_4O_5^{2-}}}+D_H\frac{v}{k_H}+D_{NH_4^+}\frac{v}{k_{NH_4^+}}\right)$$

where σ° is the initial value and u is the electrochemical mobility of each ion. This equation can also be written in the form:

$$\sigma=\sigma^\circ+\alpha\frac{v_{max}[S]}{K_M+[S]}$$

where α is the proportionality coefficient, including the diffusion coefficient of ionic species, geometric parameters, etc.

The film of thickness d and surface area A, has a resistance of:

$$R = \frac{d}{A} \frac{K_M + [S]}{\sigma^\circ K_M + [S](\sigma^\circ + \alpha v_{max})}$$

It is more advantageous to use conductance (1/R) here rather than resistance when exploiting data (Figure 13.9a):

$$\frac{1}{R} = \frac{A}{d} \frac{\sigma^\circ K_M + [S](\sigma^\circ + \alpha v_{max})}{K_M + [S]}$$

The limit values are: $\frac{1}{R_0} = \frac{A}{d}\sigma^\circ$ and $\frac{1}{R_{max}} = \frac{A}{d}\left(\sigma^\circ + \alpha\, v_{max}\right)$

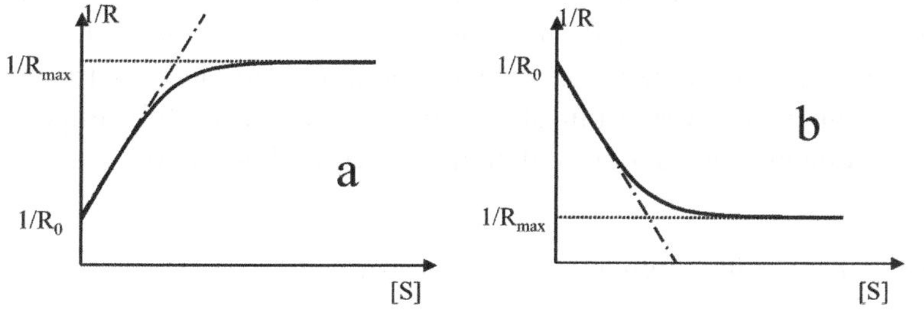

Figure 13.9. *Idealized conductimetric sensor response curves for: a) an increase; b) a decrease of conductance during enzyme reaction*

Figure 13.9 shows that if [S] is small compared to K_M, the response becomes linear.

$$\frac{1}{R} = \frac{A}{d}\left[\sigma^\circ + \frac{[S](\sigma^\circ + \alpha\, v_{max})}{K_M}\right]$$

It becomes clear that, for this representation, sensitivity depends on the species' diffusion coefficients, geometrical parameters and also v_{max}, which is proportional to the enzyme concentration in the matrix.

If ions are used in the enzyme reactions, the same relation will be true but will have a negative sign before αv_{max} as conductance will decrease with the substrate concentration (Figure 13.9b):

$$\frac{1}{R}=\frac{A}{d}\frac{\sigma^{\circ}K_M+[S](\sigma^{\circ}-\alpha\, v_{max})}{K_M+[S]}$$

13.5. Affinity sensors

DNA sensors and Immunosensors are called affinity biosensors. Their principle is based on the recognition phenomenon which happens between biomolecules (DNA/DNO, antibody/antigen, etc.). One of the two biomolecules is immobilized on the transducer and enables the recognition of the second biomolecule present in the sample. Electrochemical transduction methods are the most widespread and impedance measurement (or conductometry if the measurement is carried out at one frequency only). Only impedance measurement transduction will be discussed in this section.

13.5.1. *Use of blocking electrodes (non-faradic)*

Biodetection can be used by measuring the impedance (resistance or capacitance) of the space between the electrodes. The recognition phenomenon modifies the biofilm's electric and dielectric parameters (Figure 13.10). The immobilized species is called the probe and the species being analyzed is the target. Recognition happens according to the following equilibrium:

target + probe ⇔ target-probe complex

where the equilibrium constant is usually very large.

The more targets the analyzed solution contains, the more the space above the electrodes is covered by the target. This causes the biofilm's electric and/or dielectric properties to change.

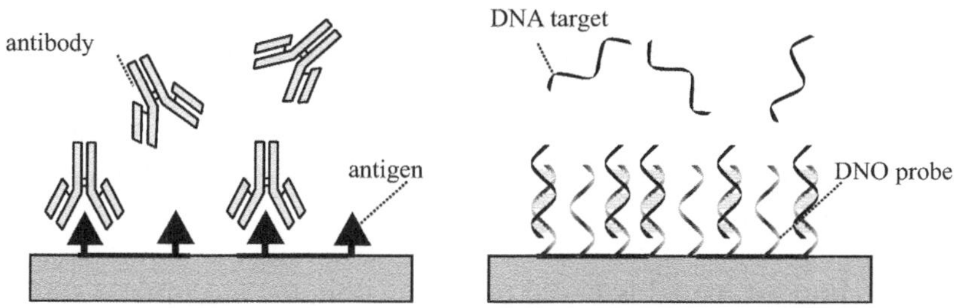

Figure 13.10. *Detection principle of affinity biosensors: a) antibody/antigen; b) DNA/DNO*

The detection method is still complex and is only used by impedance measurement specialists. It requires knowledge of interface electric modeling methods to deconvolute impedance spectra and identify the parameters influenced by recognition.

In order to trace the corresponding impedance diagram, impedance will be represented in a Nyquist-type diagram in the case of the biofilm's resistance (electric properties) being highly modified by recognition; on the other hand, if it is the capacitance (dielectric properties) that is highly modified, a Cole–Cole representation is preferred.

If only one frequency is used (conductometric measurements), a measurement frequency that is high enough to prevent the interference of the double layer capacity placed at the solution's interface should be chosen.

13.5.2. *Use of faradic electrodes (electrochemical probes)*

This type of sensor is based on the properties of one species, or of a redox couple present in the analyzed solution. The species (or couple) is called a redox-probe. It does not intervene in the recognition reaction. Usually, a ferricyanide/ferrocyanide couple, $K_3Fe(CN)_6/K_4Fe(CN)_6$, is used as the electronic transfer of the electrode's reaction is well defined in aqueous solution, and has good reproducibility.

When recognition occurs, crowding greatly alters the electrode's reaction kinetics. This is an indirect measurement as, unlike blocking electrodes, the electrochemical properties of the probe in the biofilm, and not the biofilm itself, are observed. The choice of a redox-probe is therefore critical.

The simplest model is based on passivating the electrode's surface with an overlay. For a given reaction, the redox-probe's transfer resistance is constant on the interface, the value R_{tr} varies similar to inverse functions of the electrode's available surface area for reactions. The surface area variation can be expressed as:

$$A = A°(1 - \alpha[\text{target}])$$

where coefficient α is a constant combining the recognition reaction's equilibrium constant and the covering factor of the species masking the electrode. Therefore the equation for resistance variation is:

$$R_{tr} = \frac{R_{tr}^0}{1-\alpha[\text{target}]}$$

where $R_{tr}°$ is the biosensor's transfer resistance before target recognition.

The Nyquist impedance diagram obtained is shown in Figure 13.11. It consists of a semicircle of radius equal to the probe's charge transfer resistance, and a line at 45°, called Warburg, which expresses the diffusion phenomenon from the probe to the electrode.

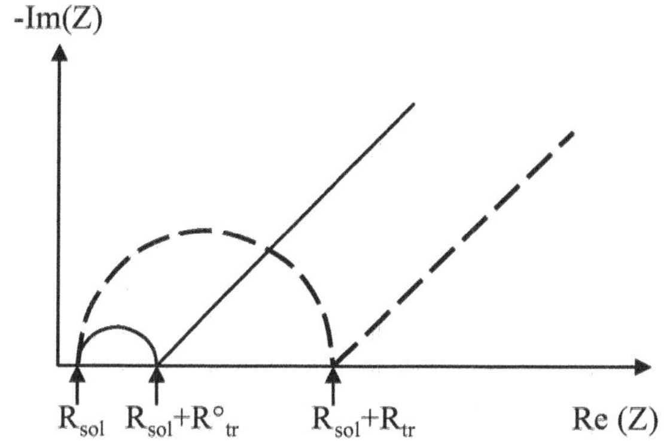

Figure 13.11. *Nyquist diagram for an affinity biosensor, in the presence of a redox-probe. R_{sol}: solution resistance; R_{tr}: biofilm transfer resistance before target recognition (solid line); R_{tr}: after target recognition (dotted line)*

If only one frequency is used (conductometric measurements), it must be in the low frequency range, just before the Warburg line. In these conditions the variation of impedance will best represent the redox-probe's charge transfer resistance.

13.6. Conclusion

There are numerous examples of commercial electrochemical transduction biosensors, whether amperometric, potentiometric or conductometric. Amperometric remains the most commonly used method.

Electrochemical transduction has several different advantages. The first is great selectivity: if we consider the amperometric method, the choice in working potential allows the selection of the redox reactions put into play. Moreover, electrochemical biosensors are compact and transportable; biosensors are measuring devices that are simple and can be miniaturized. Lastly, the cost of biosensors is low. Commercialization of microsensors is only feasible if the cost remains low.

Chapter 14

Fiber-optic Biosensors

14.1. Introduction

By definition, fiber-optic biosensors consist of a biomolecule or biological element that allows the recognition of the species to be measured, and a fiber-optic that acts as the transducer [BOR 08, WOL 08]. The biological element is essential for recognition properties and therefore for the biosensor's selectivity. The modified fiber-optic acts as a transducer by converting the recognition phase into an optic signal that is transported by the fiber. These biosensors can be miniaturized and used in portable devices. They enable remote measurements with a quick response time and are often minimally invasive. This last point is an important fundamental criterion for biomedical applications

In addition, measurements are sensitive and generally do not damage the sample. However, optic sensors can be sensitive to ambient light, depending on the detection method, which has the result of limiting their analytic performance. Numerous applications in a wide range of fields have been developed to meet society's expectations with regard to the environment, medicine and chemical

---

Chapter written by Neso SOJIC.

bioanalysis. To name a few examples, these can be the detection of pesticides, viruses, bacteria, hormones or cancer cells [MON 04].

Optic biosensors are therefore an active research area, as is testified by the vast number of publications (>1,500 in the last decade) and the commercial applications that have been derived. The subject is vast and it is not possible to discuss all aspects of it here. This chapter will limit itself to the main classes of fiber-optic biosensors and will highlight recent developments related to miniaturization, biosensor arrays and multiplexing (i.e. simultaneous detection of more than one analyte), which are the main driving forces for current research in this area. These three tendencies are in line with current requirements and are the answer to the need for analyzing complex samples, and simultaneously detecting several analytes. A large variety of applications based on optical arrays have been developed in several fields, from diagnostics to DNA biochips.

Each sensitive element (sensor, functionalized particle, cell, etc.) is spatially addressed. The creation of such miniaturized arrays benefits from progress achieved in micro/nanofabrication; this considerably increases their density and therefore the final number of analytes measured. For example, the fiber-optic detection method, which uses a CDD camera, allows information from thousands and tens of thousands of sensing elements to be collected and read very simply without the need to connect them individually.

However, remarkable progress in nanotechnology has meant that it is now possible to create an array of objects with sub-micrometric steps. The resolution limit related to light diffraction (around $\lambda/2$) is therefore an intrinsic barrier for global reading via optic paths, if one does not consider near-field scanning optical microscopy (NSOM). This field of study requires considerable instrumentation and expertise that are rarely compatible with biosensors at present.

Recognition and transduction recognition functions are closely linked and control the sensor's final performance. It is therefore essential to consider both these aspects at the design stage of new optic biosensors. For clarification, potentiometric pH electrodes and CO_2 optical sensors are not biosensors.

The same is true for molecular probes which are sometimes wrongly qualified as biosensors. Even if they are of high interest to a number of scientific domains, their relative operating principles are not based on biomolecules or biological elements, which by definition exclude them from being a biosensor.

Section 14.2 will briefly present biomolecules, or biological elements, used in the field of optical biosensors and the resulting classification. In section 14.3 we will then discuss the different types of fiber-optics used and their fundamental properties. Section 14.4 will present several applications based on different detection principles that illustrate the current changes in this multidisciplinary field.

14.2. Biomolecules and recognition

Biological elements are an essential part of biosensors. They are diverse and range from enzymes to living cells, including antibodies or DNA. One of their properties that is particularly noteworthy is that they are extremely specific.

These biological elements are immobilized on the surface of the optic fiber so that the biosensor may recognize the target to be detected. They are generally common, or close to those used in electrochemical biosensors and their description will be very brief. Given the operating mode of the biological components, biosensors can be separated into two main categories:

– *Catalytic biosensors*: mainly enzymes, as they have remarkable properties for catalysis and specificity for a given substrate or group of substrates. Living cells are also used to make biosensors but will provide a global response such as toxicity or cell viability.

– *Affinity biosensors*: based on recognition and the specific bond between the analyte and the biological element. The latter can be diverse in nature. DNA biosensors are based on recognition between two complementary oligonucleotide sequences. The specific interactions between antibodies and antigens are exploited to develop immunoassays and immunosensors. The last type of biosensor is based on the use of biological receptors.

14.3. Fiber-optics

Fiber-optic biosensors have benefited from the discoveries made in the field of fiber-optics since 1966, when low-loss fiber-optics were produced. Since then, biosensors have exploited the progressive developments of new fiber-optics even when the primary motive in this domain is telecommunications.

14.3.1. *Individual fiber-optics*

Fiber-optics are circular waveguides traditionally made from silica or plastic. Fibers consist of a core of index n_c surrounded by an outer cladding of index n_g. As the core index is higher than the fiber cladding index, light is guided along the core by total internal reflection. For glass fibers, the small index difference is obtained by doping (germanium, erbium, fluorine, etc.) the silica.

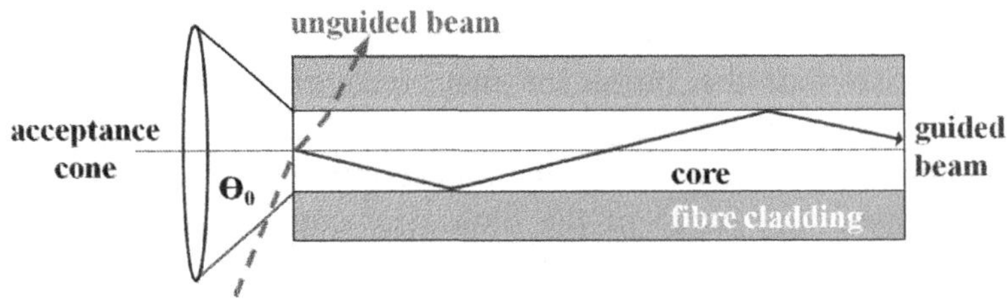

Figure 14.1. *Diagram of step-index fiber-optic*

The index difference between the core and the fiber cladding defines the acceptance cone (Figure 14.1). Light beams arriving on the fiber-optic's surface, with an incidence smaller than the limit angle Θ_0, and therefore in the acceptance cone, are guided through the fiber to its exit. Light beams arriving with an incident angle greater than Θ_0 will not be transmitted through the fiber. The value of the limit angle is linked directly to the numerical aperture and depends on the following equation:

$$sin\,\theta_0 = \frac{\left(n_c^2 - n_g^2\right)^{1/2}}{n_0} \qquad [14.1]$$

with n_0: index of air.

14.3.2. *Fiber-optic bundles*

Fiber-optic bundles consist of several thousand individual fiber-optics (or cores). Each core is surrounded by fiber cladding in order to transmit light by total internal reflection. These fibers are produced by hot-stretching an initial pre-form, as shown in Figure 14.2a.

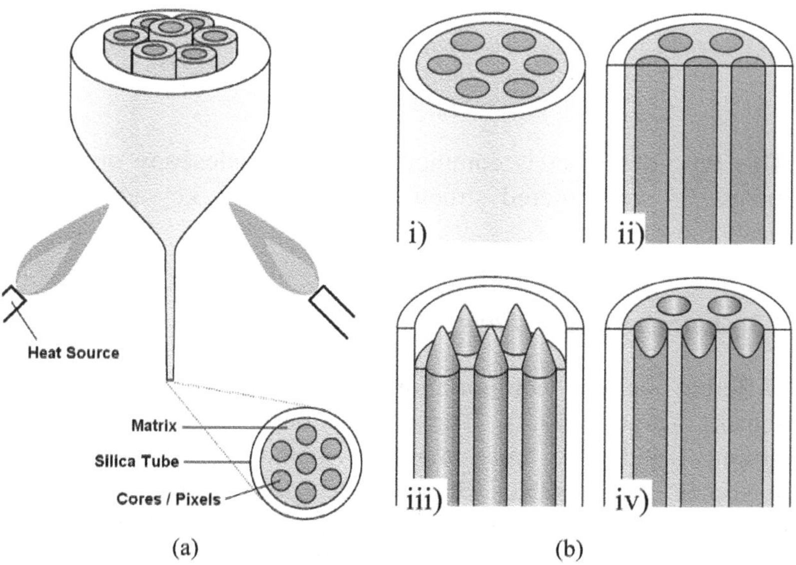

Figure 14.2. *a) Diagram of current methods for production of bundles. b) Simplified views: (i-ii) resulting bundle; tip (iii) or cavity (iv) arrays produced by selective chemical etching [DEI 10]*

When the fibers are arranged to go from one end of the bundle to the other, photons are coherently guided through each core and an imagig can be transmitted via the bundle. This is referred to as coherent bundles or image bundles. The spatial resolution depends on the distance between two adjacent cores. If the core diameters increase, a pixelization effect appears progressively. However a micrometric resolution can be reached with optimized fibers where the core radius is 1.5 µm. Figure 14.3 shows an image from such a bundle, with 6,000 cores, keeping in mind that the number of cores can reach hundreds of thousands.

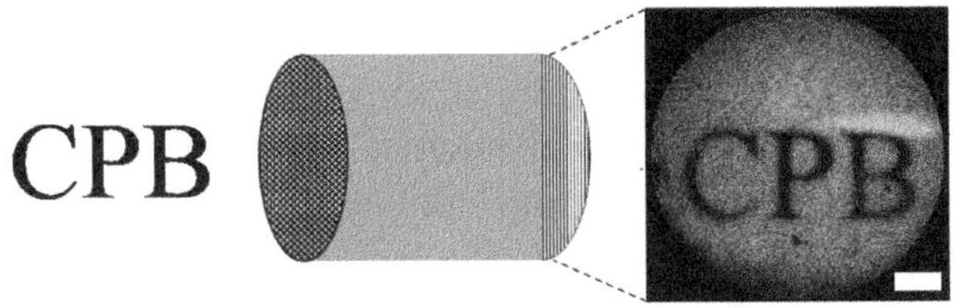

Figure 14.3. *Diagram on the imaging properties of a bundle with 6,000 individual fiber-optics. The global bundle diameter is 270 µm*

This type of object is commonly used in endoscopy or as optic elements. Its well-ordered structure constitutes a key-advantage to fabricate biosensors.

14.3.3. *Optical configurations*

Different optical configurations are used in the design of biosensors (Figure 14.4). The choice depends on experimental constraints and the measurement method used (absorption, fluorescence, time-resolved fluorescence spectroscopy, fluorescence resonance energy transfer or FRET, chemiluminescence, electrochemiluminescence, Raman spectroscopy, surface plasmon resistance or SPR, evanescent wave spectroscopy, etc.). However, fluorescence is the most commonly used method due to its intrinsic qualities for sensitivity and ease of implementation.

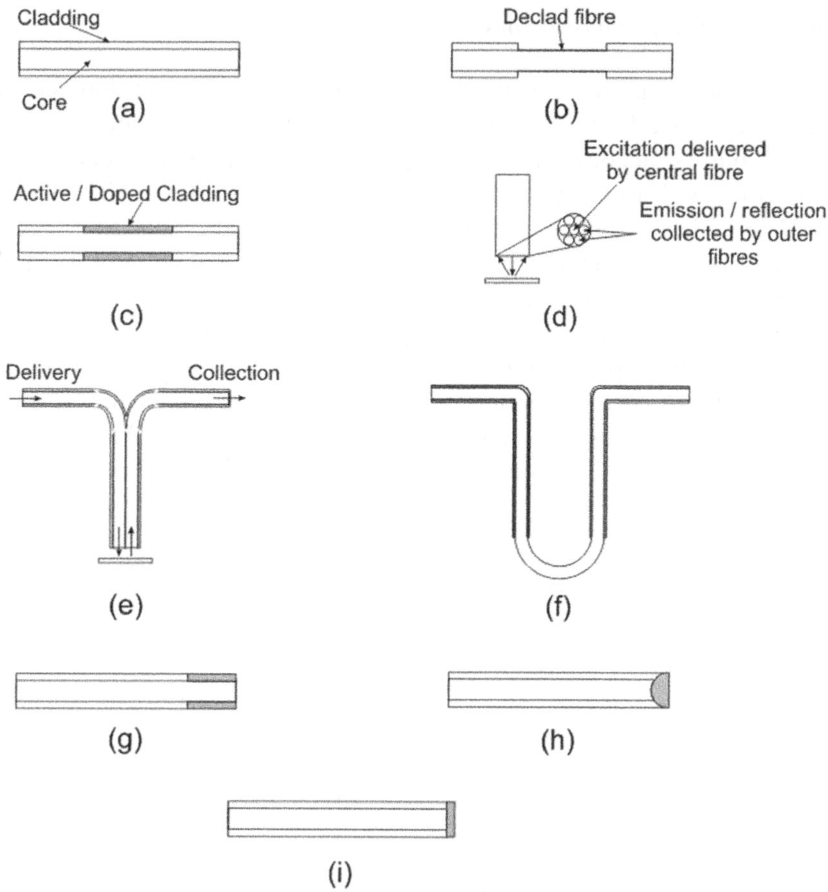

Figure 14.4. *Optical configuration traditionally used for fiber-optic biosensors: a) unmodified; b) stripped fiber cladding; c) doped fiber cladding; d) fiber bundle; e) double fiber for excitation and collection; f) U-shaped assembly; g-i) modified distal face with different geometries [MCD 08]*

14.4. Fiber-optic sensor applications

A great variety of fiber-optic biosensors have been developed using a range of detection schemes and measurement methods. The following sections will explore some typical examples.

14.4.1. *Enzymatic biosensors*

Enzymes catalyze reactions by recognizing substrate molecules with great specificity. The reaction does not directly induce a light signal but is accompanied by the consumption or production of molecules with light molecular weights, such as O_2, CO_2, NH_3, H_2O_2 or H_3O^+. These molecules interact with an optically sensitive layer. Their optic measurement can be linked to the enzyme activity and therefore to the substrate concentration. Figure 14.5 represents the outline of a simple and commonly used scheme for the design of enzymatic biosensors combined with a fiber-optic. In epifluorescent configurations, the excitation wavelength inserted into the fiber's nearest face excites the sensitive layer's fluorescence. Part of the emitted fluorescence is intercepted and guided by the fiber-optic to the photodetector.

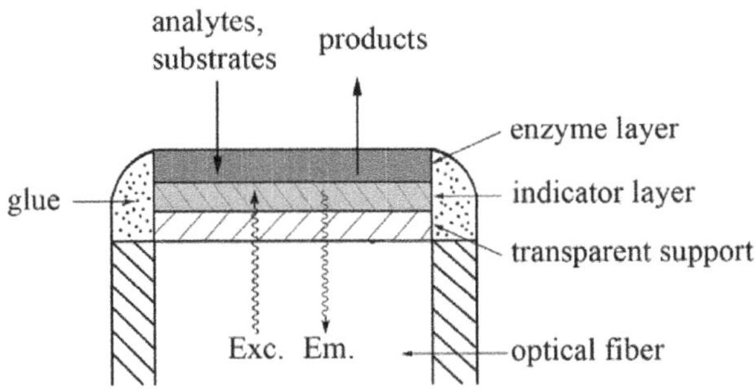

Figure 14.5. *Cross-section of an enzymatic biosensor [BOR 08, KUS 01]*

Amongst the vast number of detected compounds (lactate, cholesterol, glutamate, xanthin, penicillin, urea, ethanol, phenols, etc.), glucose is the model analyte that is measured most. The situation is similar to that of electrochemical biosensors. The measurement of glucose uses glucose oxidase which frequently catalyses β-D-glucose (reaction 1). The product D-gluconolactone then hydrolyzes to form oxidation D-gluconate (reaction 2).

$$\beta\text{-D-glucose} + O_2 \rightarrow \text{D-gluconolactone} + H_2O_2 \text{ (reaction 1)}$$

$$\text{D-gluconolactone} + H_2O \rightarrow \text{D-gluconate} + H^+ \text{ (reaction 2)}$$

Several detection paths are therefore possible:

1) Dioxygen is consumed during the enzymatic reaction. By following the decrease in O_2 with an O_2 optic sensor, it is possible to measure β-D-glucose levels indirectly. Measurement of O_2 takes place on the sensitive layer where fluorescent molecules (e.g. di-phenyl-anthracene, deca-cyclene) are immobilized. This molecule is chosen for its ability to undergo efficient quenching by dioxygen. Kopelman *et al.* have described a production of optic microsensors based on this principle that uses a ruthenium complex with phenanthroline ligands; the enzyme is immobilized in acrylamide gel via photopolymerization [ROS 96, ROS 95]. The sensor's tip diameter is 2 μm and the response time is 2 s.

2) Hydrogen peroxide, resulting from the reduction of dioxygen in enzyme reactions, is another possibility for measuring glucose. The concentration of the product is directly proportional to the enzyme activity, and therefore to the β-D-glucose concentration. Hydrogen peroxide can be measured by chemiluminescence as it reacts with luminol in the presence of a catalyst such as horseradish peroxidase (HRP). The light intensity resulting from the chemiluminescent reaction is measured via fiber-optic and gives an indirect measurement of the β-D-glucose concentration.

3) Glucose concentration can be measured by following the pH variations due to reaction 2. In this case, the sensitive layer associated with the enzyme is a proton-permeable hydrogel where pH indicators such as 8-hydroxypyrene-1,3,6-trisulfonate (pKa=7.3) are immobilized.

4) Wolfbeis *et al.* have put forward a method based on the intrinsic fluorescence of glucose oxidase, which contains a flavine adenine dinucleotide (FAD) cofactor [GRA 92]. The oxidized state fluoresces at 530 nm whereas the reduced state has a different spectrum. However, the scope for measurement is quite limited

These examples illustrate the possibilities of glucose detection. Other analytes can be detected using the same methods by changing the oxidase used, or more generally, the enzyme used.

14.4.2. *Immunosensors*

14.4.2.1. *Evanescent waves*

As described in section 14.3.1, light propagation in fibers occurs by total internal reflection. A non-propagative wave on the other hand, only exists at the core/cladding interface. This evanescent wave's amplitude decreases exponentially with distance the further it is from the core. Through localized stripping of the cladding, it is possible to immobilize antibodies on the stripped part of the wave guide. After incubation with the sample to be analyzed, the fiber is put into contact with detection antibodies that carry fluorescent markers, for example. If the sought antigen is absent from the sample, the marked antibody remains in the solution and only induces a weak signal. However, if the sought antigen is present, a sandwich is formed on the core's surface, within the zone sampled by the evanescent wave. If the marker is fluorescent, the evanescent wave excites fluorescence and is transmitted by the fiber. The marker can also be an absorbing molecule, in which case the intensity of guided light is decreased. This fluorescent marker approach is used to detect bacteria (e.g. *Salmonella typhimurium*) [ZHO 97], herbicides, drugs [WOL 08] or explosives [BAK 99].

Vo-Dinh *et al.* described the fabrication of sub-micrometric immunosensors for benzo[a]pyrenetetrol detection inside living cells [ABE 96]. The fiber-optic is stretched so that points, in the tens of nanometers range, are obtained, covered by a fine silver layer and finally modified by the corresponding antibody. The limit of detection for this fluorescent nanosensor was found to be 0.64×10^{-11}M for BPT.

Evanescent waves can also be coupled with more sophisticated optical methods such as FRET (Figure 14.6) [KO 06]. FRET's efficiency depends on the overlap between the donor's emission spectrum and the acceptor's absorption spectrum. As it is also a non-radiant energy transfer linked to dipole–dipole interaction between the energy donor and acceptor, the phenomenon is inversely proportional to the distance ($1/r^6$) between the donor and the acceptor. When the antigen to be detected is absent, the evanescent wave excites the fluorescence of a probe, referred to as the donor, and only the fluorescence wavelength λ_d is emitted. In the antigen's presence, the immobilized antibody undergoes a change in configuration that brings the donor closer to the acceptor. As a result the energy transfer and fluorescent emission mainly happens at the acceptor's wavelength λ_{ad}. The emission intensity's ratio measurement for the donor and the acceptor gives the quantity of antibodies depending on their antigen.

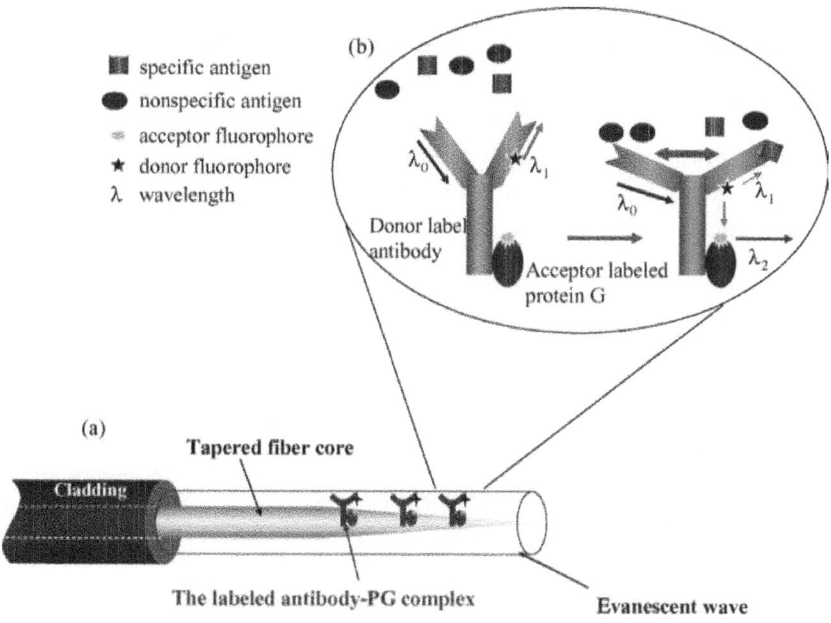

Figure 14.6. *Detection principle of a fiber-optic immunosensor associating evanescent waves with FRET [KO 06]*

14.4.2.2. *SPR*

The SPR method has two important advantages over other biosensors: firstly its high sensitivity as it is possible to reach protein concentration under a few femtomoles, and secondly the possibility of following the real-time kinetics between the ligand and the receptor without a labelling step. Configuration based on fiber-optics offers additional advantages for miniaturizing the device and carrying remote measurements. In general, single-mode fiber-optics are used to maintain polarization and reflection angle. Part of the fiber cladding is taken off and a thin gold layer is deposited in its place around the bare core. An alternative configuration is to polish the fiber's distal face with an adequate angle.

It is possible to associate plasmon effects with fiber-optics in different ways. One of these is to immobilize gold nanoparticles on the fiber's stripped surface [ABE 96]. The evanescent wave therefore excites localized surface plasmons (LSPR) and highly sensitive immunoassays can be carried out [FER 96, WAL 00].

14.4.2.3. *Chemiluminescence*

The groups led by S. Cosnier and R.S. Marks developed a new strategy to make immunosensors that combine electrochemistry and fiber-optics.

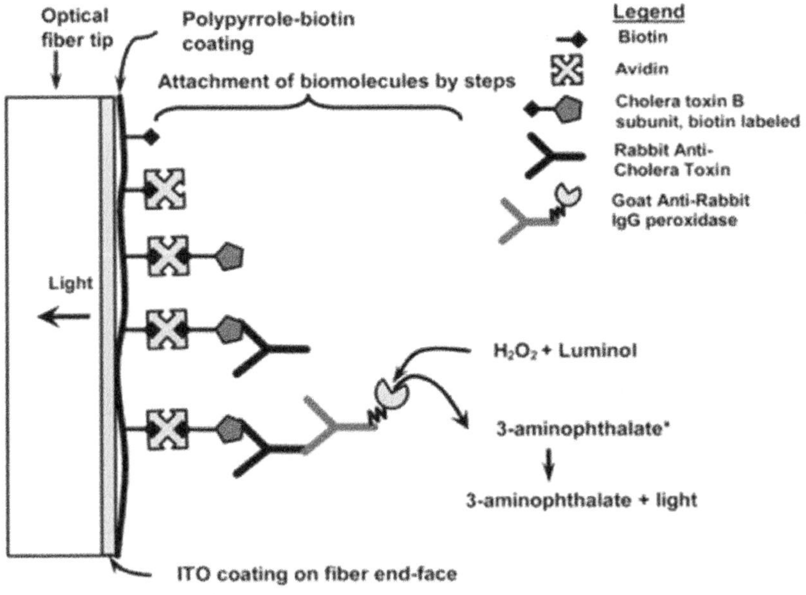

Figure 14.7. *Operating principle of chemiluminescent immunosensors for the detection of cholera toxin B [KON 03]*

Their original idea was to immobilize antibodies on the surface of a fiber-optic by electrochemistry. To do this, the fiber needs to be made conductive [KON 05, MAR 02] so that it may be used as an electrode. In order to maintain the fiber's light transmission properties, the teams used ITO, which is a transparent semiconductor. Pyrrole monomers were modified using biotin. Electropolymerization then produced a polypyrrole film containing biotin groups that can easily be functionalized by biomolecules such as antibodies. The authors have thus achieved immunoassay of cholera toxin B in sandwich format with the marker HRP [KON 03]. For the measurements, the sensor is dipped into a solution containing luminol and an oxidizing agent (Figure 14.7).

14.4.3. *DNA sensors*

The first fiber-optic biosensors, for the detection of DNA, were created at the beginning of the 1990s [GRA 92, ABE 96]. Short oligonucleotide sequences were immobilized on a fiber. The complementary strands were marked with fluorescein and the detection was carried out by evanescent waves.

Simultaneous measurement of DNA sequences were described by D. R. Walt's team with the use of an assembly of seven fiber-optics [FER 96]. Each fiber, of diameter 200 µm, was individually functionalized by a given oligonucleotide sequence. The seven fibers were then assembled to create an array and the detection was carried out using fluorescence. This approach allowed a fast (<10 minutes) and sensitive (10 nM) detection of several sequences. However the manufacturing process was an important restriction for broadening this method to the measurement of a large number of sequences.

In the year 2000, a superior alternative approach, still based on fiber-optics, was described and used commercially by the same team [WAL 00]. It consists of randomly ordered functionalized microbead arrays [EPS 02, EPS 03a, EPS 03b]. The initial material is a bundle of 50,000 fiber-optics, which have a distal face that is chemically etched (Figure 14.2) in order to obtain an array of 50,000 cavities. The process makes use of the difference in etching rates between the core and the cladding. The manufacturing process is simple, economic and highly reproducible. Beads are encoded with fluorophores so as they may be recognized depending on their functionalization. They are mixed to make sequence libraries (Figure 14.8a).

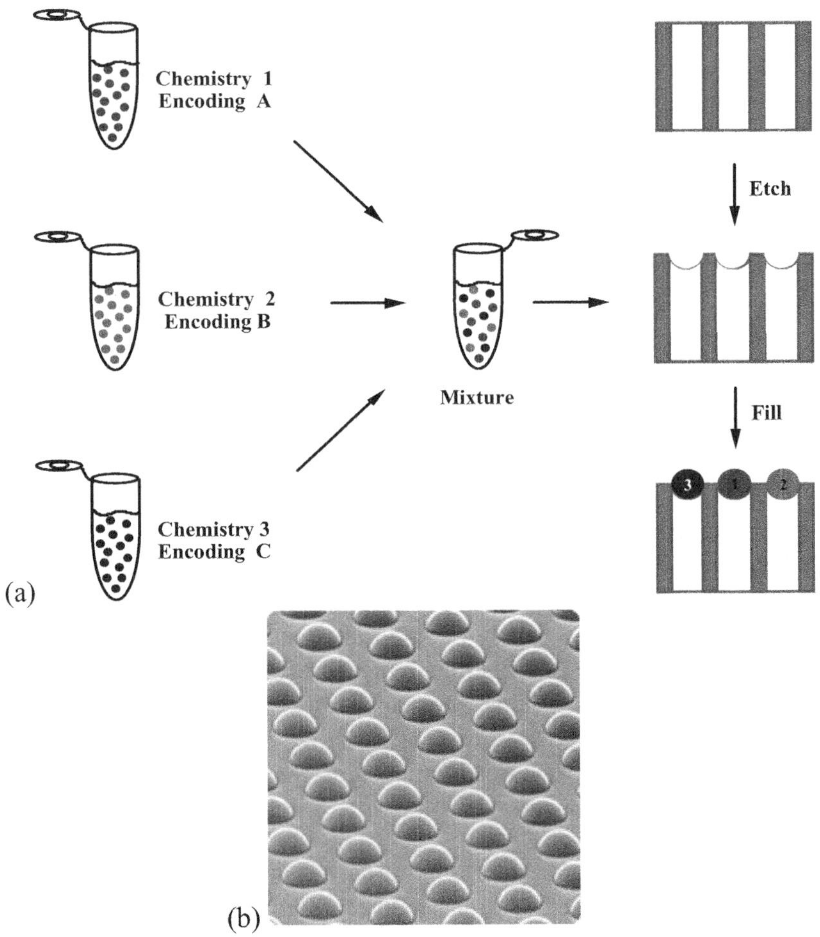

Figure 14.8. *a) Fabrication principle of the array. b) Image of part of an array of microbeads functionalized in the microcavities [LAF 08]*

After a bead library has been chosen, they are spread over the array surface and the cavities are therefore filled randomly by functionalized beads. The size of the cavity corresponds to the bead's diameter so that each cavity contains one bead only (Figure 14.8b). The position and nature of all the beads in an array are decoded in a few seconds with fluorescence. The essential advantage of this manufacturing process is that it is both simple and cheap. However, as the bead arrangement is random for each array, it is necessary to decode each one separately. In addition each nucleotide sequence must be synthesized in advance and grafted onto the beads, unlike photochemical approaches based on combinatorial synthesis. The array is then incubated with the sample and read using fluorescence again. An inherent consequence of this manufacturing process is redundancy: several cavities are occupied by the same type of bead. This redundancy increases the size of the array but guarantees that each array will contain a bead for each DNA sequence. The advantages are being able to discriminate and eliminate false positives and false negatives. This statistical aspect also enables the increase of detection sensitivity. These technological advantages are the basis for Illumina's (http://www.illumina.com) success in the field of DNA biochips.

14.5. Conclusion

Fiber-optic biosensors have undergone considerable progress over the past few years. They incorporate the discoveries made in the fields of nanotechnology, plasmonics, biology, chemistry, etc. In terms of sensitivity, the performances are progressively improving and are comparable to those of laboratory devices in a number of cases. Their miniaturization allows the sample to have a considerably smaller volume and enables measurements to be carried out for living cells. An important challenge is to be able to simultaneously measure several species in subcellular compartments to access cellular machinery. However, the development of robust and portable commercial biosensors remains the main objective and there is still much work to be done. Nanostructured fiber-optic arrays are a promising approach for the screening of new medicines and for new generation biochips.

14.6. Bibliography

[ABE 96] A.P. ABEL, M.G. WELLER, G.L. DUVENECK, M. EHRAT, H.M. WIDMER, *Anal. Chem.*, vol. 68, pp. 2905-2912, 1996.

[BAK 99] I.B. BAKALTCHEVA, F.S. LIGLER, C.H. PATTERSON, L.C. SHRIVER-LAKE, *Anal. Chim. Acta*, vol. 399, pp. 13-20, 1999.

[BOR 08] S.M. BORISOV, O.S. WOLFBEIS, *Chem. Rev.*, vol. 108, pp. 423-461,2008.

[DEI 10] F. DEISS, N. SOJIC, D. J. WHITE, P. R. STODDART, *Anal. Bioanal. Chem.*, vol. 396, pp. 53-71, 2010.

[EPS 02] J.R. EPSTEIN, M. LEE, D.R. WALT, *Anal. Chem.*, vol. 74, pp. 1836-1840, 2002.

[EPS 03a] J.R. EPSTEIN, J.A. FERGUSON, K.-H. LEE, D.R. WALT, *J. Am. Chem. Soc.*, vol. 125, pp. 13753-13759, 2003.

[EPS 03b] J.R. EPSTEIN, A.P.K. LEUNG, K.-H. LEE, D.R. WALT, *Biosens. Bioelectron.*, vol. 18, pp. 541-546, 2003.

[FER 96] J.A. FERGUSON, T.C. BOLES, C.P. ADAMS, D.R. WALT, *Nat. Biotech.*, vol. 14, pp. 1681-1684, 1996.

[GRA 92] C.R. GRAHAM, D. LESLIE, D.J. SQUIRRELL, *Biosens. Bioelectron.*, vol. 7, pp. 487-493, 1992.

[KO 06] S. KO, S.A. GRANT, *Biosens. Bioelectron.*, vol. 21, pp. 1283-1290, 2006.

[KON 03] T. KONRY, A. NOVOA, S. COSNIER, R.S. MARKS, *Anal. Chem.*, vol. 75, pp. 2633-2639, 2003.

[KON 05] T. KONRY, R.S. MARKS, *Thin Solid Films*, vol. 492, p. 313, 2005.

[KUS 01] B. KUSWANDI, R. ANDRES, R. NARAYANASWAMY, *Analyst*, vol. 126, pp. 1469-1491, 2001.

[LAF 08] C.N. LAFRATTA, D.R. WALT, *Chem. Rev.*, vol. 108, pp. 614-637, 2008.

[MAR 02] R.S. MARKS, A. NOVOA, T. KONRY, R. KRAIS, S. COSNIER, *Mat. Sc. Eng. C*, vol. 21, pp. 189-194, 2002.

[MCD 08] C. MCDONAGH, C.S. BURKE, B.D. MACCRAITH, *Chem. Rev.*, vol. 108, pp. 400-422, 2008.

[MON 04] D.J. MONK, D.R. WALT, *Anal. Bioanal. Chem.*, vol. 379, pp. 931-945, 2004.

[ROS 95] Z. ROSENZWEIG, R. KOPELMAN, *Sens. Actuat. B*, vol. 35, pp. 475-483, 1995.

[ROS 96] Z. ROSENZWEIG, R. KOPELMAN, *Anal. Chem.*, vol. 68, pp. 1408-1413, 1996.

[WAL 00] D.R. WALT, *Science*, vol. 287, pp. 451-452, 2000.

[WOL 08] O.S. WOLFBEIS, *Anal. Chem.*, vol. 80, pp. 4269-4283, 2008.

[ZHO 97] C. ZHOU, P. PIVARNIK, S. AUGER, A. RAND, S. LETCHER, *Sens. Actuat. B*, vol. 42, pp. 169-175, 1997.

Chapter 15

In Vivo Analyses with Electrochemical Microsensors

15.1. Introduction

15.1.1. *Microelectrodes in electrochemical analysis of living systems*

Electrochemical analysis of molecules of biological interest in living entities (organisms, tissues, cells) was developed during the 1970s [ADA 76]. The beginnings of this scientific field corresponded to the development of electrochemical microsensors called microelectrodes [BAR 80, DAY 80].

Microelectrodes have at least one characteristic dimension of micrometer order: typically the radius of their active surface (Figure 15.1). They are generally made from carbon fibers or metal wire (platinum, gold, etc.) and have a diameter which varies by a few dozen microns at most and a length adapted to the microelectrode's support (Figures 15.1a and b). The micron-sized diameter of the fiber or conducting wire is obtained using various techniques to reach nanometric analysis scales (Figure 15.1c) [KAW 91, ARR 04, STR 08]. The techniques are mainly based on the material's thermal etching or a chemical attack which gives it a new conical shape.

Chapter written by Stéphane ARBAULT.

Microtechnologies based on photolithography have also been used to create differently shaped supports and microelectrode networks (mainly made of metal). Designed multisensor systems can therefore be implanted in tissues, for example (Figure 15.1e) [BUR 00, SRE 96].

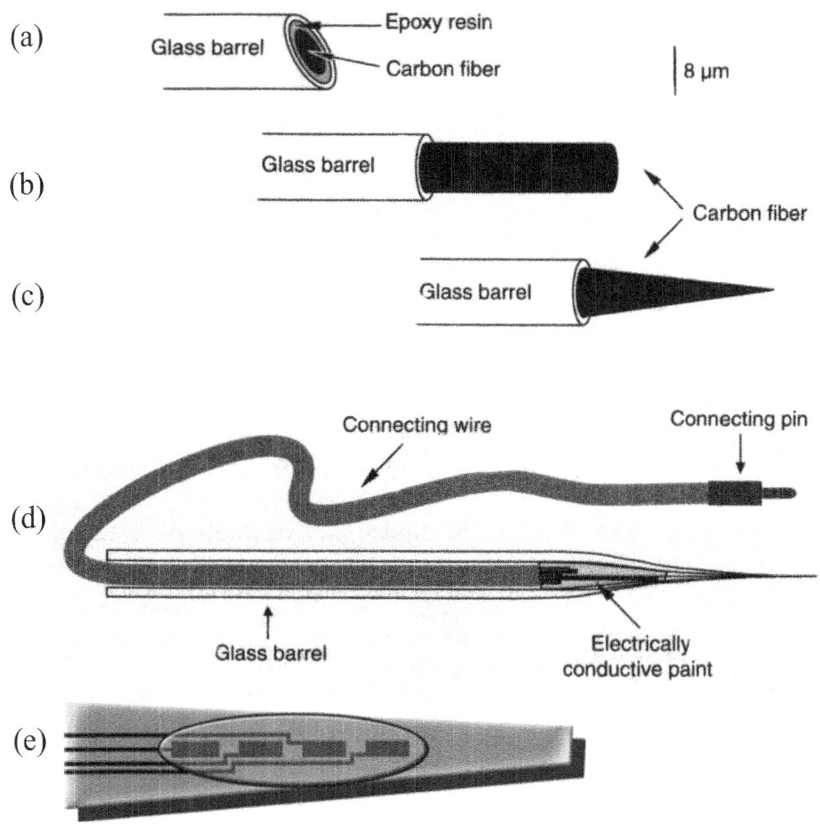

Figure 15.1. *a), b) and c) different topologies for carbon fiber or metal microwire microelectrodes for in vivo and ex vivo detection. d) The microsensor is usually integrated into a body (glass, polymeric material) of mm width for manipulation and to create a connection with the potentiostat. e) Example of a microelectrode network on a ceramic supported made via photolithographic techniques (adapted from [BUR 00])*

During the making of microelectrodes for the purpose of biology, the non-invasive nature of sensors is essential when handled or inserted into biological entities. The overall diameter of microelectrodes used for analysis in biological environments must be minimized and remain in the range of micrometers, as opposed to microelectrodes used for the study of reactions in solutions that have an external body dimension in the centimeter range. These electrodes are used as local, *in situ* sensors that cause minimal disturbance to the observed response (non-invasive). In other words, when measurements are carried out in tissues (for example, brain tissue) it is crucial to prevent tissue damage and the measurement of responses induced by analytical methods.

15.1.2. *Principles of the electrochemical analysis of living systems*

The topology and dimensionality of existing electrochemical sensors are parameters that affect analytical processes applied to biological systems. They consist of confining chemical information in a minimal analysis volume as well as using micrometric sensors to place them close to cells or tissues without causing damage (see Figure 15.2a).

Biological systems such as living cells contain quantities of messenger or reactive species that can be considered infinitesimal under stationary and homogeneous conditions. These quantities are between pico-mole and zepto-mole ranges and act in nano- to micrometric volumes, which corresponds to considerable local concentration variations, micro or millimolar and even up to molar for certain mechanisms. The concentration dissolves quickly due to the release of species being spatially targeted. Original cellular information no longer has an effect at distances similar to the source dimension (see Figure 15.2b).

In situ detection by a microsensor at living cell level mimics the biological process of information confinement [AMA 08]. Most electrochemical techniques are faradic in nature: they measure a reduction or oxidation current proportional to concentration. The confinement is therefore favorable as it amplifies the concentration variation or flow of species. When molecules of biological interest are electroactive under these conditions, electrochemical techniques are competitive compared to other methods presumed more sensitive (fluorescence spectrophotometry). Moreover, electrochemical analysis is generally carried out without the use of marked or modified biological systems unlike other fluorescence methods.

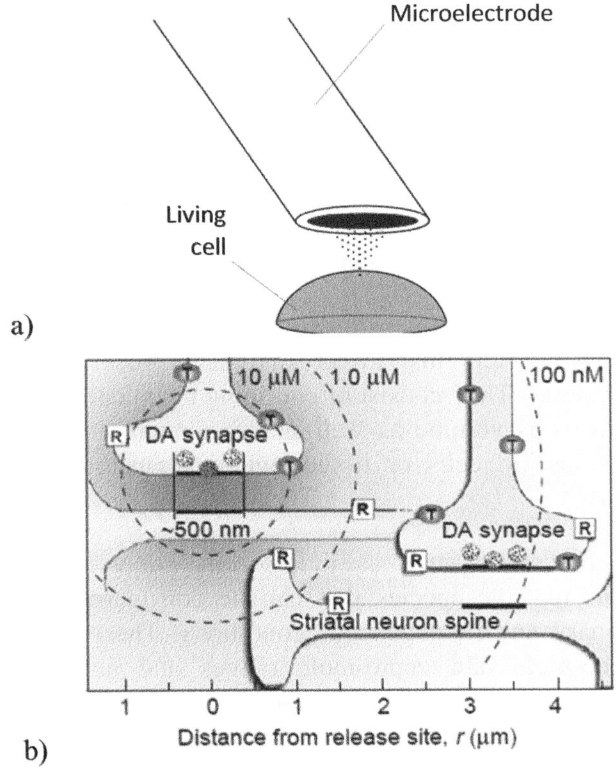

Figure 15.2. *a) Electrochemical detection at a single cell by confining information in a micrometric analysis volume between a microsensor and a biological source. b) The effect of diffusion distance on the concentration of species produced by the biological source, for example a neuron synapse (adapted from [CRA 04])*

The electrochemical analysis methods implemented are mainly: chronoamperometry, which offers the highest sensitivity to detect weak currents (flow), and cyclic voltammetry, enabling the characterization of detected redox couples (see section 15.2.1).

15.1.3. *Chemical species detected in vivo and ex vivo*

Chemical species released by living cells are often neurotransmitters or neuromodulators [AMA 08, ROB 08]. These species are either electroactive or directly detectable on electrode surfaces, namely platinum and carbon, or non-electroactive in water redox (media and biological buffers). They become detectable by biosensors based on the alteration of the electrode's surface by the recognition systems of specific enzymes, for example.

It is therefore possible to classify a vast majority of molecules detected in biological systems with following criteria – neurotransmitter or neuromodulator and electroactive or non-electroactive.

Electroactive neurotransmitters:	Electroactive neuromodulators:
– Dopamine	– Ascorbate
– Adrenalin / nor- adrenalin	– Uric Acid
– DOPAC acid	– Hydrogen peroxide
– Serotonin	– Nitrogen monoxide
– Histamine	– Nitrite
– Homovanillic Acid	– Oxygen
	– Adenosine
Non-electroactive neurotransmitters:	**Non-electroactive neuromodulators:**
– Glutamate	– Insulin
– Acetylcholine	– Glucose
– Lactate	– Adenosine tri-phosphate
– γ-aminobutyric acid (GABA)	

Table 15.1. *Example of the classification of molecules detected in biological systems*

The detection of these species is often considered to be carried out *in vivo*, even though the term is generally incorrect. Most studies are done on cell or tissue cultures, i.e. explanted from an organism and therefore *ex vivo*. The only *in vivo* studies are carried out on living animals such as rats or mice. Particular experimental conditions, like the animal's mobility, make them even more exceptional. Fascinating *in vivo* results have been reported in literature and examples will be presented later in this chapter.

15.1.4. *Experimental assemblies for in vivo and ex vivo detection*

Ex vivo studies on cell cultures and tissues are generally carried out on a microscope platform in order to combine observation (often of visible or infrared light transmitted using different phase-contrast techniques), drip or control over the biological entity's survival medium and *in situ* positioning of electrochemical sensors (Figure 15.3a).

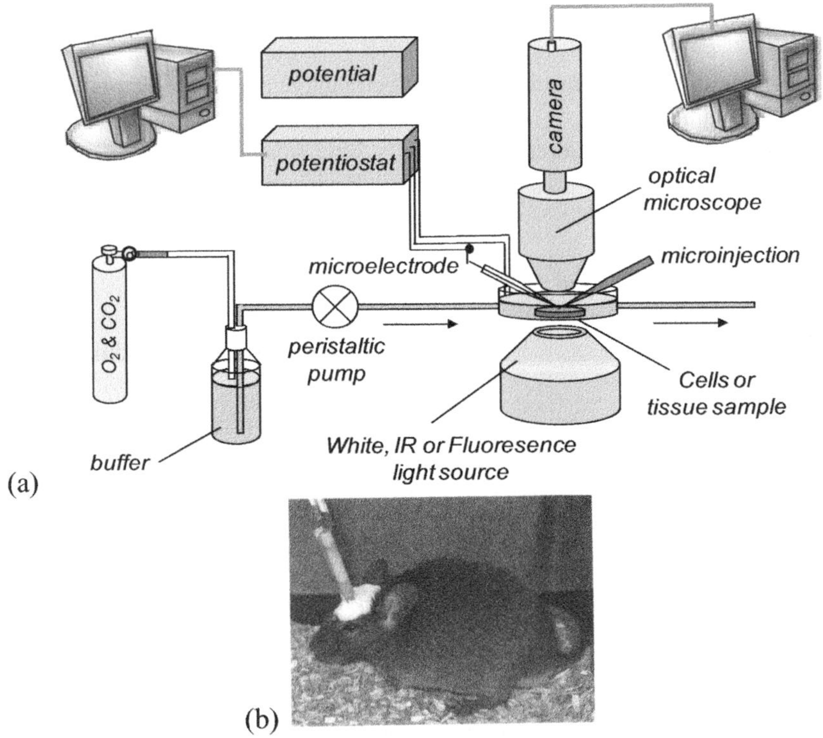

Figure 15.3. *a) Diagram of an experimental installation on a microscope platform for simultaneous observation and detection of cell cultures or tissue responses. b) In vivo neurotransmitter detection experiment on an animal by an electrochemical microsensor*

In vivo studies on animals are done by cutting off the tissue surface and inserting a microsensor deep inside the tissue (Figure 15.3b). There is usually no visual control for the positioning of microelectrodes in tissues. The location is determined through knowledge of histology and a position in z (depth of penetration). Recent research has aimed to develop telemetric systems for the transmission of information detected by electrodes implanted in animals [GAR 07]. This allows the analysis of physiological variations (often cerebral) in an animal that is free to move and subject to stimulations and restrictions.

15.2. *In vivo* electrochemical detection of neurotransmitters

The *in vivo* electrochemical detection of molecules of biological interest has often been oriented towards neurotransmitter detection. The most common neurotransmitters in cerebral tissue, besides glutamate, are amines or catecholamines, which can easily be electrochemically oxidized, i.e. at voltages under 1 V versus Ag/AgCl (Figure 15.4a).

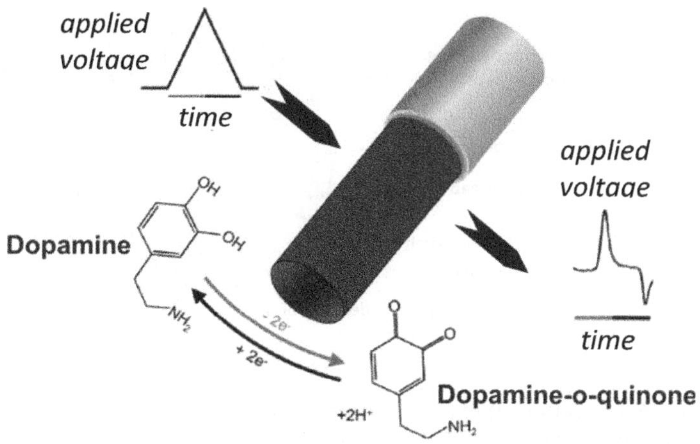

Figure 15.4. *a) Detection of catecholamine type neurotransmitter (dopamine) on a cylindrical microelectrode's surface (extract from [ROB 08])*

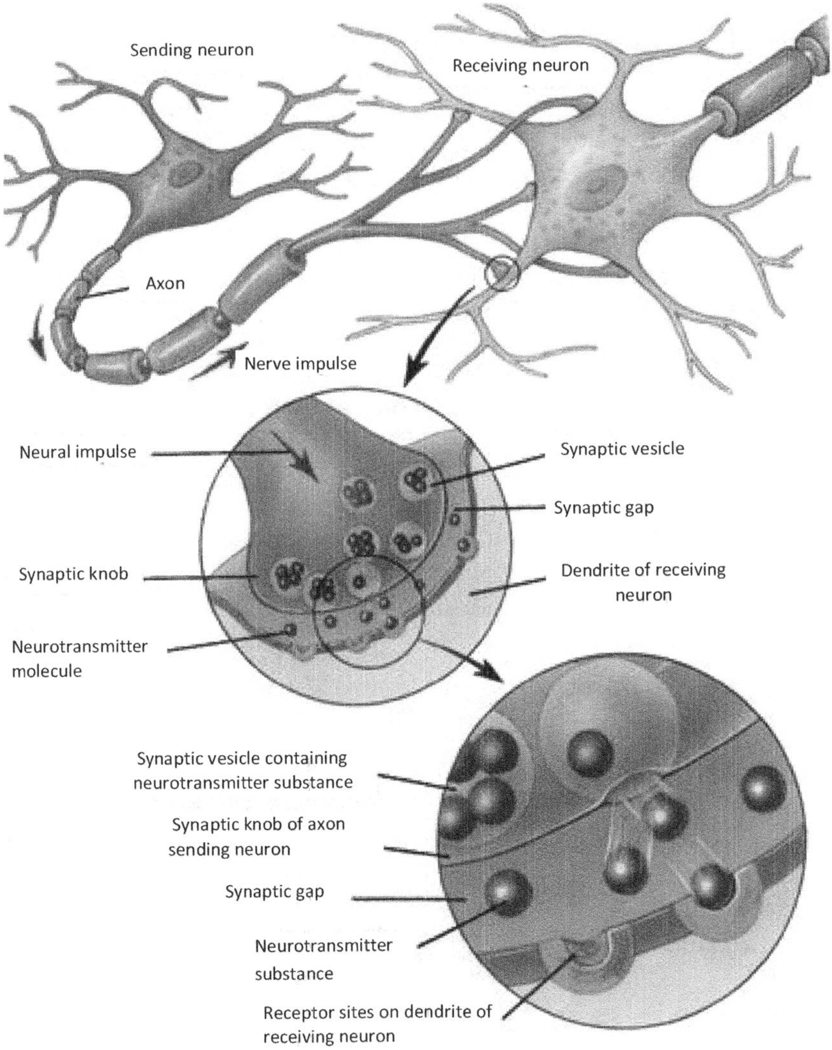

Figure 15.4. (continued) *b) Principle of neurotransmission mechanisms at the chemical synapse level between two neurons (http://www.mhhe.com /socscience/intro/ibank/set1.htm)*

The neurotransmission mechanism operates at synapse level, i.e. at the junction between neurons (Figure 15.4b). Neurotransmitters previously stocked in containers (vesicles) in the ends of emitting neurons, are released and diffuse into the synapse. They interact with receptors located on the receptor neuron's membrane surface or, more generally, diffuse around the synaptic knob. The quantity of molecules diffusing in the tissue (see Figure 15.2b) is what will be detected by the microelectrode placed near the cellular source.

Detection happens on the microelectrode's planar (often disc shaped) or volumetric (often cylindrical) surface. The former gives a local measurement of micrometric scale and the neurotransmitter's concentration, whereas the latter allows a measurement of the tissue's depth (several dozen to several hundred microns). This depth often corresponds to the thickness of the histological zone, consisting of specific neurons (e.g. dopaminergic neurons), for example.

15.2.1. *Methods for in vivo electrochemical detection of neurotransmitters*

In vivo and *ex vivo* detection is nearly exclusively faradic in nature. Studies consist of analyzing the flow of neurotransmitters emitted by one or more cells. Two faradic methods are commonly used: chronoamperometry and cyclic voltammetry. The first consists of applying one single voltage throughout the duration, i.e. oxidation (or reduction) voltage for the analyzed species. The time evolution of the oxidation (or reduction) current is therefore measured proportional to a species' *in situ* concentration in the electrode's neighboring region (in the diffusion layer). This simple method offers the highest resolution in terms of species flow as it is free of capacitive contributions. However, it does not enable us to characterize the species detected by the electrode. For this reason, research teams have developed detection poles using cyclic voltammetry (Figure 15.5).

Voltage cycles are applied to the electrode at regular intervals. The voltage limits are chosen according to the redox couples being studied and their electron transfer kinetic with the electrode. Characteristic current-tension curves are thus obtained and are used as the detected species' signature. Moreover, this method is used as a sampling method for biological response as the detection voltage is not constantly applied. In order to maintain rapid kinetic analysis, and even real-time analysis, it is necessary to setup very fast scanning speeds, typically of several hundred volts/second. This considerably increases the capacitive contribution to the total current in comparison to the faradic current of the detected species, and requires specific signal processing (averaging, subtraction, measurement of voltage evolution).

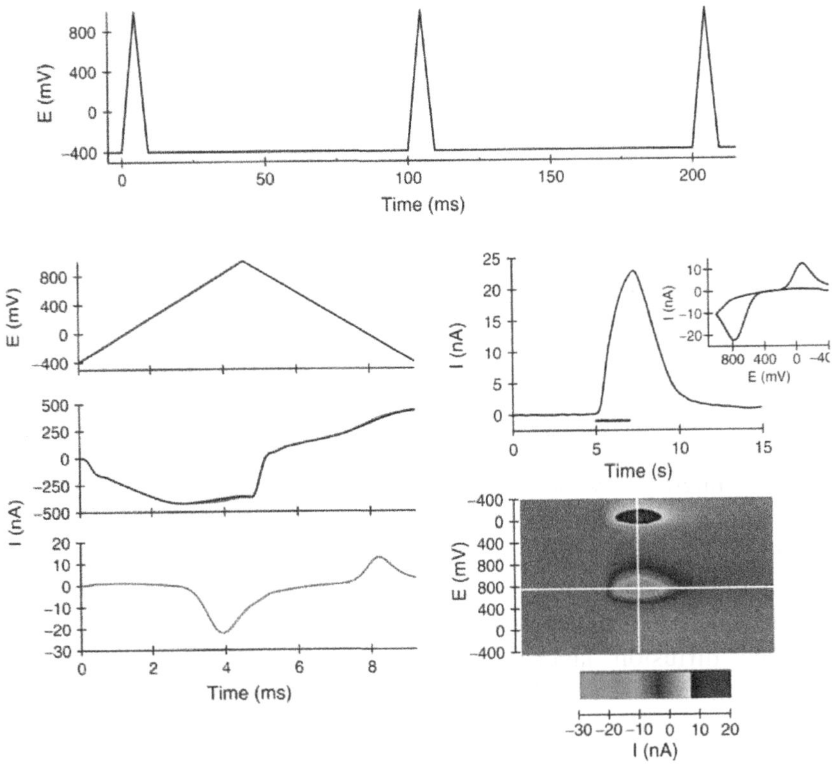

Figure 15.5. *(Above) detection protocol for the in vivo catecholamine detection [GAR 07]. (Left) for each voltage cycle, the faradic current due to dopamine by cyclic voltammetry is subtracted from the base current. (Right) the detected species is characterized by its voltametric response, and its release kinetics analyzed by the measurement of peak current evolution for each voltagram*

15.2.2. *Ex vivo and in vivo detection in rat brains*

A considerable amount of electrochemical research on rodent brains has been done using sample (thin slices 200-500 µm thick). The first electrochemical *in vivo* analyses of neurotransmitters were carried out under these conditions: analysis of an organ sample maintained active with an artificial cerebrospinal fluid. Cylindrical carbon fiber microelectrodes (diameter between 5 and 30 µm) were inserted in different zones of the brain to detect the presence of dopamine and study the release/capture mechanisms of various neurons. Even though dopamine is a rare neurotransmitter in the brain, it is found in a few specific locations and plays a role in the control of a number of functions: cognition, motor functions, motivation, rewards, sleep or memoration.

R.M. Wightman's team (University of North Carolina, USA) was a precursor in the field of *in vivo* electrochemistry by carrying out detection experiments combining *in situ* and local injection of a dopamine flow inside the tissue [KEL 87]. They highlighted the important role of transporter channels present on the surface of dopaminergic neurons, which are responsible for the elimination of excess dopamine (recapture) that has not reacted with the receptors. This mechanism controls the activation time and frequency of dopamine paths with great precision [MAY 88].

Modulation of the recapture phenomenon was then studied using electrodes implanted directly into living animals. Rats were subjected to the consumption of cocaine in order to understand the effects of the drug on dopamine activity [JON 95]. It was shown that dopamine concentrations in the brain and its release time increased with the consumption of cocaine by the animal. The modulations were linked to changes in the recapture/degradation of dopamine by neurons. These results based on *in vivo* electrochemical measurements supported various hypotheses about cocaine physiopathology. Other studies by the same team also looked into the effects of cannabinoids.

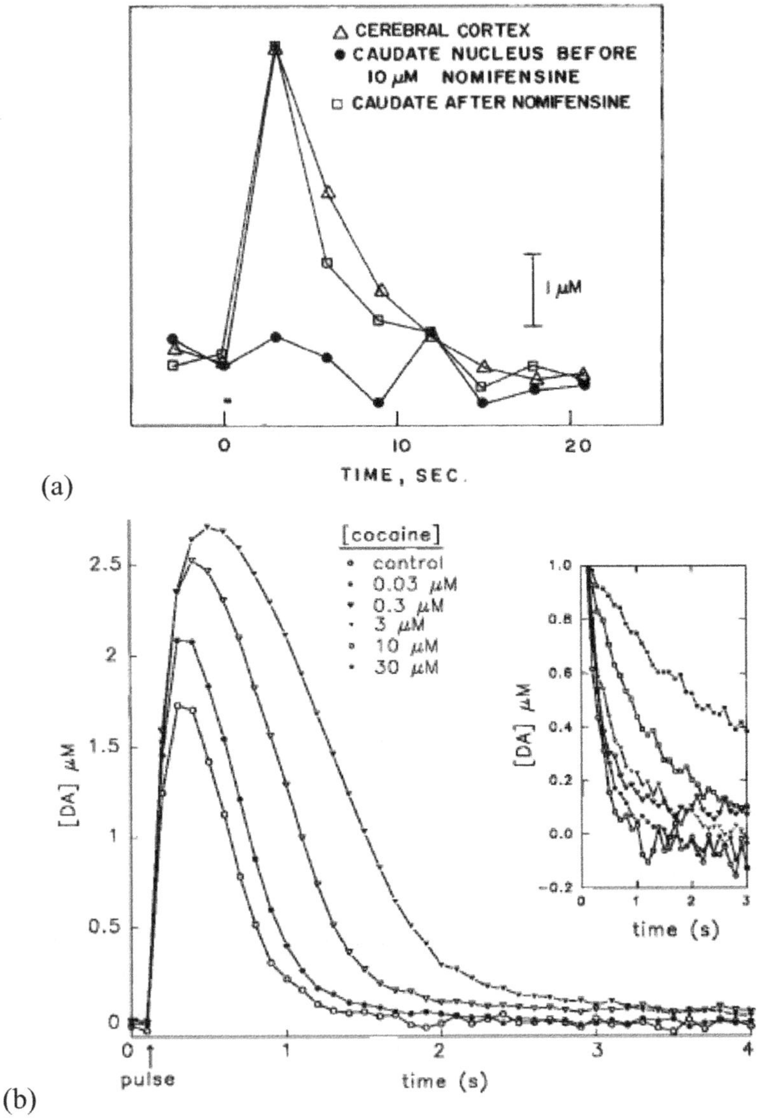

Figure 15.6. *a) Local injection by microsyringe of dopamine and its simultaneous electrochemical detection by carbon fiber electrodes inside different cerebral areas of a rat (extract from [KEL 87]). The dopamine injected is not detected in zones rich in dopamine neurons as it is recaptured by specific channels (antagonist: nomifensine). b) Effect of cocaine on the dopamine concentration and release kinetics in the brain (extract from [JON 95])*

Another series of studies consisted of characterizing the neurochemistry of animal behaviors [ROB 02]. It is known that dopamine neurons are involved in reward, behavior and pleasure. *In situ* analysis of dopamine concentrations in a rat brain (striatum) was undertaken during the encounter of a male rat with a female rat (Figure 15.7).

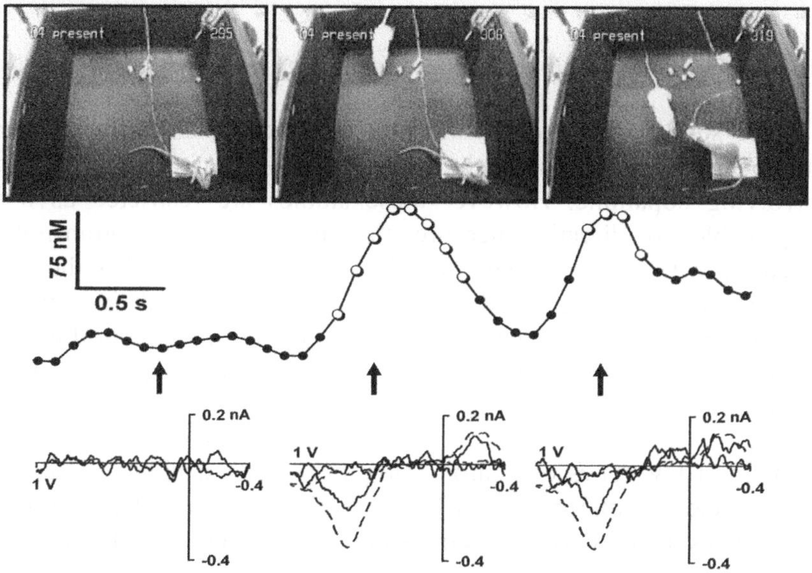

Figure 15.7. *In vivo analysis of dopamine concentration variations in the brain during the encounter of two rats of the opposite sex. Dopamine levels are compared for when the male rat is alone (left), when a female enters the cage (center) and when they are in physical contact (right). In situ variations in dopamine levels are determined using peak currents of voltammograms detected in vivo and compared to in vitro calibrations of electrodes (extract from [ROB 02])*

Low base dopamine levels were detected when the male rat was alone in his cage. However, when the female was placed in the cage, an increase in cerebral dopamine was detected even when the male could not see the female. Olfactory recognition is the reason behind the dopamine activity. The male rat then approached the female and physical contact happened, in particular via vibrissa on the nose. This was reflected by more transitory increases in dopamine levels, showing the pleasure taken by the animal during the encounter. Besides the behavioral aspects of mammals, the study was based on a precise analytical study as the continuous electrochemical analysis of the animal's brain was carried out without disturbing its behavior. The use of dopamine as a neurotransmitter in these animal responses was shown using voltammograms detected *in vivo* at each stage of the encounter between the two animals of opposite sex.

15.2.3. *Ex vivo detection of neurotransmitter release in a single cell*

The studies presented in the previous section show that *in vivo* microelectrochemical analysis reveals responses of cerebral activity involving dopamine. However these studies are restricted as it is impossible to directly measure the amount of neurotransmitter released in the synapse. The quantity of species detected is the ratio between quantities released and captured, or degraded, by specific control mechanisms. In the 1990s, several teams involved in this field of research worked towards analyzing the release mechanism of neurotransmitters in a single isolated cell [KAW 91, CHE 94].

Only a few types of neurons have been studied in isolation from tissues due to experimental difficulties of extraction and cell culture. The teams focused on neurotransmitter or hormone secreting cells present in the endocrine system. Some cells in particular could be isolated from the pancreas and adrenal glands for the study of vesicular exocytosis mechanisms at a single cell level.

Many studies have been done on chromaffin cells isolated from adrenal glands, as they are the main component of the latter's core (medulla) [LIV 84]. The cells are responsible for the release of catecholamines into the blood flow (adrenalin and noradrenalin, dopamine in smaller quantities) and various peptides during the activation of sympathetic pathways. This corresponds to the famous "adrenalin rush" induced by psychological factors in individuals. Chromaffin cells have the advantage of integrating all neurosecretion machinery, including storage vesicles for large quantities of neurotransmitter (see Figure 15.4).

Chromaffin cells can be isolated and stored in cultures for several days. A carbon microelectrode's surface (disc-shaped) is placed under a microscope by micromanipulation at a micrometric distance from the cell's surface (Figure 15.8). The detection and release of catecholamine neurotransmitters is often done using cyclic linescan voltage voltammetry or amperometry depending on the protocols developed for *in vivo* analysis described above.

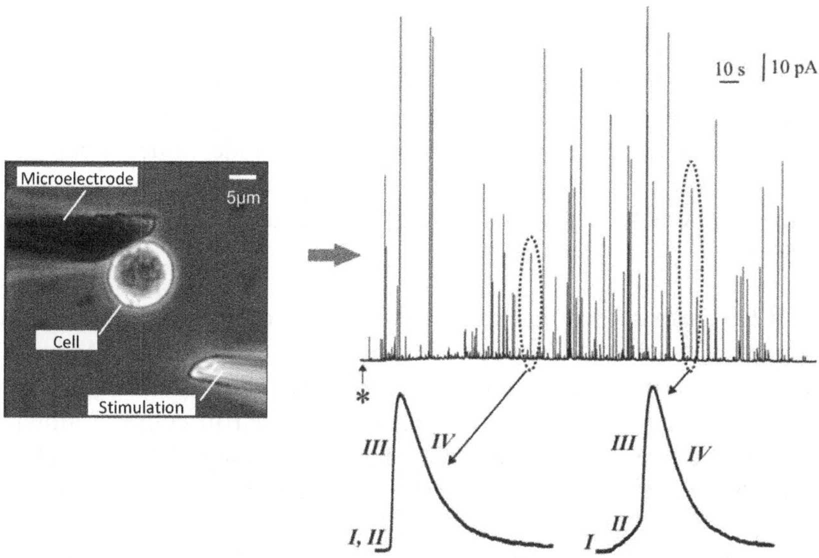

Figure 15.8. *Experiment using a carbon fiber microelectrode for the detection of catecholamine secretion. Secretion is activated by microinjection of a barium or potassium ion solution into the cell's environment (*). Released catecholamines are detected using constant voltage chronoamperometry. (+650 mV vs. Ag/AgCl)*

Figure 15.8 shows a typical amperometric response detected at an oxidation voltage of +650 mV versus Ag/AgCl of chromaffin cells after secretion stimulation by calcium channel activation. The response consists of a number of peaks, of diverse current amplitude (1 to 300 picoamperes), and rapid kinetics (lasting in the region of 100 milliseconds). The experimental data obtained show the analytic difficulties of single cell detection as the weak faradic current variations (of the picoampere order) need to be detected with extremely precise temporal resolution (less than or equal to one millisecond). The best measurement resolutions reported in literature reach values of 0.5 pA and 100 μs, i.e. the equivalent of 1,000 adrenalin or noradrenalin molecules/millisecond.

This limit can be achieved due to the effect of chemical information confinement (redox species), described previously. An infinitesimal flow of released molecules is detected before dilution in the cellular environment.

Each current peak can be kinetically and quantitatively analyzed. Analysis shows that peak kinetics are highly homogeneous despite the considerable distribution of the current's amplitude values. Each peak therefore corresponds to a well-defined dynamic release process of variable intensity. The average quantity of catecholamine detected for each peak is 3×10^6 molecules.

It has been shown that each isolated peak corresponds to a unique vesicular exocytosis process (Figure 15.9) [KAW 91, AMA 03]. Neurotransmitters are stored in secretion vesicles inside the cell (up to several tens of thousands of vesicles in chromaffin cells). Following calcium activation, vesicles are transported along the cytoskeleton towards the cell membrane so as they may interact with each other.

The reconciliation of the vesicle and cell membrane causes them to merge with each other, thus forming a connection termed fusion pore. The channel created is estimated to be several nanometers in diameter, and molecules stored inside the vesicle diffuse out towards the cell's environment. However, this phospholipid structure is unstable and quickly increases in diameter to facilitate the opening of the vesicle. The two membranes therefore fuse and the release of the intravesicular medium becomes complete.

The electrochemical analysis of the mechanism has the advantage of presenting a kinetic and quantitative signature for each step (numbered I to IV in Figures 15.8 and 15.9) [AMA 03]. Each step can be observed as soon as molecules diffuse out from the vesicular medium to the microelectrode's surface.

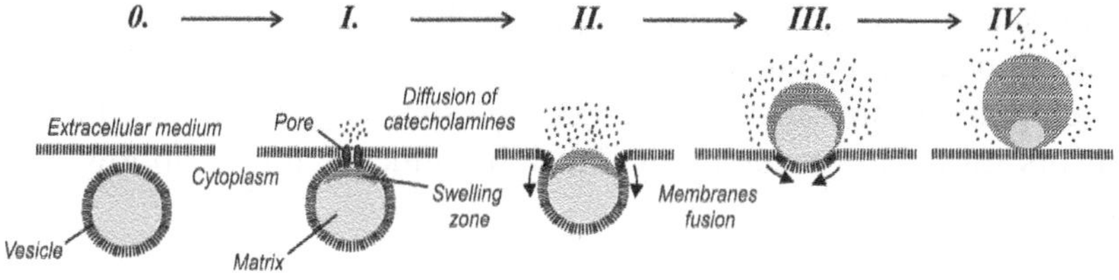

Figure 15.9. *Steps of neurotransmitter secretion by vesicular exocytosis*

The fusion pore phase in particular (phase I) has been demonstrated by the detection of peaks of low current variation (pre-peaks) before the sudden increase of the actual peak. This type of peak represents around 30% of the information detected and corresponds to vesicles having particular dimensions and properties, meant for delay-secretion and larger amplitudes [AMA 05].

Analysis of a single cell by an electrochemical microelectrode has enabled the study of a number of factors that control exocytosis mechanisms, whether they are biological or physico-chemical.

Numerous proteins and their supra-molecular structures are involved in vesicle transport and anchoring, forming fusion pores, etc. as well as in the subsequent re-forming of new vesicles using the membrane surface [HAU 11]. These studies have led to a number of publications in the field of neurobiology and biophysics. However, exocytosis also involves physico-chemical processes such as the interaction between membrane phospholipids during fusion, the destructuring of the vesicular protein matrix under osmotic gradient or even membrane tension effects [AMA 07a]. These processes were revealed by electrochemical analyses of cell secretion in experiments with modified media (ion enrichment, pH variation, etc.) or cells (pre-incubation with molecules inserting themselves in the membrane, etc.) [AMA 09, AMA 07b, AMA 06].

15.3. Conclusion

Electrochemical microsensors have been used for over 30 years for the analysis of biological activities within living systems, ranging from animals to cells. New properties have been discovered using these tools, particularly in the field of neurobiology. Industrial developments are also taking place, especially in the domain of sensors for energy metabolites, including glucose for the follow-up of glycemia.

Many new perspectives are opened by the decreasing size of microelectrodes (tending towards dimensions in the nanometer range), their increased detection sensitivity by their surface structure and the development of sensor networks for the simultaneous detection of several biological markers. Moreover, new strategies for the combination of detection principles and methods are appearing, for example, spectro-electrochemical analysis on the surface of transparent electrodes [MEU 11].

15.4. Bibliography

[ADA 76] R.N. ADAMS, *Anal. Chem.*, vol. 48, p. 1126A, 1976.

[AMA 03] C. AMATORE, S. ARBAULT, I. BONIFAS, Y. BOURET, M. ERARD, M. GUILLE, *ChemPhysChem*, vol. 4, p. 147, 2003.

[AMA 05] C. AMATORE, S. ARBAULT, I. BONIFAS, Y. BOURET, M. ERARD, A.G. EWING, L.A. SOMBERS, *Biophys. J.*, vol. 88, p. 4411, 2005.

[AMA 06] C. AMATORE, S. ARBAULT, M. GUILLE, F. LEMAÎTRE, Y. VERCHIER, *ChemBioChem*, vol. 7, p. 1998, 2006.

[AMA 07a] C. AMATORE, S. ARBAULT, M. GUILLE, F. LEMAÎTRE, *ChemPhysChem*, vol. 8, p. 1597, 2007.

[AMA 07b] C. AMATORE, S. ARBAULT, I. BONIFAS, M. GUILLE, F. LEMAÎTRE, Y. VERCHIER, *Biophys. Chem.*, vol. 129, p. 181, 2007.

[AMA 08] C. AMATORE, S. ARBAULT, M. GUILLE, F. LEMAÎTRE, *Chem. Rev.*, vol. 108, p. 2585, 2008.

[AMA 09] C. AMATORE, S. ARBAULT, I. BONIFAS, F. LEMAÎTRE, *Biophysical Chemistry*, vol. 143, p. 124, 2009.

[ARR 04] D.W. ARRIGAN, *Analyst*, vol. 129, p. 1157, 2004.

[BAR 80] A.J. BARD, L.R. FAULKNER, *Electrochemical Methods. Fundamentals and Applications*, John Wiley and Sons, Inc., New York, 1980.

[BUR 00] J.J. BURMEISTER, K. MOXON, G.A. GERHARDT, *Anal. Chem.*, vol. 72, p. 187, 2000.

[CHE 94] T.K. CHEN, G.O. LUO, A.G. EWING, *Anal. Chem.*, vol. 66, p. 3031, 1994.

[CRA 04] S.J. CRAGG, M.E. RICE, *Trends Neurosci.*, vol. 27, p. 270, 2004.

[DAY 80] M.A. DAYTON, J.C. BROWN, K.J. STUTTS, R.M. WIGHTMAN, *Anal. Chem.*, vol. 52, p. 946, 1980.

[GAR 07] P.A. GARRIS, P.G. GRECO, S.G. SANDBERG, G. HOWES, S. PONGMAYTEGUL, B.A. HEIDENREICH, J.M. CASTO, R. ENSMAN, J. POEHLMAN, A. ALEXANDER, G.V. REBEC, in A.C. MICHAEL, L.M. BORLAND (eds), *Electrochemical Methods for Neuroscience*, vol. 12, CRC Press, Boca Raton, FL, p. 233, 2007.

[HAU 11] V. HAUCKE, E. NEHER, S.J. SIGRIST, *Nat. Rev. Neurosci.*, vol. 12, p. 127, 2011.

[JON 95] S.R. JONES, P.A. GARRIS, R.M. WIGHTMAN, *J. Pharmacol. Exp. Ther.*, vol. 274, p. 396, 1995.

[KAW 91] K.T. KAWAGOE, J.A. JANKOWSKI, R.M. WIGHTMAN, *Anal. Chem.*, vol. 63, p. 1589, 1991.

[KEL 87] R.S. KELLY, R.M. WIghtman, *Brain Res.*, vol. 423, p. 79, 1987.

[LIV 84] B. G. Livett, *Physiol. Rev.*, vol. 64, p. 1103, 1984.

[MAY 88] L.J. MAY, W.G. KUHR, R.M. WIGHTMAN, *J. Neurochem.*, vol. 51, p. 1060, 1988.

[MEU 11] A. MEUNIER, O. JOUANNOT, R. FULCRAND, I. FANGET, M. BRETOU, E. KARATEKIN, S. ARBAULT, M. GUILLE, F. DARCHEN, F. LEMAITRE, C. AMATORE, *Angew. Chem. Int. Ed.*, vol. 50, p. 5081, 2011.

[ROB 02] D.L. ROBINSON, M.L. HEIEN, R.M. WIGHTMAN, *J Neurosci.*, vol. 22, p. 10477, 2002.

[ROB 08] D.L. ROBINSON, A. HERMANS, A.T. SEIPEL, R.M. WIGHTMAN, *Chem. Rev.*, vol. 108, p. 2554, 2008.

[SRE 96] G. SREENIVAS, S.S. ANG, I. FRITSCH, W.D. BROWN, G.A. GERHARDT, D.J. WOODWARD, *Anal. Chem.*, vol. 68, p. 1858, 1996.

[STR 08] A.M. STRAND, B.J. VENTON, *Anal. Chem.*, vol. 80, p. 3708. 2008.

Chapter 16

Microbial Biosensors for Environmental Applications

16.1. Pollution and environment

In order to discuss environmental pollution, it is necessary to define a few key points. The term pollution is a general term and encompasses chemical, physical (radiation, for example) and biological (microorganisms, accidentally introduced species in a biotope) pollution.

Chemical pollution consists of inorganic contaminants, mainly metals, non-metals and organic micropollutants. The ones which first come to mind occur naturally in the Earth's crust and are limited to elements in the periodic table. Pollution is therefore the result of an enrichment of contaminant in the environment, which has a more or less toxic effect on organisms. Organic pollutants are the product of human activities and their numbers are unlimited.

Over the last few years, awareness of environmental pollution has broadened from a purely scientific sphere to society as a whole, causing political institutions to consider the problem as a major risk for public health and the economy.

Chapter written by Gérald THOUAND and Marie José DURAND.

Various regulatory texts have been issued with the aim of protecting the environment by limiting and controlling both industrial and domestic waste. For example, the European Union published a list of 33 priority substances to be detected in the context of water-monitoring, in the 2001 issue of the official journal (N° 2455/2001/EC, see Table 16.1). Amongst these substances, metals, organotins, metals, polycyclic aromatic hydrocarbons (PAH) and long-chained alkanes found in various environments such as polluted seawater, are classified as "priority hazardous substances".

Legislation is becoming more specific and stricter for both local authorities, who will have to implement devices for pollution analysis, and industrial authorities, who will have to ensure that their liquid and gaseous waste is harmless. This systematic approach for waste analysis must become routine and measurements must be more frequent in order to avoid measurements being taken "between two pollution peaks". The process usually consists of real-time continuous checks. How is the demand for analysis satisfied? It must be fast, simple to use for non-specialists, cheap and serve as a warning. It must therefore be able to continuously measure pollution in aqueous environments of both "freshwater" and "seawater"-type.

Evaluating the pollution of a biotope is not straightforward as it is not simply limited to the disposal of a substance. Pollutants can be carried into the atmosphere and water and be distributed far from the emission site (Figure 16.1).

Figure 16.1. *Pollution cycle in the environment*

CAS number[1]	Name of the substance	CAS number[1]	Name of the substance
15972-60-8	Alachlor	7439-97-6	Mercury and its compounds
120-12-7	Anthracene	91-20-3	Naphthalene
1912-24-9	Atrazine	7440-02-0	Nickel and its compounds
71-43-2	Benzene	25154-52-3	Nonylphenols
n.a.[2]	Brominated diphenylethers	104-40-5	(4-(para)-nonylphenol)
7440-43-9	Cadmium and its compounds	1806-26-4	Octylphenols
85535-84-8	C_{10-13}-Chloroalkanes	140-66-9	(para-tert-octylphenol)
470-90-6	Chlorfenvinphos	608-93-5	Pentachlorobenzene
2921-88-2	Chlorpyriphos	87-86-5	Pentachlorophenol
107-06-2	1, 2-Dichloroethane	n.a.	Polyaromatic hydrocarbons
75-09-2	Dichloromethane	50-32-8	(Benzo(a)pyrene)
117-81-7	Di(2-ethylhexyl)phtalate (DEHP)	205-99-2	(Benzo(b)fluoranthene)
330-54-1	Diuron	191-24-2	(Benzo(g,h,i)perylene)
115-29-7	Endosulfan	207-08-9	(Benzo(k)fluoranthene)
959-98-8	(alpha-endosulfan)	193-39-5	(Indeno(1,2,3-cd)pyrene)
206-44-0	Fluoroanthene	122-34-9	Simazine
118-74-1	Hexachlorobenzene	688-73-3	Tributyltin compounds
87-68-3	Hexachlorobutadiene	36643-28-4	(Tributyltin-cation)
608-73-1	Hexachlorocyclohexane	12002-48-1	Trichlorobenzenes
58-89-9	(gamma-isomer, Lindane)	120-82-1	(1,2,4-Trichlorobenzene)
34123-59-6	Isoproturon	67-66-3	Trichloromethane (Chloroform)
7439-92-1	Lead and its compounds	1582-09-8	Trifluralin

Table 16.1. *Alphabetical list of the 33 priority substances in the context of water established by the European Union.*
(1) CAS: Chemical Abstract Services;
(2) na: not applicable

The toxic effect on organisms depends on the interactions between the biotope and the substances. For example, the speciation of metals depends on water pH, salinity and solubility. Waste substances can undergo chemical or biological transformations that may influence their toxic properties. Interaction between substances can also increase (synergic effect) or decrease toxicity via antagonistic effects.

Chemical analyses, although efficient, cannot take into account these phenomena and cannot reveal the bio-accessible fraction of pollutants. Biosensors, on the other hand, allow the appreciation of these interactions through the use of living organisms such as microorganisms.

16.2. Introduction to biosensors

16.2.1. *Definition, principles, classification*

16.2.1.1. *Definition*

Biosensors are measuring devices that incorporate a *biological element* (also called a bioelement or bioreceptor) connected (*intimate contact*) to a *physical measurement device* (transducer). Biosensors selectively and reversibly measure the concentration and/or presence of a chemical or biological species (called the analyte) in all sorts of samples.

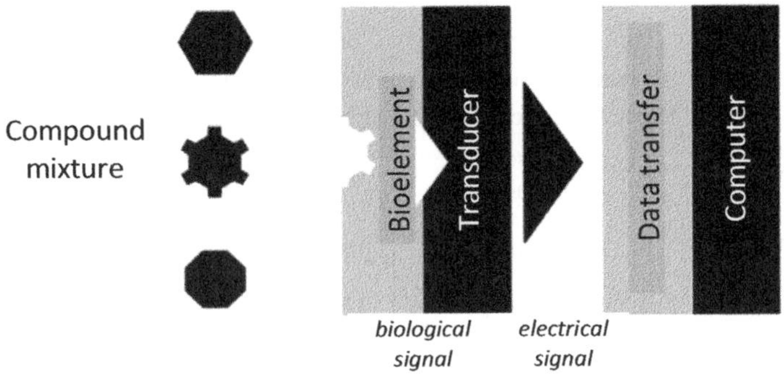

Figure 16.2. *Diagram of biosensor principle [JOU 11a]*

From an operational point of view, the biosensor's bioelement recognizes the analyte and therefore defines the system's specificity. The interaction produces a signal (pH, temperature, photons, electrons, etc.) that is converted into a measurable electrical signal by the transducer.

Biosensors – bioassay

Non-specialist literature confuses the terms bioassay and biosensor. Bioassays are distinct from biosensors as the bioelement is not in contact with (or fixed to) the transducer. Kit analysis methods are bioassays. For example, the measurement of blood glucose level is carried out in a tube by adding an enzyme (glucose oxidase) and a corresponding co-enzyme ($NADH_2$) to the blood sample. The resulting product is measured using spectrophotometry.

16.2.1.2. *Classification*

We have just seen that biosensors are composed of at least two elements. The classification of biosensors can therefore be carried out in two ways: a) classification according to transducers; b) classification according to bioelements fixed on the transducer (enzymes, microorganisms, antibodies, cells or tissues). Some authors also classify biosensors depending on their use in different domains such as for environmental, health, food-processing and defense.

In this chapter, biosensors will be presented according to the fixed bioelements. However, applications are more easily discussed if they are divided into their application domains (Table 16.2).

The two biosensor groups are:

– *Metabolic biosensors*: here, the bioelement reacts with the analyte that is transformed. Enzymes (with or without co-factors) and microorganisms (eukaryotes or prokaryotes) are taken as the biosensors' bioelement

– *Affinity biosensors*: in this category, bioelements react with the analyte that is not transformed. Bioelement/analyte reactions result in a complex that is used to reveal the analyte's presence, antibodies and nucleic acids (deoxyribonucleic acid-DNA, ribonucleic acid-RNA).

Biological element (bioelements)		Transducers		Immobilization techniques	Application domains
Cells Organelles	•Bacteria •Algae •Cyanobacteria •Yeasts •Animal (mitochondria) •Plant (chloropasts)	•Optical •Calorimetric •Electrochemical	•CCD (charge-coupled device) •Photodiode •Photomultiplier	•Physical adsorption •Inclusion in polymer matrix (agarose, Sol-gel k-Carrageenan) •Confinement	Food-processing Environmental
Molecules	•Nucleic acids •Antibodies •Enzymes	•Acoustic wave piezoelectric (quartz crystal microbalance (QCM), etc.) •Electrochemical •Optical	 •Amperometric •Potentiometric •Conductometric •Surface plasmon resonance (SPR) •Infra-red •RAMAN	•Physical adsorption •Covalent grafting	Medical Food-processing Environmental

Table 16.2. *Biosensor classification according to the bioelement used or the application*

16.2.1.3. *Transducers*

Transducers use the biochemical modification of the substrate by the bioelement to ensure energy conversion from one form into another. As a consequence, transducers cannot be partnered with any bioelement. An appropriate transducer will be chosen depending on the biochemical modification to maximize the bioelement's reaction and obtain a sensitive signal. For example, if the bioelement/analyte reaction leads to the production of photons, an amperometric electrode (measurement of electrons) cannot be used as the transducer. However a photomultiplier or a CCD (charge-coupled device) camera may be used (Table 16.2).

Transducers used for microorganisms belong to five different classes, for which the principles are summarized in Table 16.3.

Amperometric, potentiometric, conductometric and optical transducers are the most commonly used. Calorimetric transducers are rarely used.

Signal	Transducer	Principle	Example
H+, other ions	Potentiometric	Measurement of a cell's electromotive force (EMF) without current. EMF is proportional to the logarithm of the analyte's concentration (Nernst law). Measurement of the analyte's activity (concentration) against a reference electrode containing the same analyte that is kept at constant concentration.	pH electrode
Ions	Conductometry	Ion-containing solutions conduct electric current. Conductance varies with various chemical or biochemical reactions.	Conductometry electrode
e-	Amperometry	A voltage is imposed (positive or negative) between anode and cathode. This causes the analyte's oxidation or reduction. The magnitude of the peak current is proportional to the analyte's concentration.	Oxygen probe
Heat	Calorimetry	All chemical or biological processes produce or absorb heat (variation of 0.001°C) measured by thermistors. The concept is simple but application is difficult due to interference from the outside environment.	Thermistor
Light	Photomultiplier, CCD camera, photodiode, photographic film, etc.	Light emitted by microorganisms can be amplified by photomultipliers.	

Table 16.3. *Transducers used in the case of microbial biosensors*

16.2.2. *Application fields, interest, drawbacks*

16.2.2.1. *Application fields*

Biosensors have advantages that enable the development for many activities: medical, food-processing, environmental and defense.

In the medical field, biosensors are used for automatic laboratory analysis (the interest being the ability to multiply analyses for the same biosensor), analysis in intensive care units where time is critical, and autoanalysis carried out by the patient (for example, measurement of blood glucose levels for diabetics).

In the field of food-processing, biosensors are mainly dedicated to measurements for follow-up processes (for example, processes requiring continuous detection of lactate) and food quality (detection of pathogens, heavy metals, bioelectronic noses to assess the olfactive quality of wines, for example).

In environmental fields, biosensors are mainly used for the quick detection of chemical compounds (specific measurements) or their effect (toxicity measurement). One of the current difficulties encountered in this field is due to the large number of compounds to be detected (potentially toxic compound environments in comparison to medicine where 20 or so substances are usually measured).

In the field of defense/security, the military and police require portable devices capable of detecting toxic vapors (such as "sarin" gas, mustard gas, bacterial toxins present in biological bombs, etc., methyltrinitrobenzene – TNT – in explosives), and hard drugs on the operational scene.

16.2.2.2. *Interest*

In comparison to conventional methods, biosensors can also be highly sensitive (of the order of a picogram for the most sensitive) and specific (enzyme or antibody only recognizes one molecule).

They are also designed to avoid lengthy preparation before the actual measurement is carried out. The analysis time is decreased (30 min to one hour; in comparison conventional methods which have analysis times of several hours or even days, as in microbiology). Biosensors are re-usable (unlike conventional methods) and simple to use (usable by non-specialists). They can also be miniaturized to make them transportable and usable on site (autonomous). Lastly, they are not polluting as they do not require the use of extraction solvents.

16.2.2.3. *Drawbacks*

A major drawback of bioelements is their stability. This translates into a limited storage and measurement time. A bioelement's storage time ranges from several days to several weeks (however in the case of bioassays, bioelements can be kept several months when lyophilized). Biosensors can only ensure a limited number of measurements in measurement conditions. This is the case for microbial biosensors in particular and is between a few hours and a few days.

16.2.3. ***Markets, patents and legislation***

In 2003 the world market was worth 7.3 billion dollars and it is predicted to be worth 4,400 billion dollars in 2014 (source, Frost & Sulivan, http://www.sensorsmag.com). The field is expanding despite difficulties in perfecting the devices, and their acceptance in the industrial sector. The medical sectors offers the largest market (60%), followed by food-processing and the environmental sectors (defense is too confidential to estimate the market it represents).

Research and development is progressed by 20 big companies, more or less (start-ups not being properly identified). In France, less than ten small start-ups have been identified, whereas the Ministry of Industry surveyed over 800 biotechnology companies.

Around the world there are roughly one hundred research centers dedicated to the study of biosensors. The most active in the development of biosensors are located in the USA, Canada, Japan,

Germany and the UK. France is still behind for the moment despite the considerable market opportunities.

The number of patents registered since the emergence of this sector (1984) is roughly 5,014, against a total 70 million (i.e. 0.007%, query European Patent Office 2011 with key word "biosensor" in the title).

There are no standards or laws regarding the domain of biosensors. In general, the measurement methods are complementary and alternative to standardized chemical analysis measurement methods. Exceptions concern the use of genetically modified microorganisms used in microbial biosensors. In this case, commercialization and trials require authorization. In France, where legislation is strict, it is obligatory to ask authorization from the Genetic Engineering commission in order to perform trials. To commercialize a biosensor, a request must be submitted to the Commission for Molecular Biology.

Lastly, some countries accept and recognize particular biosensors more than others. For example, Germany and Japan both accept BOD (biological oxygen demand) biosensor measurements carried out in wastewater treatment plants.

16.3. Microorganisms

This part of the chapter will incorporate the aspects of microbiology that are essential for understanding microbial biosensors.

16.3.1. *Diversity of microorganisms used in microbial biosensors*

Four main types of microorganisms are used as bioelements in biosensors: heterotrophs (bacteria, yeast-mycets), microorganisms that are mainly bioluminescent before or after insertion of bioluminescence genes from marine bacteria (such as *Aliivibrio fischeri* or *harveyi*), photosynthetic organisms (algae) and, to a lesser extent, bacteriophages (Figure 16.3).

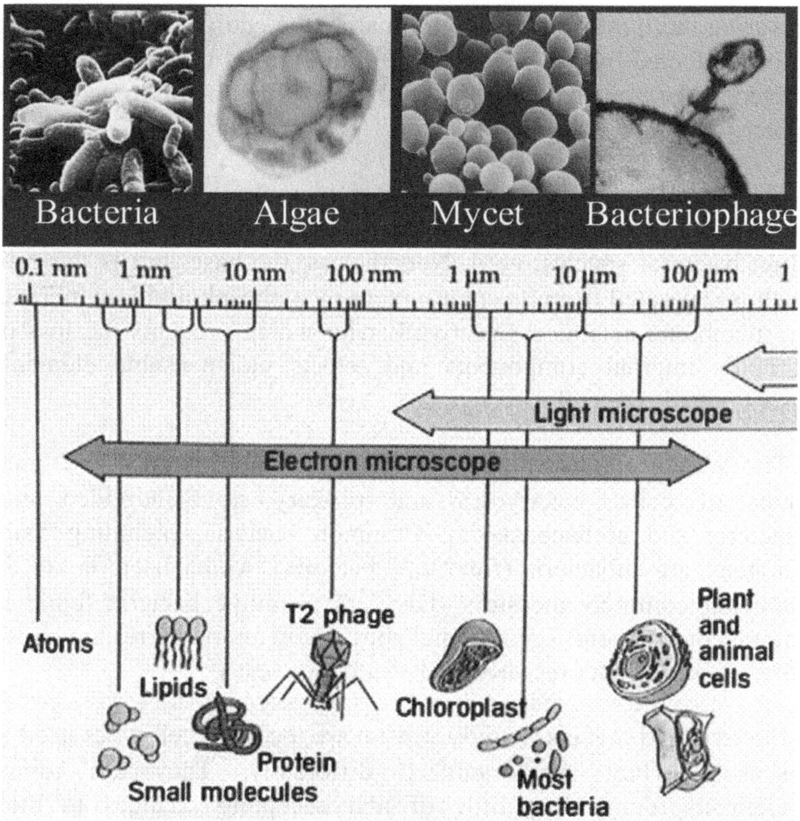

Figure 16.3. *Photographs of the four main types of microorganism. Bacteria (1-2 µm long), algae (4-10 µm), mycets (4-10 µm) and bacteriophage (several nanometers), the one in the photo has latched onto a bacteria's surface [ALB 07]*

In the case of heterotrophic microorganisms, biosensors use the process of respiration (consumption of oxygen, for example). For luminescent bacteria, the production or loss of a light signal is used for measurement. For algae, it is the change in photosynthesis that is measured. Bacteriophages are bacterial viruses that are used only occasionally. They are capable of infecting bacterial cells and either provoke its death or insert genetic material. These viruses are generally modified before use by adding bioluminescent genes, thus giving the infected bacteria the ability to produce light. In this case, measuring the light produced is ultimately the equivalent of detecting bacteria infected by the virus. This is mainly used in food-processing to detect pathogenic bacteria. This subject will not be discussed in this chapter.

The variety of cells available is considerable (as an example, the human body consists of 200 different cell types, and there are several billion bacterial species, etc.). Nonetheless, the base unit is the cell, which is separated from its environment by a phospholipid membrane. The membrane acts as a selective barrier which prevents the loss of the cell's internal components and selects indispensable elements present outside the cell.

The Woëse classification shows the existence of only two main groups of cells: eukaryotes and prokaryotes (subdivided into eubacteria and archaebacteria). Common bacteria, including main pathogens, are eubacteria (for "real" bacteria). Archaebacteria are, in effect, our common ancestors. They are primitive bacteria found in extreme environments (geothermal sites where temperatures are above 100°C, there is high pressure, i.e. several bars, etc.).

Bacteria and mycets (mainly yeasts) are the main cell types used in biosensors. They are organized differently. They are robust bioelements, demanding little of and accepting changes in their environment (unlike human cells, which require a constant temperature in the region of 37°C, pH 7, etc.).

A priori, there are no restrictions opposing the use of any of these microorganisms. However those most commonly used are:

– natural or modified strains belonging to the *Escherichia* type (mainly);

– pseudomonas;

– bacillus;

– some examples of cyanobacteria- and thermophile-type bacteria;

– mycets (*Saccharomyces cerevisiae*);

– microalgae (*Chlorella*);

– bacteriophages (Phage P22, used in *Salmonella* detection biosensors).

The choice is often made depending on the ease with which the microorganism can be grown. It may be noted that the use of marine bacteria is absent for biosensors (with the exception of bioluminescent bacteria of the *Vibrio* type).

16.3.2. *Characteristics of prokaryotes: Escherichia coli*

This is the most well-known bacteria for biologists. It is naturally present in the human intestine and is the starting point of most great biological discoveries. The study of its structure is well-known and it is a representative of other bacteria in the same group.

The bacteria will be described from inside to outside.

Escherichia coli is a small stick (called "rod-shaped bacillus" or simply "bacillus"). Its weight is estimated to be around one pictogram, i.e. 10^{-12} g and 50% of it consists of proteins. It does not exceed 2 µm in length and has a diameter of 0.5 µm. It requires very little: an organic carbon source (glucose, etc.), a nitrogen source (ammonium), a phosphate source, mineral elements (k, Mg, etc.) and most importantly an electron acceptor: oxygen. With these base sources, bacteria divide three times every hour and synthesize all the essential elements (proteins, nucleic acids, lipids, etc.) that they require during their lifespan (Figure 16.3).

Prokaryotes (eubacteria and archaebacteria), represented by *Escherichia coli*, lack compartments like the nucleus. All molecules (proteins, nucleic acids, lipids, carbohydrates) essential for the cell's proper functioning float in an aqueous, viscous liquid called cytosol. Bacteria consist of more than 2,000 different types of molecules. Genetic information is carried by one single chromosome, DNA (deoxyribonucleic acid). In comparison, our cells have 23 pairs of chromosomes.

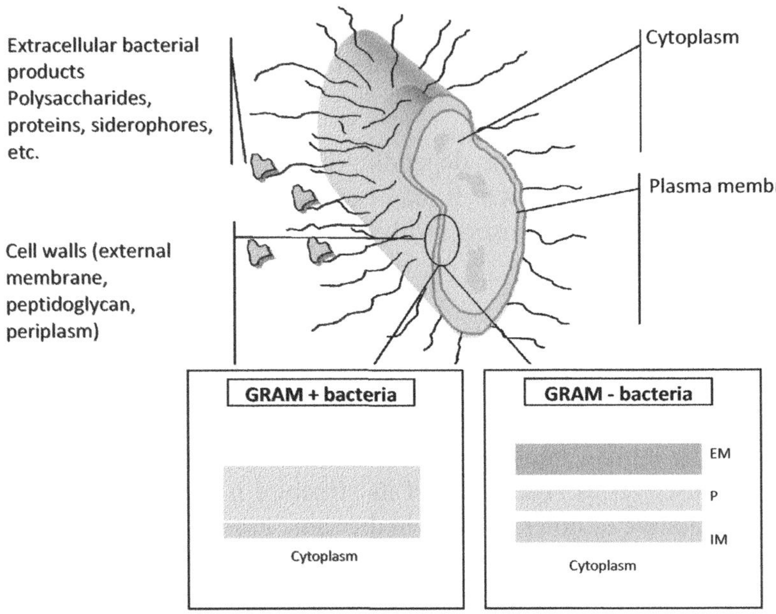

Figure 16.4. *Diagram of bacterial structure (of the Escherichia coli type)*

Cytosol is surrounded by a membrane, termed "internal or plasma", consisting of lipids and proteins, and which play an important role as a selective barrier.

The plasma membrane is itself surrounded by a "mesh" that protects the bacteria from osmotic pressure imbalances that could cause it to swell and burst. This "mesh" is called the "peptidoglycan".

Above the peptidoglycan, a second membrane envelops the cell and is called the "external membrane". This membrane is similar to the internal membrane as it also consists of lipids and proteins. beware: This structure is not present in all bacteria.

Finally, protein structures are anchored on the bacteria's surface. They are:

– flagella, which ensure bacteria's locomotion. It can move at a speed of 50 μm/S, i.e. 180 m/h or 25 times its size/s;

– pili, still called fimbriae, which help the transfer of information "by conjugation" (the term sexual pili is still used) between two bacteria or in surface adherence phenomena (teeth, etc.).

All these details are necessary to understand that reactions used in biosensors between bacteria and their analyte happens increasingly frequently in the cell (in the cytosol). In other words, this reaction is limited by all the structures mentioned above.

16.3.3. *Eukaryotes*

Eukaryotes (from the Greek meaning "true") include protista, mycets (yeasts used in biosensors), plants and animals.

They have a nucleus and other organelles, all confined within a membrane that is different to that of prokaryotes. Eukaryotic cells are much larger and more complex than prokaryotes and their growth/division rate is much slower (one division per day on average).

16.4. Microbial biosensors

16.4.1. *Brief history*

Microorganisms were first observed in the 17th Century by Antoine van Leeuwenhoek and Louis Joblot. This was the start of modern day microbiology. It is estimated that 50% of the earth's biomass is microbial in nature (parasites, diatoms, algae, fungi, bacteria and viruses) and that viruses alone (living parasitic organisms that depend on the cells they infect) reach the astronomic number of 10^{31}. Human beings and animals within their environment are therefore dependent on the activities of microorganisms that are the critical keystones for all geochemical cycles.

Microorganisms can be seen in both a negative and positive light depending on the observer. Doctors consider them to be harmful and medical history supports this point of view. Biotechnologists on the other hand, try to understand them and try to use them in processes as old as bread-making, or as promising as molecule biosynthesis for therapeutic reasons.

Microorganisms (mainly bacteria) have been used in the field of medicine in order to evaluate the effect of chemotherapeutics initially discovered by Paul Ehrlich in 1904. Antibiograms where the first bioassay developed between the years of 1940 and 1945, in order to determine the activities of newly-discovered antibiotics with respect to bacteria, tests which are now termed “sensitive bioelements”.

Considerable scientific and medical progress was achieved and from the 1960s onwards, scientists were concerned with the effects of chemical substances on the environment, capable of modifying major geochemical cycles. Ecotoxicology therefore developed a considerable number of tests using representative microbial bioelements to assess the impact of chemical pollutants on their activity, and to deduce the possible consequences on microbial ecosystems. One of the main criteria for a bioassay’s success lies in its simplicity of use. Acknowledgement must be given to the research carried out by Kempner and Hanson, followed by those of Bulich, Isenberg and Ultizur in the 1980s, as they ensured the success of a simple bioassay based on the bioluminescence of marine bacteria *Vibrio fischeri* (renamed *Aliivibrio fischeri*) upon contact with toxic substances. The test was commercialized under the name of Microtox™ and is an international reference. It most certainly contributed to the development of modern microbial bioassays.

In parallel to research on the effects of toxic substances on the environment, studies published in the 1960s put forward the biodegrading properties of organic substances by microorganisms, leading the way for the first articles, in 1963, on the topic of biodegradation measurement. The aim was to assess the biodegradation of chemical substances before their commercialization through the means of microorganisms in closed reactors, and to estimate their persistence in the environment. This problem is still pertinent today, especially after the implementation of the EU regulation REACH 1907/2006 (Registration, Evaluation and Authorization of Chemicals), where Microbial biosensors are fully justified.

Modern analysis using microorganisms is intimately linked to sensor technology. With the advent of oxygen sensors from as early as the 1960s, Isao Karube, following the research of Clark and Lyons in 1962 on enzyme biosensors, succeeded in 1977 in immobilizing bacteria on an oxygen sensor for the measurement of biological oxygen demand (BOD), which now appears in all known generations of microbial biosensors. The simple but revolutionary idea was to convert biological signals (in this case the consumption of oxygen by bacteria due to the consumption of organic matter) into electrical signals through the use of an oxygen sensor. At the time, it was much easier to measure the signal than to dose the organic matter itself.

The year 1982 was a turning point as far as microbial biosensors are concerned: two teams, led by Baldwin and Engebrecht, simultaneously published their findings on the subject of cloning genes for bioluminescence present in *Vibrio harveyi* and *Aliivibrio fischeri*, in *Escherichia coli*. The latter therefore became bioluminescent. From the 1960s onwards, this scientific skill, which is now common, enabled the development of bacteria that emit light in the presence of specific targets (metals, organic or metal organic chemical substances). The advantage is the measurement of pollutant concentrations termed bioavailable (those in contact with the microorganism) unlike chemical analysis which measures the total concentration. Chemical and biological analysis are therefore complementary [THO 11].

16.4.2. *Summary of the state of the art*

A special 2011 journal issue on the use of microorganisms for analysis gives a comprehensive overview of the current knowledge of biosensors using microorganisms [THO 11]. Microbial biosensors find their place in the environmental field as they are able to give an account of an effluent's or a chemical molecule's toxicity (non-specific biosensor) or else act as specific biosensors for the measurement of specific pollutants (metals, organotins, pesticides, etc.). Microorganisms are sensitive in all cases to the pollutant's bioavailable part, i.e. the concentration of the pollutant acting on a microorganisms particular mechanism (respiration, synthesis, DNA, etc.).

Conversely, chemical analysis, which often extracts pollutants using a solvent, gives an account of the total concentration. Chemical and biological analyses are not competing but complimentary.

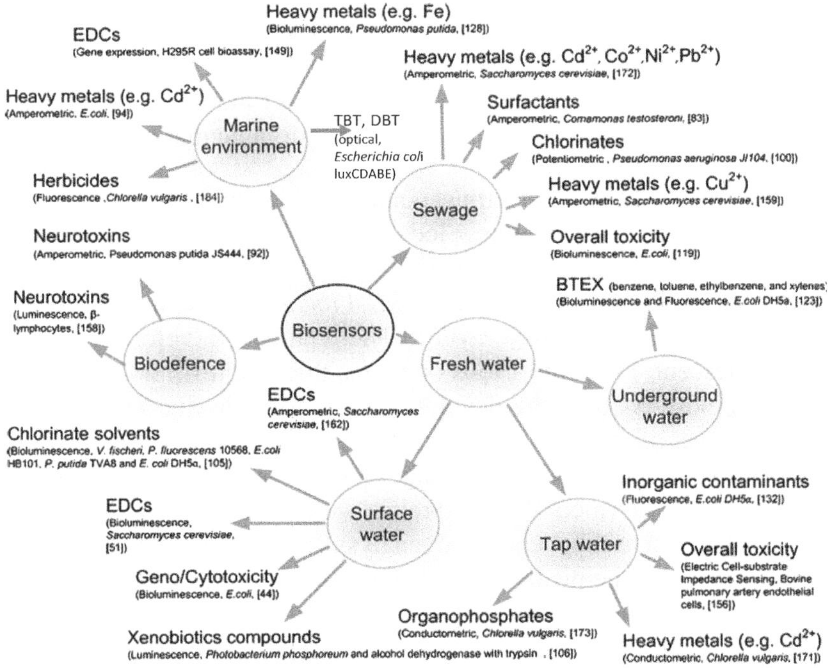

Figure 16.5. *Summary of various state-of-the-art microbial biosensors used to detect pollutants in different environmental areas [ELT 11]*

Figure 16.5 shows chemical pollutants from the list of priority substances detected by microbial biosensors over the last 20 years for different applications (underground water, etc.). The signals emitted by the microorganism upon contact with the pollutant can be processed through two different transduction modes: optical (luminescence, fluorescence) or electrochemical (amperometric, conductometric, potentiometric). The limits of detection vary greatly and depend on both pollutant and microorganisms. The lowest limit is most certainly that obtained for dibutyltin, a derivative of TBT (0.03 μg.L^{-1}) or mercury (1 pM-*Escherichia coli* K12 pBMer*lux* [JOU 11a, JOU 11b]). The limit of detection lies on average between μg and mg/L. The detection time varies between 20 minutes and two hours.

However, specific pollutant detection is not historically the first application. Microorganisms were initially used to measure global pollution, called biological oxygen demand (BOD). This abstract term means the quantity of oxygen required by microorganisms to oxidize organic pollutions present in water (this measurement is of interest mainly for sewage treatment plants). The measurement of an effluent or a molecule's toxicity is a major advantage of microorganisms. This effect needs to be taken into account when measuring the loss of a particular biochemical function (light emission, respiration, photosynthesis, etc.).

Microorganisms are capable of specifically measuring an analyte (metal, pesticide, etc.) due to its modified genetic heritage. The specific element is a promoter (small DNA sequence) that reacts more or less specifically with the pollutant. Detection of the analyte/promoter association is highlighted by reporter genes. These are either genes for bioluminescence (therefore the transducer is optical) or genes producing electrochemically active substrates (transducer will then be electrochemical).

The microorganism constantly produces light (or constantly uses oxygen, light for algae, etc.). Any alteration to the light production process will be a sign of toxicity on the process.

In this case, the respiration process (consumption of oxygen or production of CO_2) is used to measure the consumption (biodegradation) of the carbon present in the effluent. This type of biosensor is widespread and is termed the "BOD biosensor" (see section 16.5.1).

16.4.3. *Immobilization*

Immobilization consists of maintaining a bioelement against the transducer for the duration of the analysis and during cleaning phases. It must ensure a favorable environment for reaction (pH, salinity, transparency if necessary, etc.). The downside is that the immobilized element limits the transfer of the analyte to the bioelement (for example, oxygen will have difficulty diffusing to the matrix, thus slowing down its progression towards the bacteria).

There is no standard immobilization method which caters for all applications. Only main principles exist. Multiple methods are often tried before the best compromise is found in order to maintain the bioelement's stability and to best respond to measurements (for more details, please refer to [PER 09]).

Immobilization	Method	Support	Scheme
Superficial	Adhesion	Direct link	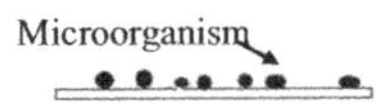
		Indirect link	
	Adsorption	Membrane	
		Porous support	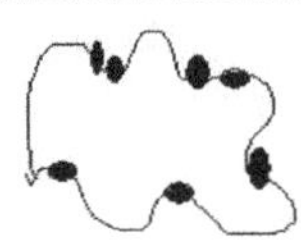
Inclusion	Encapsulation	Matrix (agarose, etc.)	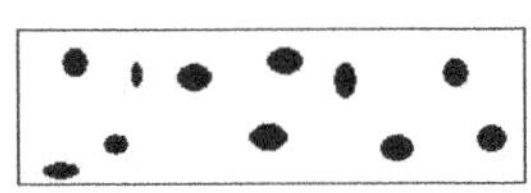

Table 16.4. *Different immobilization principles of microorganisms (according to [PER 09])*

The methods used for microbial biosensors are inclusion/encapsulation of natural polymers (agarose, alginate, k-carrageenan) or synthetic (PVA – polyvinyl alcohol, latex, sol-gel, etc.) in gels (Tables 16.4, 16.5 and 16.6).

Alginate
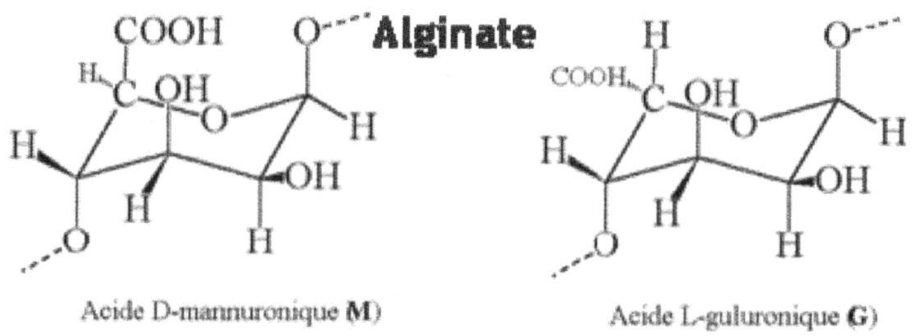
Agarose
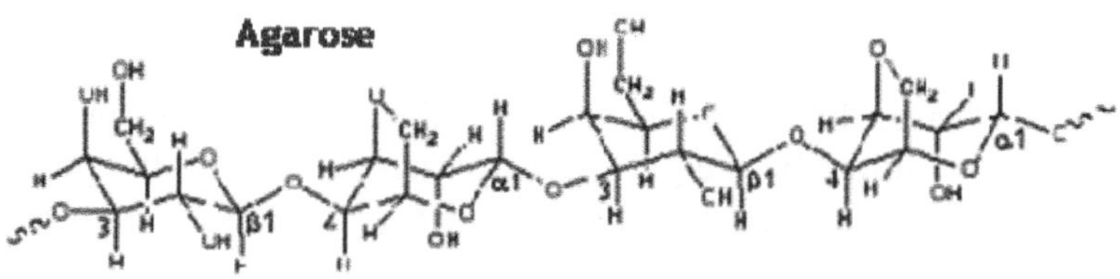
k-carrageenan
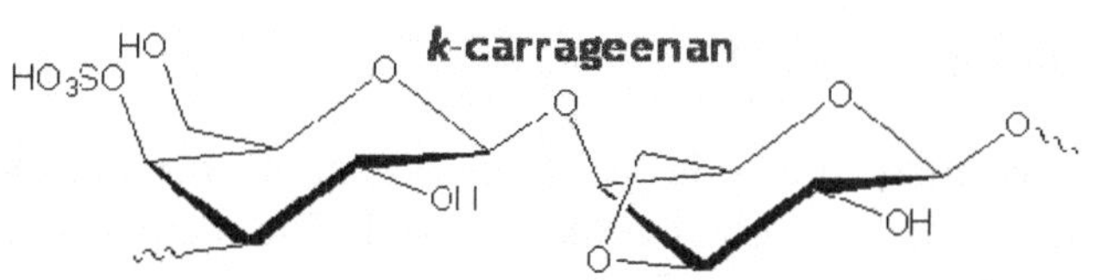

Table 16.5. *Natural gel structures: alginate, agarose and k-carrageenan (according to [PER 09])*

PVA	Polyacrylamide
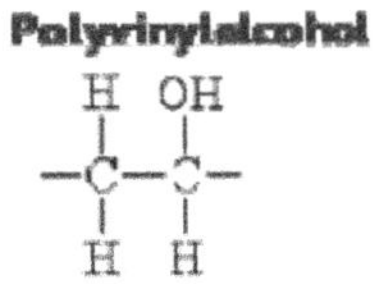	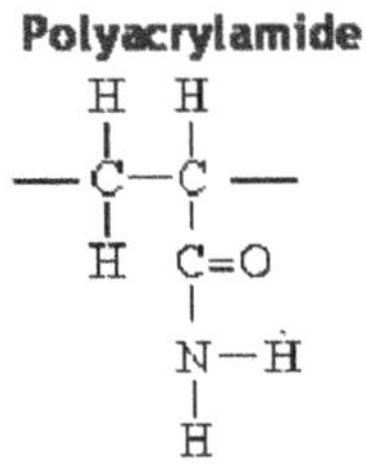
PUR	**PCS**
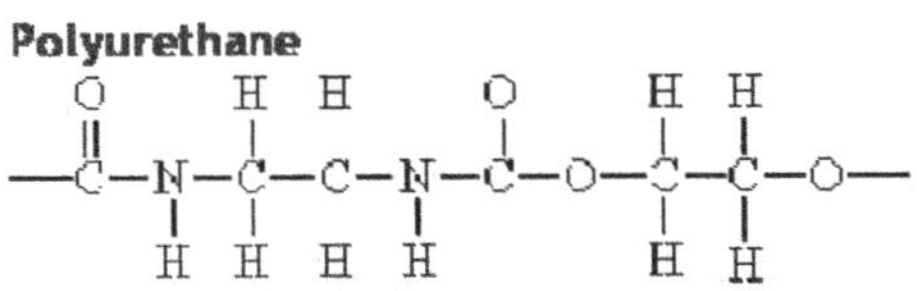	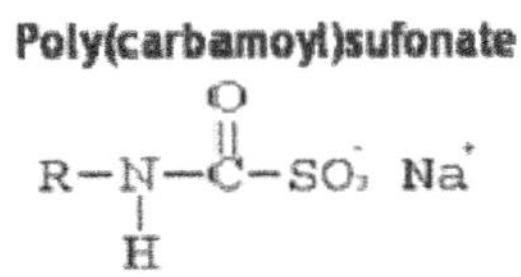

Table 16.6. *Gel structure of polyvinyl alcohol (PVA), polyacrylamide, polyurethane (PUR) and poly(carbamoyl)sulfonate (PCS) (according to [PER 09])*

In this case, bioelements are free in the solution but their movement is limited by the gel's trellis structure. The trellis' porosity is controlled to avoid losing the bioelement and allow the movement of nutrients and release of products.

Some systems use an additional sandwich technique that consists of trapping the polymer gel + microorganism between two nylon membranes, cellulose acetate, etc., in order to prevent the loss of microorganisms.

More marginal methods or immobilization on acetate paper, 0.22 μm filters or fiber glass could be used. The method requires the microbial suspension to be filtered on the filter or paper that retains the cells. The filter is then placed on the transducer. This method has the disadvantage of allowing microorganisms to escape more easily.

In addition to problems related to the material and heat diffusion within microbial gels, there is an additional drawback which is the gel's instability during the use of microorganisms. The latter divide and are ultimately released to colonize adjacent wells as well as the system's entire fluidics.

16.4.4. *Performance*

The creation of a new biosensor requires the determination of characteristics allowing it to be compared to both existing biosensors and traditional methods. The following criteria must be determined.

Selectivity

A biosensor must be specific: this is the reason for its creation. Its specificity is determined by its biological elements. These can be:

– exclusive (limited) to a molecule with a particular configuration (for example D-glucose recognized and not L-glucose);

– large with respect to a molecular group (all benzene derivatives including toluene, xylene, etc.);

– non-specific (non-specific bacteria that will indicate an effluent liquid's global toxicity).

The scope of measurement and the limit of detection

It is necessary to know the analyte's concentration range which is linearly measurable by the biosensor. In general, biosensors must be able to measure an analyte with a concentration varying between two orders of magnitude.

The limit of detection (LOD) often represents an indicative value. It is defined as the smallest concentration detected by the biosensor. Microbial biosensors sometimes have a higher LOD than conventional chemical analysis methods (mg/L and even µg/L).

Signal and measurement reproducibility and repeatability

Every measurement must be reproducible. This is even more important for biosensors that can have a deviation signal due to the biological detection element (therefore are unstable with respect to time).

Repeatability (measurements with the same bacteria membrane) and reproducibility (measurements with different microorganism membranes) are usually between 90 and 95%.

Response time

A biosensor's response time is the time taken for the sensors to reach a stationary state where variations in signal are no longer perceived [TRA 91]. In practice the response time varies with the amount of microorganisms and the thickness of the diffusion layer. Conversion of chemical data into an electrical signal by the transducer also requires a certain amount of time, which varies for each transducer.

The response time (which is a global measurement) therefore corresponds to the sum of different phenomena in need of optimization. For biosensors using microorganisms, the response time is usually between 20 and 60 minutes (for example, some enzyme biosensors have response times under one minute). Finally, even if the response time is particularly short, the time between two analyses (the recovery time) must be the shortest possible in order to increase the frequency of measurements.

Lifetime

One of the major disadvantages of biosensors is the short lifetime of their bioelements. As a consequence, the lifetime of the biosensor depends on the bioelement and quality of the immobilization (hydrophilic region for enzymes and bacteria, etc.).

Two types of lifetime need to be determined:

– the biosensor's operational lifetime (several days on average);

– the biosensor's lifetime under storage conditions (several weeks to several months on average).

16.5. Examples of microbial biosensors in environmental applications

This section aims to present the technological developments of various microbial biosensors for measurement of non-specific pollution (biosensors measuring the biological oxygen demand) or specific pollutants (heavy metal detection), through the presentation of specific examples.

Biosensors measuring biological oxygen demand (BOD), BOD sensors, are devices which measure overall pollution. This differentiates them from the first biosensor characteristic which is specificity. This "failure to respect" the rule is easily explained. Hundreds of different molecules are in solutions or adsorbed on colloids in the environment. For cost-related reasons, it would be impossible to create biosensors specific to each molecule. Here, an overall response is wanted through biodegradation of organic matter.

16.5.1. *BOD principle*

Biological, or biochemical, oxygen demand indicates the amount of biodegradable organic matter present in different environmental compartments (river water, seawater, sewage). Microorganisms (X) oxidize organic matter (MO) as shown below:

$$MO + O_2 + X_{initial} \rightarrow CO_2 + X_{multiplied} + H_2O.$$

The more oxygen is used by microorganisms, the more organic matter (carbon source) is biodegraded. The traditional and standardized method takes five days, in which microorganisms (from the environment) are mixed to different effluent dilutions that are to be tested. This method is long and explains why various studies have looked into developing a biosensor to shorten this analysis time.

16.5.2. *Principle of BOD biosensors*

Bacteria, often *Pseudomonas putida* or *fluorescens*, are immobilized on an oxygen probe. The more the bacteria "respire" (use oxygen), the more biodegradable dissolved carbon they will have used (Figure 16.6). The bacteria's oxygen consumption leads to a decrease in the electric current measured by the oxygen probe.

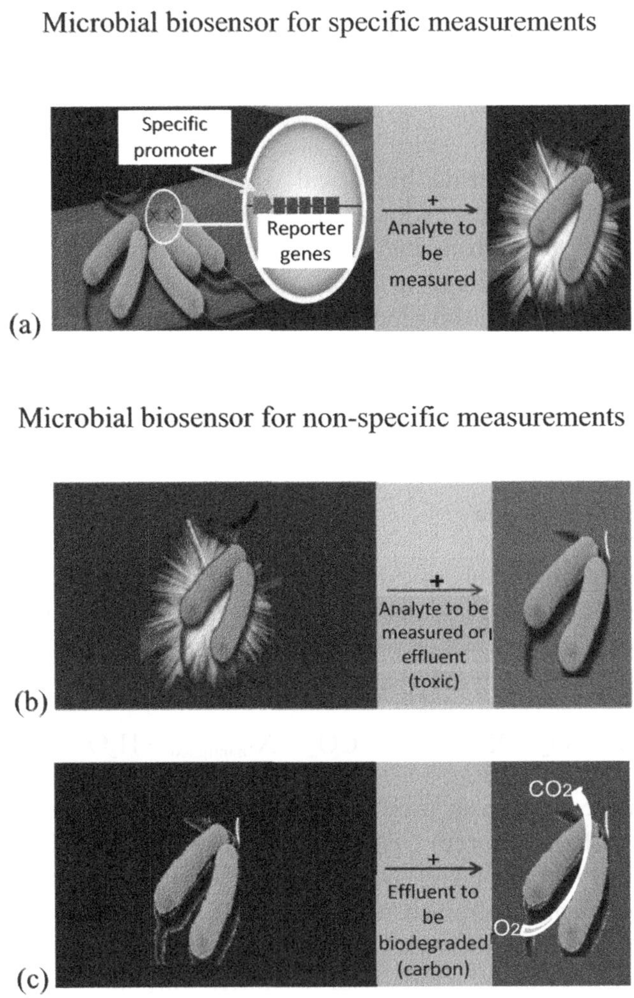

Figure 16.6. *Two main application principles of microbial biosensors detection of heavy metals by genetically modified* Escherichia coli *bacteria (example from research carried out in CBAC, UMR CNRS GEPEA, IUT laboratories at La Roche sur Yon)*

Measuring a heavy metal by conventional methods requires sophisticated apparatus and qualified personnel. The cost for each analysis is around 30 €/metal.

16.5.3. *Principle of alternative analysis methods*

Metal detection technologies rely on the measurement of bioluminescence generated by contact between a bioluminescent bacteria and the metals to be detected. When the metal is present, bacteria produce light at a frequency between 490 and 590 nm (visible) (Figure 16.7).

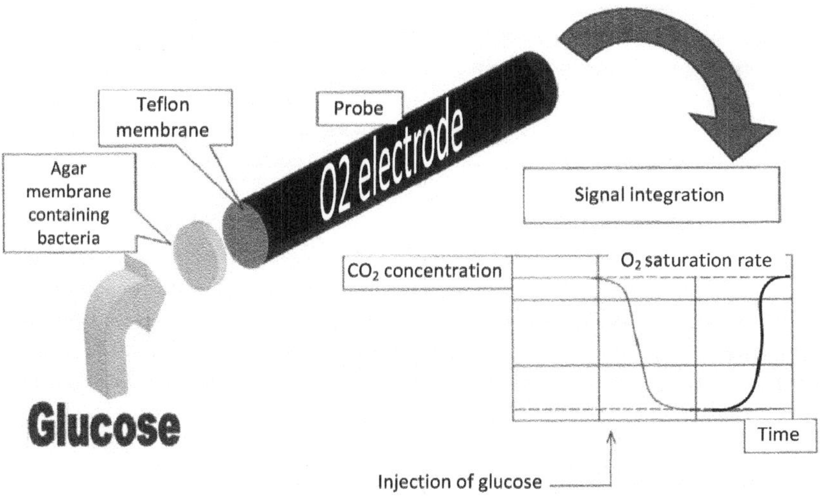

Figure 16.7. *Scheme of microbial biosensor for the measurement of BOD*

Genetic modification requires a promoter isolated from natural metal resistance mechanisms to be partnered with a bioluminescence operon. The latter taken from marine bacteria, such as *Alliibrio fischeri*, is either complete (*lux* CDABE) or truncated (*lux*AB). *LuxAB* genes code for luciferase, whereas CDE codes for the synthesis and recycling of a long-chained aldehyde (tetradecanal). The enzyme requires oxygen for the oxidation of aldehyde in fatty acid and of $FMNH_2$ produced by the bacteria's metabolism.

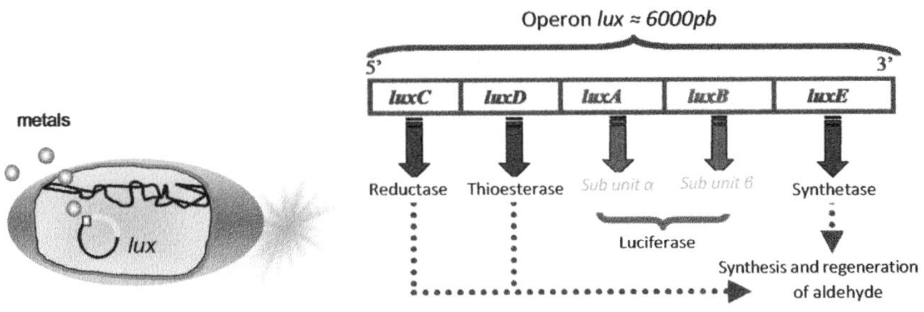

Figure 16.8. *Metal measurement principles for bioluminescent Escherichia coli bacterium*

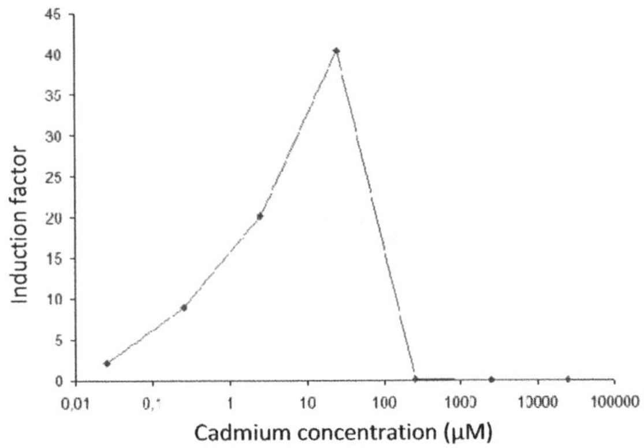

Figure 16.9. *Production of bioluminescence by Escherichia coli pBZntlux strain with the sample's cadmium concentration. The induction factor is defined as the relationship between luminescence recorded after cadmium exposure and that recorded in a cadmium free medium*

Figure 16.9 represents the response of *Escherichia coli pBZntlux* when exposed to an increasing cadmium concentration. Luminescence is measured after one hour. The figure shows that bioluminescence increases for concentrations ranging from 0.025 µM (cadmium limit of detection) to 25 µM (bacteria's induction maximum). A decrease in luminescence is observed for higher concentrations and is linked to the metal's toxic effect.

16.5.4. *Example of three metal detecting bacterial biosensors*

The proposed systems differ from a technological point-of-view but use the same bacterial strains.

Biosensor in liquid phase: Lumisens I

Lumisens I is a patented biosensor and was the first system developed in a laboratory and published by [THO 03] and [HOR 04] (Figure 16.9). The bacterial bioluminescent bioelements in this system are maintained in a stable physiological state, in a liquid medium, in a continuous growing reactor managed by sensors (temperature, dissolved oxygen, pH). The cellular density is automatically measured in real-time by the absorbance measurement (Beer-Lambert law). A laser diode is placed below the growing reactor and emits constant light radiation through the growing medium. Unabsorbed light is measured by a photodiode. The level of bioluminescence produced by the bacteria is measured with a photomultiplier.

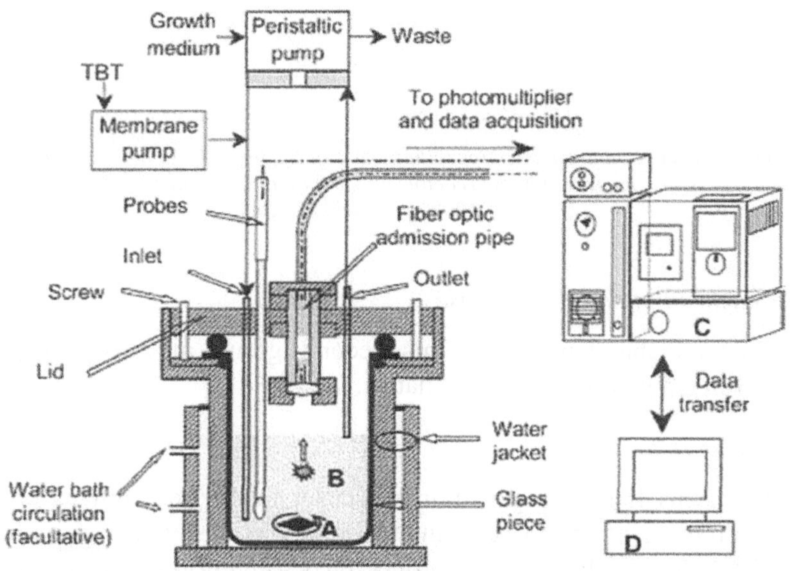

Figure 16.10. *Lumisens I – diagram of the biosensor according to [HOR 04]. A: stirrer; B: bioluminescent microorganisms cultivated in bioreactor closed by screw-down stainless-steel lid; C: photomultiplier and acquisition/regulation device; D graphical interface [THO 03]*

Figure 16.11 illustrates three successive cadmium pollution events that translate into an increase in bioluminescence induced by the bacteria in the reactor, directly linked to the sample's metal concentration. This increase in emitted light intensity is sensed by the photomultiplier and is represented graphically by a considerable peak.

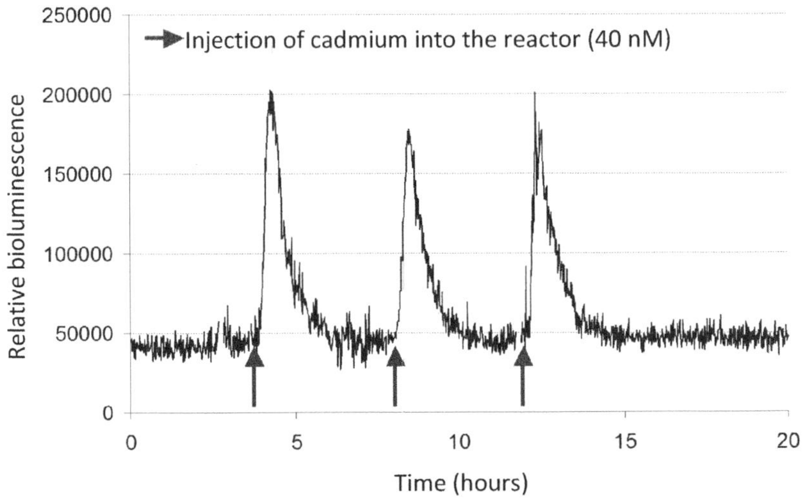

Figure 16.11. *Cadmium detection with the use of Lumisens I. Bioluminescence is measured in the reactor by a photomultiplier after the bacteria's contact with pollutant*

Biosensor in immobilized phase

Bacterial bioelements in this second group of biosensors are immobilized in a solid agarose matrix.

Lumisens II

Lumisens II is a single-strain biosensor in the immobilized phase (measurement of only one pollutant) and was designed and built in 2007 [HOR 07]. The patented concept (N°2889591) is schematically described in Figure 16.12.

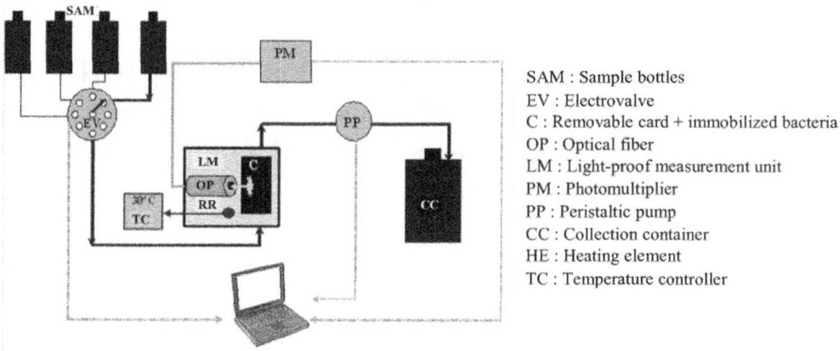

Figure 16.12. *Diagram of the biosensor*

A multiposition electrovalve allows the selection of the sample to be put into contact with the bacteria. The chosen sample goes through a measurement unit containing a removable card in which bacteria are immobilized. A peristaltic pump forces the fluid along the biosensor's channels. The measurement unit and the card are temperature controlled to the desired temperature by a heating element, regulated by manual relay and a thermocouple placed in contact with the unit. The bioluminescent signal is caught by a bundle of fiber-optics linked to a photon-counting photomultiplier. Light signals are then processed using a computer interface. However, unlike its predecessor, there is no control system for dissolved oxygen and pH in the system (Figure 16.13).

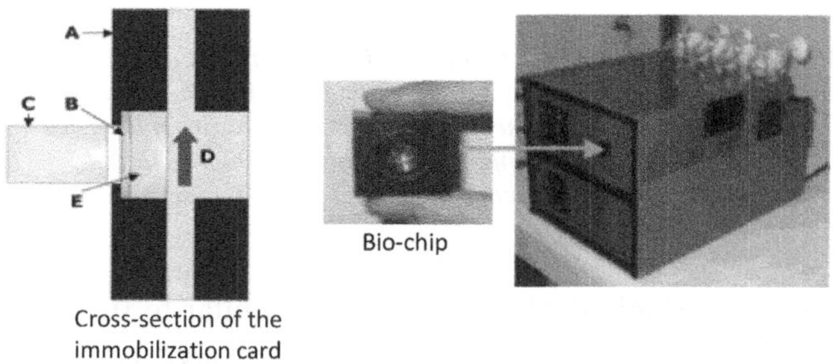

Figure 16.13. *Photograph of Lumisens II and the immobilization card (support for bioluminescent bacteria) and cross-section. Immobilized bacteria in transparent agarose (2%) touch the liquid to promote exchanges.* ***A****: Card o Bio-Chip made of black acetal;* ***B****: viewing window;* ***C****: fiber-optic;* ***D****: flow direction of the liquid;* ***E****: agarose matrix containing bioluminescent bacteria*

Lumisens III

The biosensor Lumiens III was published in 2011 by [CHA 11] in collaboration with the University of Maine. This biosensor is the most evolved version, having been directly derived from Lumisens II. Its general outline of operation is shown in Figure 16.14.

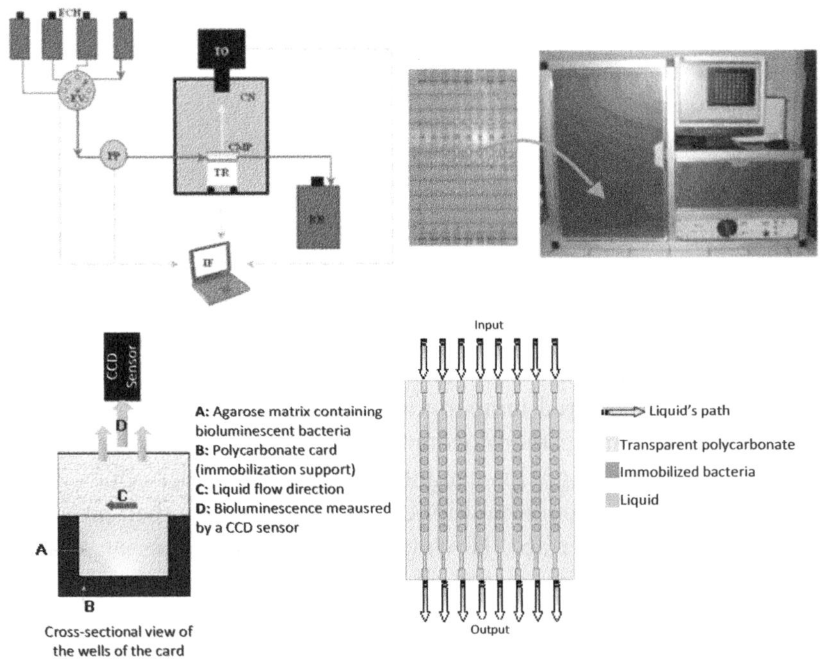

Figure 16.14. *Multi-well card for Lumisens III (general diagram and cross-section). Fluid flows in the card were optimized by [AFF 09].* ***A****: Agarose matrix containing bioluminescent bacteria;* ***B****: Polycarbonate card (immobilization support);* ***C****: Flow direction of liquid;* ***D****: Bioluminescence measured by a CCD sensor*

The general principal of fluidic flow is similar to that of biosensor Lumisens II: flow and sample selection is ensured respectively by a peristaltic pump and an electrovalve. The main difference with this system is the measurement card, which contains the immobilized bioluminescent bacteria. In this case a multi-well and multi-channel card is used which allows simultaneous exposure of a sample to several bacterial strains of different sensitivity (several sought compounds). This is what differentiates it from the two previous systems.

The use of this card implies a number of modifications of the Lumisens II system. The temperature control by means of a heating element in Lumisens is replaced by a Peltier module which optimizes thermal transfers. The configuration of the card containing bacteria no longer allows the measurement of bioluminescence emitted by each well, using a photomultiplier. In addition, the optic transducer is replaced by a CCD (charge coupled-device) camera, which allows the overall simultaneous capture of the wells. After computer processing, the levels of bioluminescence produced by each well is therefore individually obtained.

In the example illustrated in Figure 16.15, daily mercury pollution (500 nM) has been simulated. Each of the graph's peaks corresponds to an exposure to the pollutant (mercury at 500 nM). Using the biosensor Lumisens III, we are able to highlight the presence of the specific pollutant in the analyzed sample. However, a certain instability in the signal is apparent during the first three days and beyond the seventh day. This instability is linked to physiological modifications in the bacterial bioelements, which therefore limits the biosensor to a five day use.

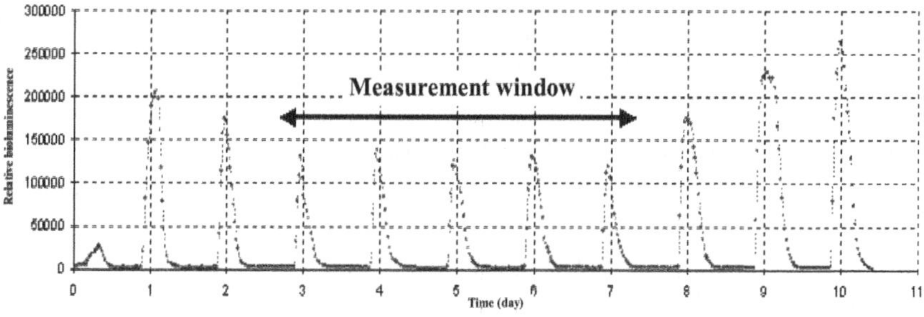

Figure 16.15. *Example of pollution detection in effluents with the use of Lumisens III. Case of daily mercury exposure*

16.6. Bibliography

[AFF 09] M. AFFI, S. CAMILLE, P. LEGENTILHOMME, J. COMITI, J. LEGRAND, G. THOUAND, "Numerical design of a card and related physicochemical phenomena occurring inside agarose-immobilized bacteria: a valuable tool for increasing our knowledge of biosensors", *Sens. Actu. B*, vol. 138, pp. 310-317, 2009.

[ALB 07] B. ALBERTS, J. ALEXANDER, L. JULIAN, R. MARTIN, R. KEITH, P. WALTER, *Molecular Biology of the Cell*, Garland Science, 2007.

[CHA 11] T. CHARRIER, C. CHAPEAU, L. BENDRIA, P. PICART, P. DANIEL, G. THOUAND, "A multi-channel bioluminescent bacterial biosensor for the on-line detection of metals and toxicityPart II: Technical development and validation of the biosensor", *Anal. Bional. Chem.*, vol. 400, pp. 1061-1070, 2011.

[ELT 11] E. ELTZOV, R.S. MARKS, "Whole-cell aquatic biosensors", *Analytical and Bioanalytical Chemistry*, vol. 400, pp. 895-913, 2011

[HOR 04] H. HORRY, M.J. DURAND, P. PICART, L. BENDRIAA, P.H. DANIEL, G. THOUAND, "Development of a biosensor for the detection of organotin compounds", *Environmental Toxicology*, vol. 19, pp. 342-345, 2004.

[HOR 07] H. HORRY, T. CHARRIER, M.J. DURAND, B. VRIGNAUD, P. PICART, P.H. DANIEL, G. THOUAND, "Technological conception of an optical biosensor with a disposable card for use with bioluminescent bacteria", *Sensor and Actuator B*, vol. 122, pp. 527-534, 2007.

[JOU 11a] S. JOUANNEAU, M.J. DURAND, G. THOUAND, "Développement de biocapteurs bactériens optiques pour la détection de métaux dans l'environnement", *REE (Revue de l'électricité et de l'électronique)*, vol. 1, pp. 24-33, 2011.

[JOU 11b] S. JOUANNEAU, M.J. DURAND, P. COURCOUX, T. BLUSSEAU, G. THOUAND, "Improvement of the identification of four heavy metals in environmental samples by using predictive decision tree models coupled with a set of five bioluminescent bacteria", *Environmental Science Technology*, vol. 45, pp. 2925-2931, 2011.

[PER 09] M. PERNETTI, D. PONCELET, G. THOUAND, "Properties and choice of material used for microbial biosensor", in R. COMEAUX, P. NOVOTNY (eds), *Biosensors: Properties, Materials and Applications*, Nova Science Publisher, New York, 2009.

[THO 03] G. THOUAND, H. HORRY, M.J. DURAND, P. PICART, L. BENDRIAA, P.H. DANIEL, M.S. DUBOW, "Development of a biosensor for on line detection of organotin compounds with a recombinant bioluminescent *Escherichia coli* strain", *Applied Microbiology and Biotechnology*, vol. 62, pp. 218-225, 2003.

[THO 07] G. THOUAND, H. HORRY, B. VRIGNAUD, Biocapteur en phase immobilisée, N°2889591, submitted 2 August 2005, delivered 9 February 2007.

[THO 11] G. THOUAND, "Microorganisms for analysis", *Analytical Bioanalytical Chemistry*, vol. 400, pp. 893-894, 2011.

[TRA 91] M.-C. TRAN, *Les biocapteurs – Principes, construction et applications*, Masson, 1991.

Chapter 17

Biofuel Cells

17.1. Presentation and risks

Modern society's demand for energy is increasing and is linked to the medium-term depletion of fossil fuels and the increasing impact of fossil fuel combustion on the environment. It is therefore necessary to find alternative short-term methods for energy production. Amongst these new, clean energy sources (no emission of greenhouse gases or environmental pollution), energy production via electrochemical methods, such as fuel cells, has increased particularly in Europe where it can be used to decentralize energy production to power stations generating different amounts of energy.

Biofuel cells are a subcategory of fuel cells. They are prominent in world research as they convert chemical energy into electrical energy by catalytic reaction of enzymes or living organisms. The concept behind biofuel cells is the use of redox enzymes for the specific targeted oxidation of fuels (alcohols, hydrogen, sugars such as glucose, fructose, lactose or cellobiose) at the anode and the reduction of oxidizers (O_2, H_2O_2) at the cathode to generate electric power.

Chapter written by Serge COSNIER.

Amongst the alcohols likely to be used as a fuel for biofuel cells, methanol, ethanol and glycerol are considered as renewable fuels that can be produced by fermentation of biomass. The design of biofuel cells does not necessarily require bioanodes and biocathodes to be separated, unlike fuel cells, as enzymes are very selective and highly active.

In addition, fuel cells need catalysts based on precious metals (platinum, rhodium, ruthenium or palladium) or alloys (Raney nickel) for the anode, and precious metals or transition metals (nickel, gold, silver or chrome) for the cathode.

The inherent ecological aspect of biofuel cells must be emphasized: they do not require noble materials (platinum, nickel, etc.) unlike fuel cells. Materials, fuels and products used in the creation and operation of biofuel cells are all biodegradable. Biofuel cells are also free of major economic risks linked to metals: mining resources for non-ferrous (aluminum, nickel, zinc, lead, etc.) and precious (gold, silver, platinum, palladium, rhodium, iridium, ruthenium, osmium) metals are becoming rarer due to the exponential demand for strategic metals by high-technology industries and for metal alloys by aeronautics or the automobile industry.

The scientific challenge is elaborating devices with power and size compatibility for use as nomadic energy sources (low power miniature generator for mobile phones and portable GPS). Biosystems could also ensure the recharging of equipment's batteries and standby modes when associated with conventional batteries.

The majority of biofuel cells produce electric energy from the electro-enzymatic degradation of glucose and O_2, two substrates found in physiological liquids (blood, extracellular liquids). The main motivation for the development of biofuel cells concerns their use in the human body as an autonomous energy source for implanted medical devices. As the fuels for biofuel cells are found in the organism, these generators will be completely autonomous and will be able to supply energy to pacemakers, micro-machines, micro-pumps, sensors, etc. Even though lithium batteries are currently the preferred energy source for electronic medical implants, research work is focused on the development of alternative energy sources that are able to operate independently without needing external refueling.

Several strategies based on harnessing the body's energy are being explored to fuel medical implants. However these systems, such as those converting the body's mechanical energy into electrical energy, or those using thermoelectric effects, have limited power generation that cannot exceed 100 μW at best. In this context, the potential societal impact of glucose biofuel cells would be tremendous as there are currently only batteries for pacemakers or insulin pumps that use little power (10 μW) and no device for powers of several hundred μW.

Even though the first biofuel cell was created in 1964 (consisting of various enzymatic bioanodes associated with an oxygen cathode), the research field remained unexplored until the end of the 1990s [YAH 64]. Since 2000, research on clean electricity production methods using renewable fuels has promoted the theme of "biofuel cells", which is receiving renewed attention illustrated by an exponential progression in terms of scientific publications, as shown by the graph in Figure 17.1.

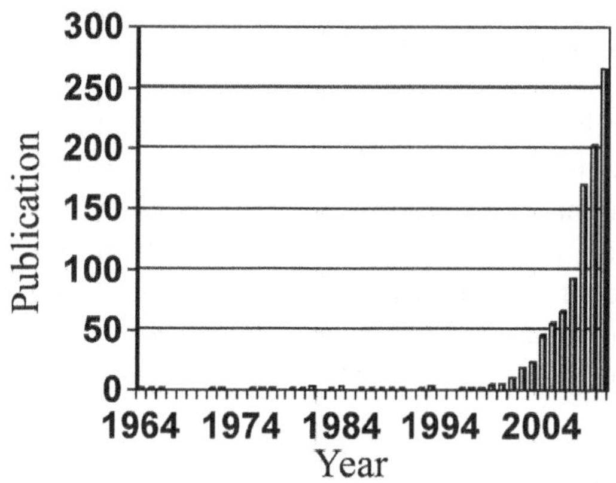

Figure 17.1. *Exponential development of this theme illustrated by the number of publications published from 1964-2010*

This trend is also explained by the fact that technological obstacles related to the development of biofuel cells have recently been circumvent thanks to the commercial availability of a wide range of purified enzymes, the design of new biomaterials for electrodes and new electrode configurations allowing the immobilization of a high enzyme density and their electrical connection. Technological progress in the domain of biosensors, in particular between 1990-2000 created considerable potential in the domain of biofuels. Examples of progress are the quick development of different immobilization possibilities for enzymes and microorganisms and strategies based on redox mediators with the aim of establishing an electrical transfer between the electrode and biomolecules.

A number of improvements have been made in the domain of enzymatic biofuel cells in the past 20 years and have been summarized in various reviews [BAR 04, ATA 07, CRA 08, BIL 08, WIL 09, IVA 10]. The use of whole organisms, with enzymatic paths resulting in the complete metabolism of sugar or alcohol substrates (microbes, bacteria, mircroorganism cultures, whole cells and even biofilms), has also given rise to a series of "microbial" fuel cells. These do not require isolation or enzyme purification, and allow multi-enzymatic systems to function in conditions similar to their natural environment [OSM 10].

The research domain is therefore in full expansion and is currently a particularly important strategic topic, as illustrated by the high number of patents registered throughout the world. The geographical distribution of patents and scientific publications highlights the predominance of the main actors in international research, with the USA ahead of Japan and China, in joint second position, followed by France, Germany and South Korea.

17.2. Principle and characterization of biofuel cells

Most biofuel cells are based on electro-enzymatic oxidation of glucose by glucose oxidase (GOX) and oxygen reduction by laccase or, infrequently, bilirubine oxidase (Figure 17.2).

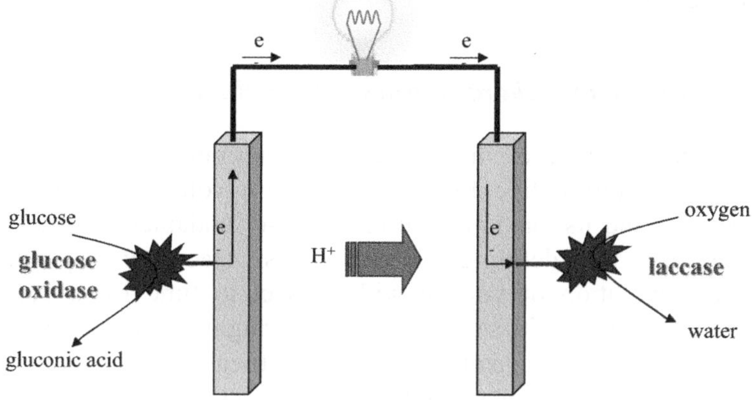

Figure 17.2. *Diagram of how a biofuel based on the electrical connection of laccase and glucose oxidase works*

Two redox mediator couples are usually involved in a biofuel cell: one to establish the electrical connection between the reduced form of flavin adenine dinucleotide, the prosthetic center of GOX, and the other to transfer electrons to the copper centers of laccase allowing the oxygen reduction. The power of biofuel cells is defined by electro-enzymatic reactions occurring on each of the two electrodes. In addition to high catalytic activity, the involved redox enzymes generally have high selectivity with respect to their substrate, thus simplify a biofuel cell's design. It is therefore conceivable to develop biofuel cells without a separating ion conduction membrane between the anode and cathode, as glucose can be mixed with oxygen.

Protein fragility, with respect to temperature and incompatible pH, and the limited access to their redox centers for electric exchange are the main disadvantages of using enzymes as electro-catalysts. Traditionally, operational conditions for biofuel cells include pH ranges from 3-8 and temperatures between 10 and 50°C. As for the prosthetic centers, they are often deeply buried inside the protein and a small mobile electronic relay is required in order to connect them. However, enzymes such as peroxidase or bilirubine oxidase either have active centers located near the protein shell or a flexible configuration, allowing direct electrical communication with the electrode surface.

17.2.1. *Performance characteristics of biofuel cells*

The performance of biofuel cells is determined in terms of the maximum electromotive force or maximum voltage output of the biofuel cell and its maximum current. These fundamental parameters are determined by defining a biofuel cell's capacity and operational stability (evolution of voltage with respect to time for a constant current discharge) or its storage stability during intermittent operation. The biofuel cell's capacity or total electrical charge supplied is determined by recording a current with constant intensity (I) delivered by the biofuel cell as a function of time (t) and electric charge (Q) produced such that: Q= I x t.

The performances of biofuel cells are often presented as a power/voltage or current/voltage curve (Figure 17.3). The shape of the power/voltage curve is similar to a bell curve, with limits being a maximum voltage output measurement related to zero current, and a maximum current measurement recorded between the bioanode and biocathode that have been electrically connected without the use of a resistor to the external circuit.

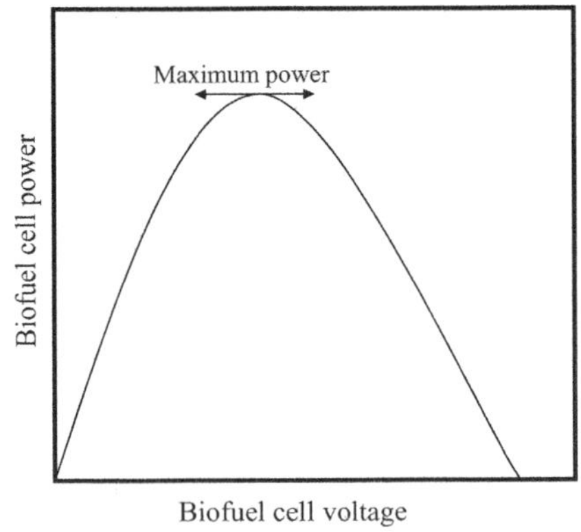

Figure 17.3. *Diagram of a biofuel cell's performance, presented as a curve: power as a function of the output voltage*

Ideally, the biofuel cell's maximum voltage can be approximated using the difference between the redox potentials of the two redox couples associated with enzymes, forming the biocathode and bioanode. It is therefore necessary to electrochemically characterize each bioelectrode with polarization curves that allow the determination of the potential value corresponding to the appearance of a catalytic current and even the maximum current value for each bioelectrode. The maximum voltage output of the biofuel cell is thus determined by the difference between these potentials. Maximum power, on the other hand, corresponds to a compromise between the limit values of the current and the voltage. Unlike fuel cells that present powers of kilowatts, biofuel cells have maximum power densities in tens and hundreds of microwatts/cm^2, recently even several milliwatts/cm^2 have been reached. This comparison must still be modulated as biofuel cells operate at ambient pressure and temperature.

One of the target applications for biofuel cells involves portable electric supply. It is imperative to consider the size of the proposed device. Keeping this goal in mind, the following parameters can be determined: power density corresponding to the biofuel cells' power delivery divided by the electrode's surface area (cm^2), and power density corresponding to the power divided by the electrode's volume (cm^3).

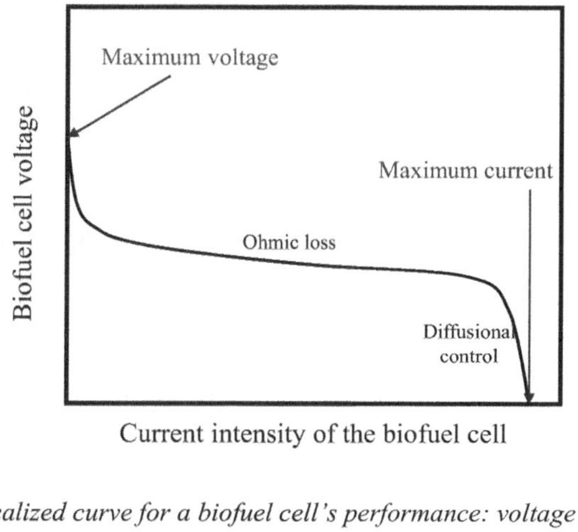

Figure 17.4. *Idealized curve for a biofuel cell's performance: voltage versus current*

17.3. Principle and characterization of a biofuel cell

Biofuel cells can be classified into different categories depending on the type of catalyst used for their creation.

17.3.1. *Biofuel cells based on mediated electron transfer*

Redox centers in enzymes are often deeply buried within the proteins and have diameters ranging from 4 to 15 nm. The spatial separation between this prosthetic site and the electrode's surface acts as a steric barrier for electron transfer. Redox mediators are used as they allow shorter electron transfer distances and enable the setup of an electronic connection between protein redox centers and electrodes, either by their diffusion or by electron hopping between them. Redox mediators should ideally have the following properties:

– a redox potential that is slightly different from the enzyme's potential in order to facilitate electron transfer, while maintaining the biofuel cell's electromotive force close to the potential difference of prosthetic centers of the enzymes used between the anode and cathode;

– a high diffusion coefficient in the electrolyte support;

– fast electron transfer kinetics between the enzyme and electrode;

– strong chemical stability that allows repeated oxido-reduction cycles;

– excellent biocompatibility so as not to denature the enzyme.

The redox mediator is often a small organic molecule or a metal complex (such as thionine, quinine, osmium) or an iron complex (such as ferrocyanide) [MAN 06, KAV 08, COS 10, NÖL 11]. The latter can be either free in a solution or connected by a long spacer arm to the enzymes themselves, or a material which captures them. It is necessary to include a membrane to prevent the diffusion of this electron relay outside the biofuel cell if redox mediators are present in the solution.

17.3.2. *Biofuel cells based on direct electron transfer*

The ideal configuration for biofuel cells resides in the possibility of establishing a direct transfer of electrons between the electrode and the active site of redox enzymes, thus reaching the enzyme's full catalytic potential and maximizing the rate of electron transfer. This direct transfer enables the thermodynamic potential, corresponding to oxygen reduction and glucose reduction at neutral pH, to be reached and leads to a theoretical voltage of 1.22V for the biofuel cell [ATA 07]. Direct electron transfer between the enzyme and electrode can also eliminate interferences often caused by undesirable redox reactions of solution components or secondary electrocatalytic reactions between redox mediators and electroactive components in the solution.

However, direct electron transfer requires prosthetic sites or electroactive relays to be located close to the protein's surface envelope. Enzymes with a deeply buried redox center cannot easily communicate with the surface of electrodes. Even if the enzyme's active center is located near the protein envelope or even on the surface, the enzyme's correct orientation, the proximity of the electrode's surface and good accessibility via substrate diffusion are pre-requisites for high speed electron transfer [BUL 06].

The practice of modifying the surface electrodes has often been used to facilitate ordered immobilization and orientation of enzymes and therefore their direct connection via the formation of self-assembled single layers of thiols functionalized on gold, which makes spatially controlled immobilization of proteins easier. Conducting polymers, polyelectrolytes, ionic liquids and nanoparticles are equally used to establish this direct communication. Direct electron transfer between the enzyme and electrode can also be created using different carbon-based materials. Carbon nanotubes are choice materials for the development of nanostructured deposits with a direct connection with adsorbed, entrapped or covalently bonded enzymes due to their high conductivity and large specific surface areas (up to 1,000 $m^2 g^{-1}$) in combination with their mechanical resistance. For example, multi-walled carbon nanotube deposits enable the adsorption and direct connection of bilirubine oxidase which catalyzes oxygen reduction in water [TAS 11, GÖB 10].

17.3.3. *Hybrid biofuel cells*

Hybrid biofuel cells are the association of half-elements with half-bioelement using metal catalysts or zinc anodes (oxidation reaction of Zn into Zn^{2+} replaces biological glucose oxidation for example). This hybrid setup allows the biofuel cell's lifetime to be increased by decreasing the number of proteins involved in reactions.

In comparison to biofuel cells, the advantage is that these mono-enzyme hybrid biofuel cells are able to work at the involved enzyme's optimum pH. Moreover, the theoretical electromotive force related to zinc oxidation and oxygen reduction results in a higher value (1.65 V) than that of an all-enzyme system based on glucose and oxygen (1.22 V) [ATA 07].

For example, a ceramic electrode modified with carbon and bilirubine oxidase was associated with a zinc electrode, resulting in a hybrid biofuel cell with a maximum tension of 1.095 V and a maximum power density of 41.3 $\mu W/cm^2$ [NOG 10]. On the other hand, one of the electrodes operates as a battery and will have to be replaced or recharged. The association of a zinc anode with a biocathode that reduces oxygen with laccase also eliminates the problems with the electric connection of glucose oxidase in the presence of oxygen, and its deactivation in the presence of enzymatically produced H_2O_2 by unconnected GOX. The presence of oxygen is necessary for the biocathode's operation and is also GOX's natural cofactor. It can, therefore, interfere with the reduced form of GOX's prosthetic center ($FADH_2$) and reduce to H_2O_2.

17.3.4. *Bio-inspired and biomimetic fuel cells*

As for bio-inspired fuel cells, catalytic oxidation of glucose is associated with oxygen reduction and is carried out by transition metal macrocycles: such as iron phthalocyanine or cobalt porphyrins that mimic the structure or activity of an enzyme's prosthetic sites. Despite the observed power loss, the complete removal of the system's fragile elements, enzymes, results in long-life bio-inspired fuel cells. If the performance of these abiotic devices is still low, their durability has been recently demonstrated. In 2008, Felix Von Stetten's group obtained performances of several $\mu W/cm^2$, and a lifetime of over seven months [KER 08].

17.4. Biofuel cell examples

The amplification of the electrical connection of immobilized enzymes in a host material (polymer, clays, etc.) has been carried out by incorporating graphite particles or carbon nanotubes in order to improve the intrinsic conductivity of composite materials. For example, the improvement of the conductivity of redox clays, lamellar double hydroxides functionalized by quinine redox mediators, with oxidized single-walled carbon nanotubes has resulted in a maximum catalytic current produced by electrically connected laccase that is 3 times higher (from 28 to 77.6 $\mu A\ cm^{-2}$) [DIN 10].

A new design concept for bioelectrodes has been developed using these different conductive particulate materials, such as graphite particle flakes. Previous conventional methods consisted of modifying the surface electrodes. The new method is based on high pressure compression of graphite particles, enzymes and redox mediators that directly leads to the creation of a disc-shaped electrode. In addition, it enables the use of redox mediators that are insoluble in aqueous media thus increasing the choice of possible mediators and the bioelectrode's stability in aqueous media. Graphite discs for the bioanode and biocathode were thus created. Those for the bioanode contain catalase associated with GOX that has been electrically connected by ferrocene. Those for the biocathode contain laccase that has been electrically connected by the 2,2 '-azino-bis (3-ethylbenzothiazoline-6-sulfonate) (ABTS) ammonium salt. The disc surfaces were then covered with different membranes that prevented the release of either enzymes or redox mediators soluble in aqueous media, ABTS in this example. The membranes which have been studied include chemically reticulated proteins, Nafion and poly-lysine films. Some biofuel cells that associate two compressed discs, 1.3 cm in diameter, have output powers of 23 μW cm^{-2} at 0.33 V and a voltage and maximum current of 0.63 V and 0.17 mA cm^{-2} respectively [COS 10]. A power of 200 μW cm^{-2} has been achieved by optimizing different parameters: graphite/enzyme ratio, graphite/redox mediator ratio, compression pressures (3,000-10,000kg cm^{-2}).

Depending on the manufacturing concept of biofuel cells by compression of an enzyme, redox mediator and graphite particle mixture, biocathodes based on polyphenol oxidase (PPO) have been developed as an alternative to biocathodes based on laccase or bilirubine oxidase for oxygen reduction. PPO is an enzyme that catalyzes the oxidation of phenols and dihydroxybenzenes into quinoid derivatives where the co-substrate, oxygen, is reduced to water. It has been shown that PPO is capable of operating *in vivo* in cerebral tissues, unlike other enzymes used in the cathodes of biofuel cells described until now.

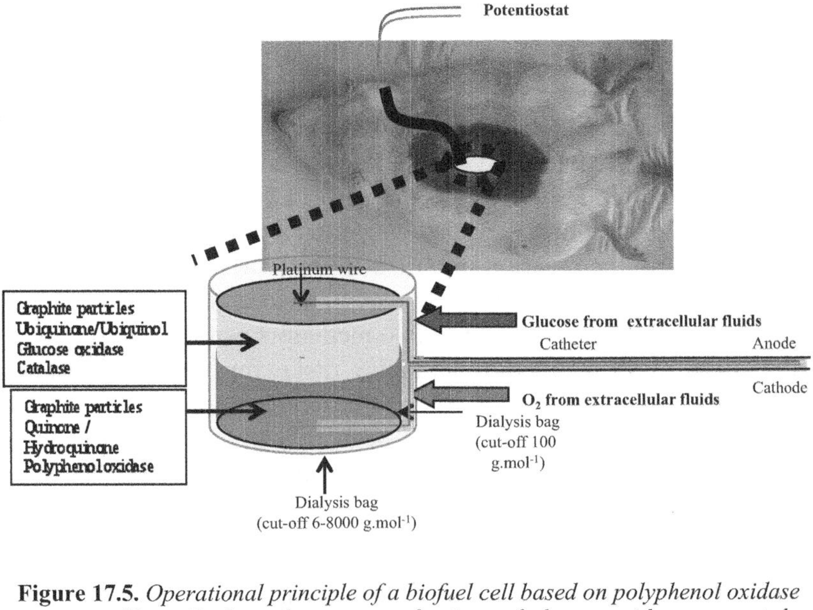

Figure 17.5. *Operational principle of a biofuel cell based on polyphenol oxidase connected by quihydrone for oxygen reduction and glucose oxidase connected by ubiquinone for glucose oxidation and location of the implant*

Biofuel cells with a bioanode based on the connection of GOX using ubiquinone and its association with catalase for the elimination of hydrogen peroxide, and a biocathode based on the connection of PPO by a hydroquinone/quinone couple, were developed in a PTFE (PolyTetraFluroroEthylene) membrane, which is required for *in vivo* experiments. Each bioelectrode was enveloped in a dialysis bag. The biofuel cell was then surgically inserted into an anesthetized male Wistar rat's (514 g in weight) retro-peritoneal space. The first trials showed the possibility of continuously generating 6.77 μW mL^{-1} for 11 days in a rat that was free of its movements from glucose and oxygen present in extracellular fluids [CIN 10].

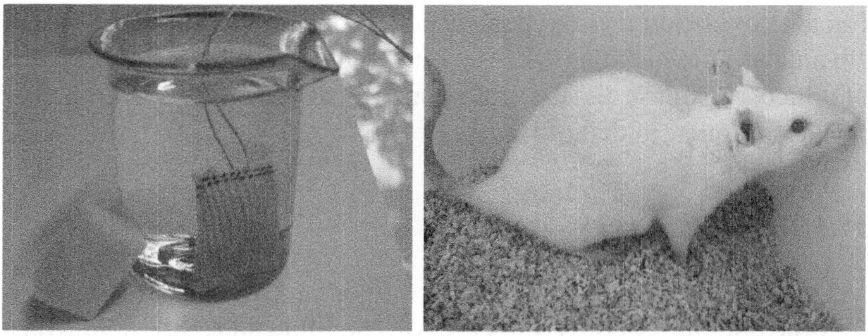

Figure 17.6. *Biofuel cell encapsulated in a PTFE membrane and implanted in a rat*

Nishizawa and collaborators have recently put forward an original manufacturing process for enzyme electrodes. Films based on carbon nanotube "forests" were used to create a hybrid assembly of enzymes by adsorption and contraction of nanotubes induced by an immersion and a drying step [MIY 11].

The resulting film has a thickness between 4 and 12 µm and consists of long (~1 mm) single-walled carbon nanotubes spaced at 16 nm intervals. The film's specific surface area is assessed via the Brunauer-Emmett-Teller (BET) method, based on adsorption isotherms of nitrogen, and is estimated to be 1,300 m^2 g^{-1}. When enzyme-containing liquids are inserted into the film and dried, a compact near-hexagonal nanotube structure traps the enzymes and the authors describe the automatic control of the intra-nanospace, of similar size to that of the enzymes, by a contraction phenomenon which amplifies the enzymes' electrical connection. This involves the direct transfer of electrons between nanotubes and enzymes.

Films based on carbon nanotube forests are used to create an anode with fructose dehydrogenase and a cathode with laccase. Electrocatalytic oxidation of a D-fructose solution (200 mM) results in very high oxidation current densities: 16 mA cm^{-2}. As for the biofuel cell, it produces power densities which reach 1.8 mW cm^{-2} at 0.45 V in stirred fructose solutions maintained in an oxygen atmosphere and buffered to pH 5. The biofuel cell also has a maximum zero-current voltage of 0.77 V, which is in good accord with the difference in fructose oxidation and oxygen reduction potentials. There is a strong advantage in associating laccase with fructose dehydrogenase instead of glucose oxidase as the optimum pH of fructose dehydrogenase (pH 4) is similar to that of laccase.

17.5. Bibliography

[ATA 07] P. Atanassov, C. Apblett, S. Banta, S. Brozik, S. Calabrese Barton, M. Cooney, B. Y. Liaw, S. Mukerjee, S. D. Minteer, *Interface*, vol. 16, p. 28, 2007.

[BAR 04] S.C. Barton, J. Gallaway, P. Atanassov, *Chem. Rev.*, vol. 104, p. 4867, 2004.

[BIL 08] M. Z. Bilha, I. Willner, *Electroanalysis*, vol. 20, p. 583, 2008.

[BUL 06] R.A. Bullen, T.C. Arnot, J.B. Lakeman, F.C. Walsh, *Biosens. Bioelectron.*, vol. 21, pp. 2015-2045, 2006.

[CIN 10] P. Cinquin, C. Gondran, F. Giroud, S. Mazabrard, A. Pellissier, F. Boucher, J.-P. Alcaraz, K. Gorgy, F. Lenouvel, S. Mathé, P. Porcu, S. Cosnier, *PLoS ONE*, vol. 5, e10476, 2010.

[COS 10] S. Cosnier, D. Shan, S. Ding, *Electrochem. Commun.*, vol. 12, pp. 266-269, 2010.

[CRA 08] J.A. Cracknell, K.A. Vincent, F.A. Armstrong, *Chem. Rev.*, vol. 108, p. 2439, 2008.

[DIN 10] S.-N. Ding, M. Holzinger, C. Mousty, S. Cosnier, *J. Power Source*, vol. 195, pp. 4714-4717, 2010.

[GÖB 10] G. GÖBEL, T. DIETZ, F. LISDAT, *Electroanalysis*, vol. 22, pp. 1581-1585, 2010.

[IVA 10] I. IVANOV, T. VIDAKOVIC-KOCH, K. SUNDMACHER, *Energies*, vol. 3, p. 803, 2010.

[KAV 08] P. KAVANAGH, P. JENKINS, D. LEECH, *Eltrochem. Commun.*, vol. 10, pp. 970-972, 2008.

[KER 08] S. KERZENMACHER, J. DUCRÉE, R. ZENGERLE, F. VON STETTEN, *J. Power Sources*, vol. 182, p. 66, 2008.

[MAN 06] N. MANO, V. SOUKHAREV, A. HELLER., *J. Phys. Chem.*, vol. 110, pp. 11180-11187, 2006.

[MIY 11] T. MIYAKE, S. YOSHINO, T. YAMADA, K. HATA, M. NISHIZAWA, *J. Am. Chem. Soc.*, vol. 133, pp. 5129-513, 2011.

[NOG 10] W. NOGALA, A. CELEBANSKA, G. WITTSTOCK, M. OPALLO, *Fuel Cells*, vol. 10, pp. 1157, 2010.

[NÖL 11] T. NÖLL, G. NÖLL, *Chem. Soc. Rev.*, vol. 40, no. 7, pp. 3564-3576, July 2011.

[OSM 10] M.H. OSMAN, A.A. SHAH, F.C. WALSH, *Biosens. Bioelectron*, vol. 26, p. 953, 2010.

[TAS 11] F. TASCA, W. HARREITHER, R. LUDWIG, J.J. GOODING, L. GORTON, *Anal. Chem.*, vol. 83, pp. 3042-3049, 2011.

[WIL 09] I. WILLNER, Y.-M. YAN, B. WILLNER, R. TEL-VERED, *Fuel Cells*, vol. 9, 2009

[YAH 64] A.T. YAHIRO, S.M. LEE, D.O. KIMBLE, *Biochim. Biophys. Acta*, vol. 88, p. 375, 1964.

List of Authors

Khalifa AGUIR
University of Aix Marseille
France

Stéphane ARBAULT
University of Bordeaux
France

Marcel BOUVET
Institut de Chimie Moléculaire (ICMUB)
University of Burgundy
France

Philippe BREUIL
Ecole Nationale Supérieure des Mines de Saint-Etienne
France

Serge COSNIER
Joseph Fourier University
Grenoble
France

Isabelle DUFOUR
University of Bordeaux
France

Marie José DURAND
IUT La Roche sur Yon
France

Chantal GONDRAN
Joseph Fourier University
Grenoble
France

Nicole JAFFREZIC-RENAULT
University of Lyon I
France

René LALAUZE
Ecole Nationale Supérieure des Mines de Saint-Etienne
France

Philippe MENINI
Paul Sabatier University
Toulouse
France

Hubert PERROT
UPMC
Paris
France

Christophe PIJOLAT
Ecole Nationale Supérieure
des Mines de Saint-Etienne
France

Louis RENAUD
University of Lyon
France

Neso SOJIC
University of Bordeaux
France

Pierre TEMPLE-BOYER
LAAS-CNRS
Toulouse
France

Gérald THOUAND
IUT La Roche sur Yon
France

Jean-Paul VIRICELLE
Ecole Nationale Supérieure
des Mines de Saint-Etienne
France

Index

E

F, G

H, I, L

M

O, Q

R

S

T

V

Printed and bound by CPI Group (UK) Ltd, Croydon, CR0 4YY
13/09/2026
14961075-0007